POLLUTION PREVENTION:

METHODOLOGY, TECHNOLOGIES AND PRACTICES

by

Kenneth L. Mulholland and James A. Dyer

Published by the American Institute of Chemical Engineers
3 Park Avenue, New York, NY 10016-5901

3 Park Avenue, New York, New York 10016-5901

Library of Congress Catalog Card Number 98-40828
ISBN: 0-8169-0782-X

Terry A. Baulch, Cover Design and Typesetting

Pollution prevention: methodology, technologies, and practices by K.L. Mulholland and J.A. Dyer.
p. cm.

Includes bibliographical references.
1. Factory and trade waste - - Management. 2. Pollution prevention.
II. Title.
TD897.5.M85 1999 98-40828
363.73'7--dc21 CIP

To our wives
Cathy and Yvette

Table of Contents

Foreword

For better or worse, engineers rarely make the headlines. When we do, it is a fair bet that it is the result of something gone wrong. Technology, when it works well, is (and should be) invisible. And as often is the case, so are the people who create it.

Consider an oil refinery. Few people, driving past it on their way to work, pause to wonder about the engineering feat of turning crude into the plastics that comprise much of their car, the gasoline that powers it, and the asphalt beneath the tires (which, incidentally, also had their origins in that same refinery, or one remarkably like it). In fact, it is likely that the majority of people driving past don't give the refinery a second thought at all.

Which, perhaps, is as it should be. But, on surprisingly rare occasion, something will happen to bring the inner workings of technology into the public spotlight.

One of the most interesting recent examples of this phenomenon is the so-called Year 2000 ("Y2K") problem. The problem stems from the way that early computer programmers recorded date information. Faced with the need to keep both code and data as compact as possible, programmers often truncated dates to record only the last two digits of the year. In other words, the year 1998 (the year that this book was originally published) would be recorded as "98." Obviously, this poses a potential problem for software that must track dates past the year 2000. Does a date entry of "02" stand for the year 2002, or 1902?

Public reaction to this rather arcane issue of software programming has been surprising. Reactions range from mild concern to dire predictions of the imminent demise of civilization. In the worst-case scenarios, banks will fail, the government will come grinding to a halt, and airplanes will fall from the sky. Of course, this assumes that they are still flying — as of this writing several airlines have announced that, as a precautionary measure, they will suspend all flights when the final hour arrives.

The truncation that seemed an expedient solution thirty years ago has "suddenly" introduced the one thing that computer programmer's try so hard to avoid — ambiguity.

Like many engineering problems, the Y2K problem was well understood by the people who created it. But (again, like many engineering problems) the decision that led to the problem was part of a conscious trade-off. Faced with the challenge of getting software to function within constraints of available memory and processor speed, it was an expedient solution to a problem whose ultimate solution was assumed to lie in the future — when improved technology would make it possible to implement a more complete solution.

The problem, of course, is that the future is now.

In ways that are strikingly similar, the process industries are facing their own version of the Y2K problem, in the form of environmental wastes. Like the Y2K problem, process waste is often the result of engineering trade-offs that were made between efficiency and capital cost, and between the state of existing technology and the need to get product to market. One result of these trade-offs is waste, which in many cases can no longer be tolerated either economically or environmentally.

Think of a chemical plant as a computer program for turning raw materials into products, with the instructions (code) written in steel, rather than bytes. Like the software engineers who must examine every line of code to find a persistent bug, we are faced with the challenge of methodically "opening up" each process, and eliminating the sources of waste and pollution that threaten the environment, and threaten the viability of the industry.

This book is an instruction manual for debugging the chemical process industries.

This book will help the chemical engineer recognize pollution prevention as nothing more than a reformulation of traditional chemical engineering problem solving, and in the process provide new applications for your skills. The same reaction engineering skills that optimized processes for maximum throughput and minimum capital investment can also be applied to the challenge of re-optimizing for high selectivity and maximum efficiency. Sometimes, for example, when dealing with persistent and bioaccumulative materials, the questions may need to be rephrased in somewhat different ways — "can we live without this intermediate in our product or process?" but ultimately, it still becomes a problem amenable to a mixture of creative problem solving and fundamental technical skills learned in school and on the job.

For the non-chemical engineer, the book demonstrates in understandable terms the application of the chemical engineer's toolkit — reaction engineering, transport phenomena, equipment design, and engineering economics

— to the challenges of preventing waste in the process industries.

Finally, for all readers, the generous use of case studies and examples from process plants will help you gain an understanding of how pollution prevention techniques can be used to reduce or eliminate plant bottlenecks, improve product quality and yield, and improve overall profitability of today's process plant.

Scott Butner
Senior Research Scientist,
Environmental Technology Division
Pacific Northwest National Laboratory
Seattle, WA

Preface

In 1994, the DuPont Company established "The DuPont Commitment" for Safety, Health, and the Environment. As part of this Commitment, the company resolved to drive toward zero waste generation at the source and zero emissions. Having spent the previous 5 years delivering on corporate goals to reduce air toxic and air carcinogen emissions, we realized the value that pollution prevention could play in improving the environment and the manufacturing process, while also making money for the business. What a concept! As a result of our work on end-of-pipe treatment of waste streams, we were also armed with valuable information on the cost of abating emissions at the tail end of the process. We quickly realized that the cost of end-of-pipe treatment could be a catalyst to help drive pollution prevention in the businesses. What was missing, however, was the ability to quickly identify feasible pollution-prevention technologies and practices that applied across multiple manufacturing processes and industries. Being chemical engineers, we envisioned a "unit operations" approach to pollution prevention.

Since 1994, we have responded to DuPont's corporate commitment to drive toward "zero waste and emissions." With the help and support of many colleagues at DuPont, particularly Robert W. Sylvester, we invented a structured methodology to identify the best pollution-prevention solutions for any manufacturing process, as well as a unique set of pollution technologies and practices that are based on chemical engineering know-how. In addition, we took full advantage of the important end-of-pipe treatment and economic evaluation skills and knowledge that were available to us at DuPont. We did not waste any time in leveraging this knowledge across the globe. By being proactive rather than reactive, we have helped DuPont reduce total waste generated by 300 million pounds per year, and have helped position the corporation to meet its goal of "zero waste and emissions," while also gaining competitive advantage through faster, more effective, revenue-producing pollution-prevention programs.

The novel methodology described in this book uses a structured brainstorming process that requires minimum resources to identify chemistry and engineering changes that can be made to existing and new processes. These process improvement ideas not only reduce the amount of waste being generated, but they also make money for the business. We have participated in or led over two dozen brainstorming sessions to identify the best waste reduction. In this sense, the methodology is proven and of practical use to engineers responsible for reducing waste generation in both new and existing processes.

To support the brainstorming of ideas, we transformed an extensive database of process-specific pollution-prevention case histories within DuPont and the open literature into fundamental pollution-prevention knowledge that is applicable to all businesses around the world. Examples of this pollution-prevention knowledge include equipment and parts cleaning, optimizing the use of water, pollution prevention in batch operations, pollution prevention through reactor design and operation, pH control as a pollution-prevention tool, high-value wastes, organic solvent selection, minimizing equipment leaks, and greener separations.

This extensive knowledge was then further simplified to a set of pollution-prevention "wisdoms" that are easily understood and broadly applicable. This fundamental knowledge allows plant operations personnel to have just as important a role in pollution prevention as do the engineers and chemists.

In the process, we developed a course based on the pollution-prevention methodology, technologies, and practices described in this book, and taught it to over 250 DuPont professionals in the United States and Europe. The course was also given in Poland as part of a World Environment Center initiative, and was taught at the University of Delaware as part of their Environmental Certificate Program and Chemical Engineering Graduate Program. In addition, we helped design and participated in three pollution-prevention workshops—two sponsored by the American Institute of Chemical Engineers' Center for Waste Reduction Technology and one organized by the Delaware Chamber of Commerce. Lastly, the work was recognized with a DuPont Safety, Health, and Environmental Excellence Award, one of only 13 achievements chosen to receive this award in 1998 by a corporate selection committee comprising DuPont professionals and external representatives from leading environmental organizations.

In summary, this book offers many practical tools and methodologies for establishing and implementing a pollution-prevention program at your site, whether large or small. Our approach combines a proven, well-structured methodology with simplified tools and techniques to speed your journey through the process. If nothing else, we hope that we have dispelled the myth that pollution prevention is process-specific.

Kenneth L. Mulholland
James A. Dyer

Acknowledgments

The authors wish to acknowledge the valuable contributions of a number of their colleagues and close friends in DuPont. Specifically:

Robert W. Sylvester for his inspiration, enthusiastic leadership, and guidance on the development and implementation of the "practice" of pollution prevention. When it comes to brainstorming, Bob is one of the best!

Wayne C. Taylor for his humor, but also for his significant contributions to the principles of engineering evaluations, the 10-Step Method, and the economics of biological wastewater treatment.

Robert A. Keller for his vital work on the economics and selection criteria for halogenated VOC air emissions and the prevention of pollution in batch operations.

Bruce M. Vrana and *Jay R. Balder* for the development of and permission to use the "Shortcut NPV Method."

R. Bertrum Diemer for his contributions to the Dyelate process case study in Chapter 6.

Bradford F. Dunn for supporting this work and allowing it to be shared much more broadly outside DuPont.

And last, but not least, a large number of colleagues who contributed in various ways to the development of the pollution-prevention technologies and practices in this book: *Ashok S. Chetty, Bryan C. Fritzler, Noel C. Scrivner, Wilford Shamlin, Wilfred K. Whitcraft, Thomas A. Kittleman, David G. R. Short, Ross E. Kendall, Stephen T. Breske, Mervin E. Meckley,* and *William W. Goudie.*

"If you always do
What you always did,
You will always get
What you always got."

Joe Juran

"It is not the answer that illuminates,
it is the question."

Playwright Eugene Ionesco

CHAPTER 1

Why Pollution Prevention?

1.1 Introduction

"Pollution prevention" has become the environmental buzzword of the 1990s. No matter what one chooses to call the task or technology of reducing waste and emissions from a chemical process—pollution prevention, waste minimization, source reduction, clean technology, green manufacturing—the challenge of implementing process changes that actually reduce waste generation is often formidable. Engineers and scientists faced with developing and implementing a pollution-prevention program for a business or a manufacturing site face many obstacles—technological, economic, and societal. Some of these obstacles are real, while many others are only perceived to be real.

In this book, we present the reader with a proven methodology, together with preferred engineering practices and technologies, for minimizing waste generation at the source, and hence for preventing pollution. The focus of the methodology is on identifying pollution-prevention engineering technologies and practices that will change what is happening *inside* the pipes and vessels of the manufacturing process, rather than just on simple procedural or cosmetic changes. In addition, the book provides techniques and tools for overcoming obstacles to implementation of solutions and for establishing an ongoing, continuous pollution-prevention program at the manufacturing level. The methodology has been and continues to be successfully practiced inside the DuPont Company. The pollution-prevention engineering technologies and practices, developed by reviewing hundreds of process-specific case histories from a variety of industries, nicely complement the methodology and provide a useful knowledge base for quickly identifying possible process changes that reduce waste generation and emissions.

1.2 Waste as Pollution

A waste is defined as an unwanted byproduct or damaged, defective, or superfluous material of a manufacturing process. Most often, in its current state, it has or is perceived to have no value. It may or may not be harmful or toxic if released to the environment. Pollution is any release of waste to the environment (i.e., any routine or accidental emission, effluent, spill, discharge, or disposal to the air, land, or water) that contaminates or degrades the environment.

Figure 1-1 depicts a typical manufacturing facility. Inputs to the facility include raw materials to produce the salable product(s), water, air, solvents, catalysts, energy,

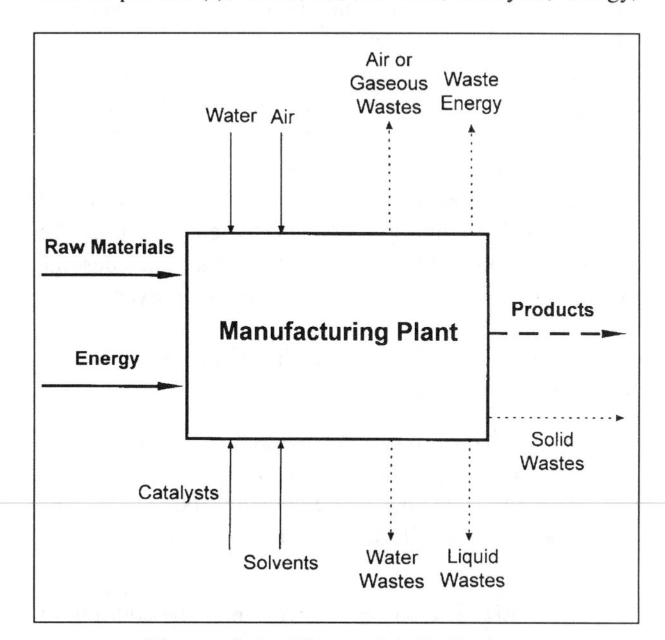

Figure 1-1. Plant with Pollution

and so on. Outputs from the facility are the salable product(s), waste energy, as well as gaseous, liquid, water, and solid wastes.

In contrast, a manufacturing facility with an absolute minimum or "zero" amount of waste being generated is shown in Figure 1-2. Inputs to the facility include only the raw materials to make the salable product(s) and energy. The only outputs are salable products.

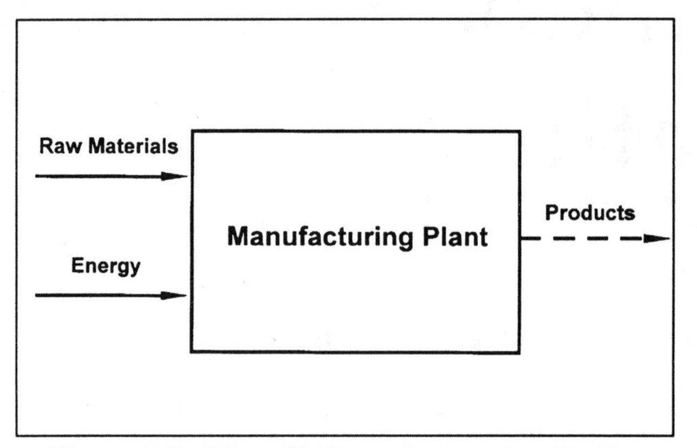

Figure 1-2. "Zero" Waste Generation Facility

1.3 How Is Pollution Prevention Defined?

In this book, we define pollution prevention fairly broadly as any action that prevents the release of harmful materials to the environment. This definition manifests itself in the form of the pollution-prevention hierarchy shown in Figure 1-3. In this hierarchy, safe disposal forms the base of the pyramid, while minimizing the generation of waste at the source is at the peak.

In contrast, the U.S. Environmental Protection Agency (EPA) definition of pollution prevention recognizes only source reduction, which encompasses only the upper two tiers in the hierarchy: minimize generation and minimize introduction.[1] The U.S. EPA defines the hierarchy shown in Figure 1-3 as Environment Management Options. The European Community, on the other hand, includes the entire hierarchy in its definition of pollution prevention, as is done in this book.

A definition of each tier in the pollution prevention hierarchy is given below:

- **Minimize Generation.** Reduce to a minimum the formation of nonsalable byproducts in chemical-reaction steps and waste constituents, such as tars and fines, in all chemical and physical separation steps.
- **Minimize Introduction.** Minimize the addition of materials to the process that pass through the system unreacted or that are transformed to make waste. This implies minimizing the introduction of materials that are not essential ingredients in making the final product. Examples of introducing nonessential ingredients include: (1) using water as a solvent when one of the reactants, intermediates, or products could serve the same function, and (2) adding large volumes of nitrogen gas because of the use of air as an oxygen source, heat sink, diluent, or conveying gas.
- **Segregate and Reuse.** Avoid combining waste streams together without giving consideration to the impact on toxicity or the cost of treatment. For example, it may make sense to segregate a low-volume, high-toxicity wastewater stream from several high-volume, low-toxicity wastewater streams. Examine each waste stream at the source and determine which ones are candidates for reuse in the process or can be transformed or reclassified as a valuable coproduct.
- **Recycle.** A large number of manufacturing facilities, especially chemical plants, have internal recycle streams that are considered part of the process. In this case, recycle refers to the external recycle of materials, such as polyester film and bottles, Tyvek envelopes, paper, and spent solvents.
- **Recover Energy Value in Waste.** This step is a step of last resort. Examples include burning spent organic liquids, gaseous streams containing volatile organic compounds, and hydrogen gas for their fuel value. The reality is that the value of energy and resources required to make the original compounds is often much greater than that which can be

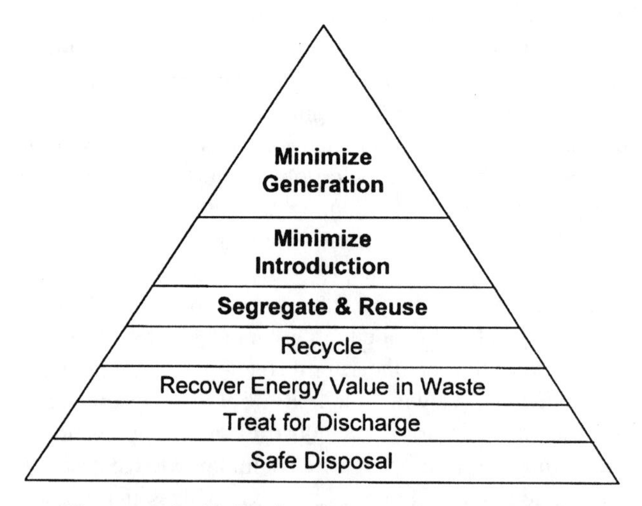

Figure 1-3. Pollution Prevention Hierarchy

recovered by burning the waste streams for their fuel value.

- **Treat for Discharge.** This involves lowering the toxicity, turbidity, global-warming potential, pathogen content, and so forth, of the waste stream before discharging it to the environment. Examples include biological wastewater treatment, carbon adsorption, filtration, and chemical oxidation.
- **Safe Disposal.** Waste streams are rendered completely harmless or safe so that they do not adversely impact the environment. In this book, we define this as total conversion of waste constituents to carbon dioxide, water, and nontoxic minerals. An example would be subsequent treatment of a wastewater-treatment-plant effluent in a private wetlands. So-called "secure landfills" would not fall within this category unless the waste is totally encapsulated in granite.

In this book, we focus on the upper three tiers of the pollution-prevention hierarchy; that is, minimize generation, minimize introduction, and segregate and reuse. This is where the real opportunity exists for reducing waste and emissions while also improving the business' bottom line.

1.4 Drivers for Pollution Prevention

Since the early 1960s, the number of federal environmental laws and regulations has been increasing at a rate three times that of the United States population. In 1960, there were only three federal environmental laws on the books; now there are more than 30. This does not even include the much larger number of state environmental laws. Figure 1-4 shows both the population growth in the United States and the number of federal environmental laws and regulations as a function of time. The reality is that laws and regulations use command and control to force industry to comply.

Toward the end of the 1980s, however, a much larger number of industries was beginning to turn to pollution prevention as a means of avoiding the installation of expensive end-of-the-pipe treatment systems. It was becoming clear to many that, with time, the succession of increasingly stringent regulations would ultimately lead to a complex, expensive series of treatment devices at the end of a manufacturing process—each with its own set of

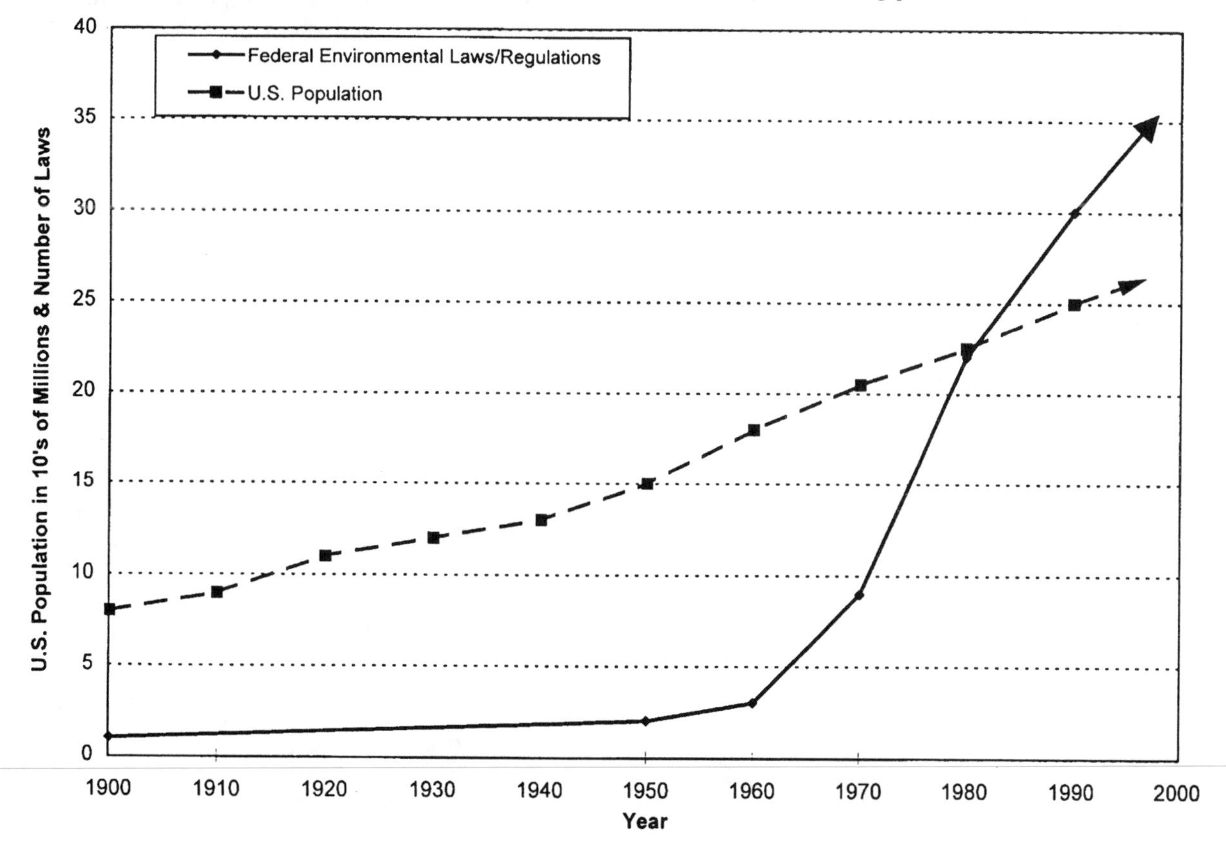

Figure 1-4.
Comparison Between the Increase in Federal Environmental Laws and the U.S. Population with Time

maintenance and performance issues.

Those industries and businesses that began to accept and implement pollution-prevention solutions instead of treatment found that they not only reduced waste generation but that they also made money. As a result of these experiences, various governmental agencies began to incorporate pollution-prevention requirements into new environmental laws. Congress recognized that "source reduction is fundamentally different and more desirable than waste management and pollution control,"[2] therefore, it passed the Pollution Prevention Act in 1990.

From our own experience in DuPont, the five major drivers for pollution prevention are:

1. The increasing number and scope of environmental regulations and laws.
2. The rising cost of waste treatment.
3. Greater government oversight and control of business operations.
4. More awareness by corporations of the value of pollution prevention to the business bottom line and to the customer.
5. The heightened awareness in society of the need for sustainability of the planet.

The first and second major drivers for pollution prevention, as just described, are regulations and laws and the cost of waste treatment. Extrapolation of the two curves in Figure 1-4 would imply that future laws and regulations will be even more stringent and, if solved by end-of-pipe treatment, even more costly.

Figure 1-5 shows conceptually the cost incurred by the business to generate waste versus the amount of waste produced by a manufacturing process. As indicated on the figure, most processes currently operate in the upper righthand portion of the graph. The figure also shows that these same processes could lower waste generation and costs with little or no capital investment by shifting downward along the solid line. The "Economic Zero," as indicated by the vertical dotted line, is the point where the slope of the curve reverses itself and normally becomes very steep. Further reducing waste generation, then, requires a significantly greater capital investment, for example, replacing large piece(s) of equipment or unit operations. Instead, to further reduce the level of waste being generated while simultaneously reducing the cost to generate this waste, new chemistry or new engineering technology is required (i.e., a new process). This is indicated by the dashed curve on Figure 1-5.

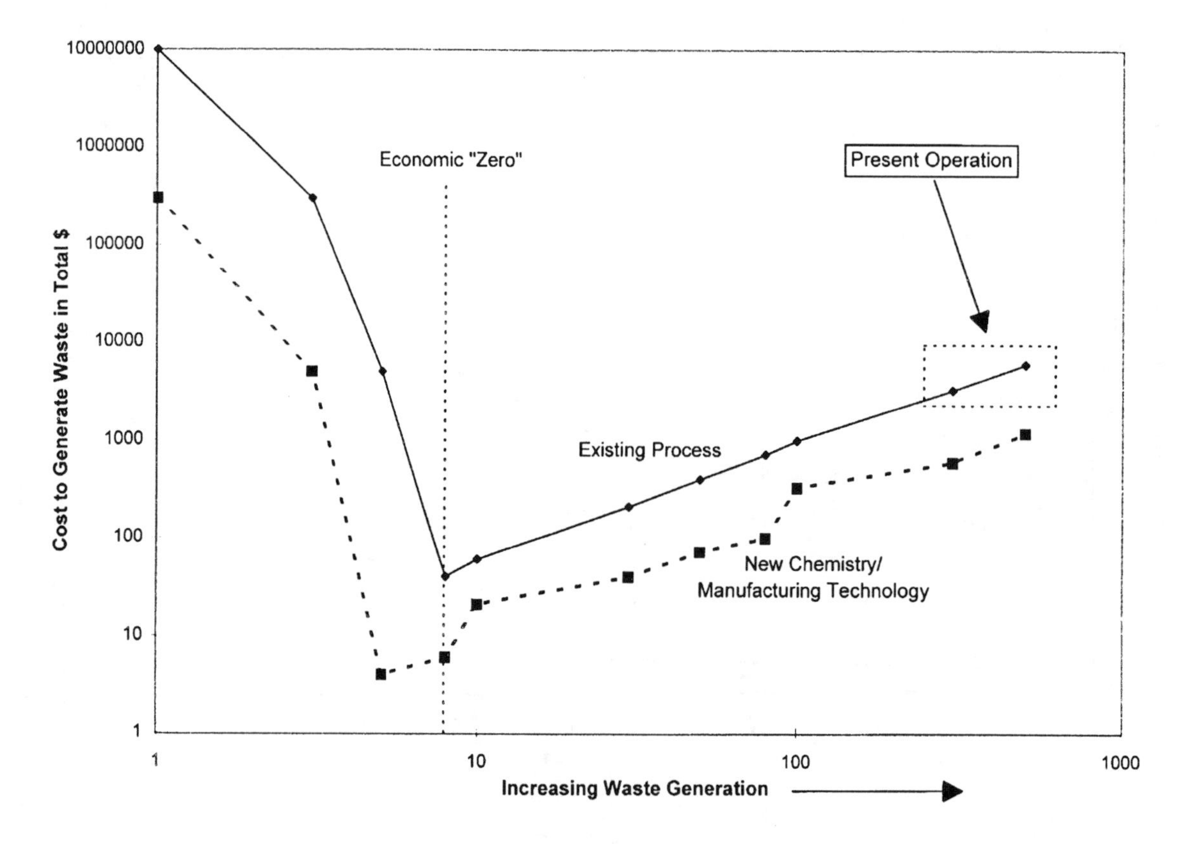

Figure 1-5. Waste Generation vs. Business Cost

Federal, state, and local governments are demanding more and more information from manufacturers—not only the size, composition, and properties of waste streams that are generated, but also what chemicals are added to the process to manufacture the final product, and descriptive information on how these chemicals are used within the process. The third major driver for pollution prevention, then, becomes control of the business. When a business does not make any waste or is below a *de minimus* level, then only a minimum amount of information is required by the governing bodies; hence, business information is conserved.

Figure 1-6 depicts the degree of control business leadership has over a business versus governmental control as a function of the amount of waste being generated by a process. Normally, there will be a *de minimus* level of waste generation below which the regulations require only minimal governmental oversight; that is, the business controls its own destiny. As the level of waste generation increases, however, so too does the amount of governmental oversight. As a result, business leadership has less control of their business and is less able to respond to various business factors that might improve their bottom line. The *de minimus* point for a Regulatory "Zero" is normally below that for the Economic "Zero" yet a business still might voluntarily choose to spend additional capital investment to increase control.

Recognizing the value of pollution prevention to the business and the customer, progressive companies are developing corporate goals to motivate their employees to reduce the amount of waste being produced. Examples include the 3M Corporate Environmental Conservation Policy[3] and the DuPont Company's Safety, Health and the Environment Commitment of zero waste generation and emissions, which is shown in Figure 1-7. The environmental group Grassroots Recycling Network is developing a Zero Waste Policy Paper for consumer products. The net result is that society is beginning to expect that the processes of the future will not generate waste and that the products are recyclable or biodegradable.

For the businesses that have implemented pollution-prevention programs, the amount saved or earned has been quite dramatic. For example, in the 3M Company, the Pollution Prevention Pays (3P) Program netted $350 million for their U.S. plants from 1976 through 1987 while reducing waste generation by more than 425,000 tons per year. A second example is the joint EPA/DuPont Chambers Works Waste Minimization Project, which resulted in a savings of $15 million per year for only $6.3 million in capital investment and a 52% reduction in waste generation.[4]

The DuPont Company has also instituted a corporate

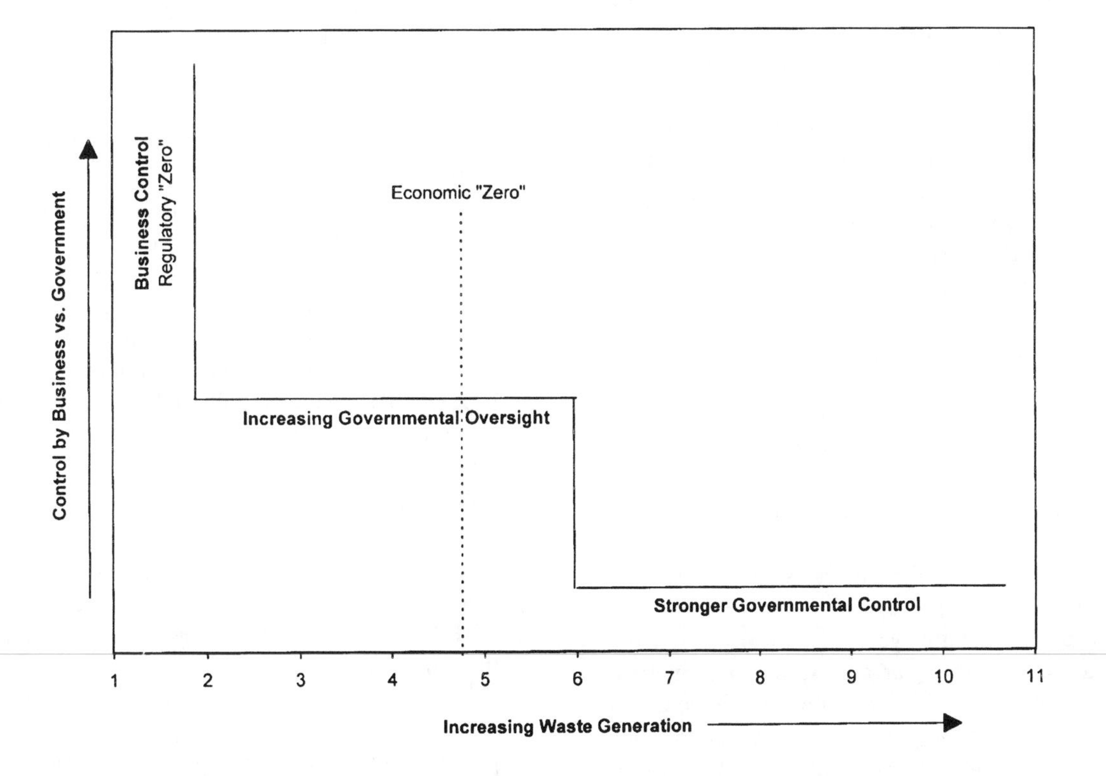

Figure 1-6. Waste Generation vs. Business Control

Environmental Excellence Award program. Of the typical 550 submissions per year, approximately 70 pass the first screening and 12 are finally selected as winners. For the years 1994 to 1996, more than $200 million per year positive return and $320 million in avoided capital expenditures were realized for the 210 programs that passed the first screening.

The fifth main driver for pollution prevention, which is growing in importance, is sustainability, that is, building a sustainable global economy or an economy that the planet is capable of supporting indefinitely.[5] Pollution prevention is one of three ways that a company can move toward sustainability. A second way is product stewardship, where a manufactured product has minimal impact on the environment during its full manufacturing life cycle. A third step toward sustainability is through clean technology; that is, technology that has a minimum impact on the environment. Examples include (1) avoiding the use of chlorine and its attendant toxic, persistent, or bioaccumulative compounds, and (2) replacing high-temperature and high-pressure processes with biotechnology routes, which can manufacture products at ambient conditions.

1.5 Pollution-Prevention Wisdom

A wealth of information is available on pollution-prevention successes across many industries; however, it is primarily packaged in the form of process-specific case histories.[1,4,6] As a result, the information is not organized in a sufficiently generalized way so as to allow the rapid transfer of knowledge from one type of industry to another. To help the practitioners of pollution prevention—engineers and scientists—to more quickly generate ideas, this process- or industry-specific information has been transformed into generalized knowledge that can be more easily implemented by project teams and existing manufacturing facilities. The information is organized in a "unit operations" format to facilitate its widespread use across different processes and industries. One might imagine a *Chemical Engineers' Handbook* of preferred pollution-prevention technologies and practices.

In DuPont, pollution prevention has evolved over the last five years

- from the compilation of *data* in the form of case studies,
- to the organization of these case studies into *information* on specific processes,
- to the development of generalized *knowledge* from the case study data and process-specific information,
- into the basic *wisdom* of pollution prevention; that is, the insight that *properly defining and parsing the problem ultimately leads to the best pollution-prevention solutions.*

Figures 1-1 and 1-2 depict a manufacturing plant with and without waste generation, respectively. The manufacturing facility depicted in Figure 1-1 produces many different waste streams and adds many different feed materials to the process. Two techniques developed and practiced within DuPont—"Waste Stream Analysis" and "Process Analysis"—help to parse or divide the waste streams and overall process into their important components or parts, so as to focus the creative imagination on generating pollution-prevention ideas. Each of these techniques is briefly introduced below, because they represent key aspects of this pollution-prevention wisdom.

1.5.1 Waste Stream Analysis

The best pollution-prevention options cannot be implemented unless they are identified. To uncover the best options, each waste-stream analysis should follow four steps:

1. List all components in the waste stream, along with any key parameters. For instance, for a wastewater stream these could be water, organic compounds, inorganic compounds (both dissolved and suspended), pH.
2. Identify the compounds triggering the concern, for example, hazardous air pollutants (HAPs), carcinogenic compounds, and compounds regulated under the Resource Conservation and Recovery Act (RCRA). Determine the sources of these compounds within the process, then develop pollution-prevention options to minimize or eliminate the generation of these compounds.
3. Identify the highest volume materials (often these are diluents, such as water, air, a carrier gas, or a solvent), because these materials or diluents often control the investment and operating costs associated with end-of-pipe treatment of the waste streams. Determine the sources of these diluents within the process. Then develop pollution prevention options to reduce their volume or eliminate them.
4. If the compounds identified in Step 2 are successfully minimized or eliminated, identify the next set of compounds that has a large impact on investment and operating costs in end-of-pipe treatment. For example, if the aqueous waste stream was originally a hazardous waste and was incinerated, eliminating the hazardous compound(s) may permit the stream to be sent to the wastewater treatment facility. However, this may overload the biochemical oxygen demand (BOD) capacity of the existing wastewater treatment facility. If so, it may be necessary

The DuPont Commitment

Safety , Health and the Environment

We affirm to all our stakeholders, including our employees, customers, shareholders and the public, that we will conduct our business with respect and care for the environment. We will implement those strategies that build successful businesses and achieve the greatest benefit for all our stakeholders without compromising the ability of future generations to meet their needs.

We will continuously improve our practices in light of advances in technology and new understandings in safety, health and environmental science. We will make consistent, measurable progress in implementing this Commitment throughout our worldwide operations. DuPont supports the chemical industry's Responsible Care® and the oil industry's Strategies for Today's Environmental Partnership as key programs to achieve this Commitment.

Highest Standards of Performance, Business Excellence
We will adhere to the highest standards for the safe operation of facilities and the protection of our environment, our employees, our customers and the people of the communities in which we do business.

We will strengthen our businesses by making safety, health and environmental issues an integral part of all business activities and by continuously striving to align our businesses with public expectations.

Goal of Zero Injuries, Illnesses and Incidents
We believe that all injuries and occupational illnesses, as well as safety and environmental incidents, are preventable, and our goal for all of them is zero. We will promote off-the-job safety for our employees.

We will assess the environmental impact of each facility we propose to construct and will design, build, operate and maintain all our facilities and transportation equipment so they are safe and acceptable to local communities and protect the environment.

We will be prepared for emergencies and will provide leadership to assist our local communities to improve their emergency preparedness.

Goal of Zero Waste and Emissions
We will drive toward zero waste generation at the source. Materials will be reused and recycled to minimize the need for treatment or disposal and to conserve resources. Where waste is generated, it will be handled and disposed of safely and responsibly.

We will drive toward zero emissions, giving priority to those that may present the greatest potential risk to health or the environment.

Where past practices have created conditions that require correction, we will responsibly correct them.

Conservation of Energy and Natural Resources, Habitat Enhancement
We will excel in the efficient use of coal, oil, natural gas, water, minerals and other natural resources.

We will manage our land to enhance habitats for wildlife.

Continuously Improving Processes, Practices and Products
We will extract, make, use, handle, package, transport and dispose of our materials safely and in an environmentally responsible manner.

We will continuously analyze and improve our practices, processes and products to reduce their risk and impact through the product life cycle. We will develop new products and processes that have increasing margins of safety for both human health and the environment.

We will work with our suppliers, carriers, distributors and customers to achieve similar product stewardship, and we will provide information and assistance to support their efforts to do so.

Open and Public Discussion, Influence on Public Policy
We will promote open discussion with our stakeholders about the materials we make, use and transport and the impacts of our activities on their safety, health and environments.

We will build alliances with governments, policy makers, businesses and advocacy groups to develop sound policies, laws, regulations and practices that improve safety, health and the environment.

Management and Employee Commitment, Accountability
The Board of Directors, including the Chief Executive Officer, will be informed about pertinent safety, health and environmental issues and will ensure that policies are in place and actions taken to achieve this Commitment.

Compliance with this Commitment and applicable laws is the responsibility of every employee and contractor acting on our behalf and a condition of their employment or contract. Management in each business is responsible to educate, train and motivate employees to understand and comply with this Commitment and applicable laws.

We will deploy our resources, including research, development and capital, to meet this Commitment and will do so in a manner that strengthens our businesses.

We will measure and regularly report to the public our global progress in meeting this Commitment.

Figure 1-7. The DuPont Commitment to Safety, Health and the Environment

to identify options to reduce organic load in the aqueous waste stream.

1.5.2 Process Analysis

The manufacturing facility in Figure 1-2 represents a case where all of the materials added to or removed from the process are valuable to the business. Therefore, to help frame the problem for a real manufacturing facility, a process analysis should be completed.

For either a new or existing process, the following steps are taken:

1. List all raw materials reacting to salable products, any intermediates, and all salable products. This is "List 1."
2. List all other materials in the process, such as nonsalable byproducts, solvents, water, air, nitrogen, acids, and bases. This is "List 2."
3. For each compound in List 2, ask "How can I use a material from List 1 to do the same function of the compound in List 2?" or "How can I modify the process to eliminate the need for the material in List 2?"
4. For those materials in List 2 that are the result of producing nonsalable products (i.e., waste byproducts), ask "How can the chemistry or process be modified to minimize or eliminate the wastes (for example, 100% reaction selectivity to a desired product)?"

Analyzing the process in these ways and then applying fundamental engineering and chemistry practices will often result in a technology plan for driving toward a minimum waste-generation process. Other key ingredients for a successful pollution-prevention program are a proven methodology and the ingenuity of a savvy group of people to generate the options.

1.6 Scope of this Book

The chapters that follow present practical information on how to develop and implement a pollution-prevention program, how to identify pollution-prevention options, and how to evaluate which options are the best for the business. Chapters 2 and 3 present the methodology for developing and sustaining a pollution-prevention program on either a large or small scale. This methodology has been used on more than 35 processes in the DuPont Company over the last 8 years. Chapter 4 discusses the economics of pollution prevention—how to quickly quantify the incentive for pollution prevention using end-of-pipe treatment costs and lost-product value, and how to evaluate a viable set of pollution-prevention alternatives. Proven tools and techniques are presented, such as a short-cut method for calculating net present value and cost curves for various end-of-pipe treatment technologies. Next, Chapter 5 walks the reader through an actual DuPont case study of a chemical manufacturing process. Finally, Chapters 6 through 18 describe preferred pollution-prevention engineering technologies and practices for reducing waste generation and emissions. This unit operations approach to pollution prevention has been very valuable within DuPont for leveraging generalized practices across different businesses. In this way, pollution prevention moves from the intimidating world of process-specific case histories to the successful world of improved processes and products.

Literature Cited

1. U. S. Environmental Protection Agency. May 1992. *Facility Pollution Prevention Guide*. EPA/600/R-92/088. Washington, DC: U.S. EPA, Office of Research and Development.
2. Thurber, J., and P. Sherman. 1995. "Pollution Prevention Requirements in United States Environmental Laws." *Industrial Pollution Prevention Handbook*, H. M. Freeman Ed. :27–49. New York: McGraw-Hill.
3. Bringer, R. P. September 1989. "Pollution Prevention Program Saves Environment and Money." *Adhesives Age*, 32:33–36.
4. U.S. Environmental Protection Agency. November 1993. *DuPont Chambers Works Waste Minimization Project*. EPA/600/R-93/203. Washington, DC: U.S. EPA, Office of Research and Development.
5. Hart, S. L. January-February 1997. "Beyond Greening: Strategies for a Sustainable World." *Harvard Business Review*, 75(1):66–76.
6. Chemical Manufacturers Association. May 1993. *Designing Pollution Prevention into the Process: Research, Development and Engineering*. Chemical Manufacturers Association.

CHAPTER 2

The Path to Pollution Prevention

2.1 Introduction

The traditional approach to process design has been to first engineer the process and then to engineer the treatment and disposal of waste streams. With increasing regulatory and societal pressures to eliminate emissions to the environment, however, disposal and treatment costs have escalated exponentially. As a result, capital investment and operating costs for disposal and treatment have become a larger fraction of the total cost of any manufacturing process. For this reason, the *total system* must now be analyzed simultaneously (process plus treatment) to find the minimum economic option.

Experience in all industries teaches that processes that minimize waste generation at the source are the most economical. For existing plants, the problem is even more acute. Even so, experience has shown that waste generation in existing facilities can be significantly reduced (greater than 30% on average), while at the same time reducing operating costs and new capital investment.

In this chapter, we present a broad overview of the path to an effective pollution-prevention program. The phases and individual steps of this proven methodology are applicable to both large-scale and small-scale problems. In fact, many of the techniques and tools that support the methodology can be easily applied by chemists, process engineers, and project engineers to individual waste streams within a process or facility. In Chapter 3, we will discuss each of the steps in this methodology in much greater detail. The goal of this chapter is to simply establish the framework for the remainder of the book.

2.2 The Recipe for Success

After participating in over 75 waste-reduction or treatment programs, one thing has become clear to us—there is a recipe for success. We have found that successful pollution-prevention programs are characterized by the following four success factors:

1. Commitment by business leadership to support change and provide resources.
2. Early involvement of all stakeholders in the process.
3. Quick definition of the cost for end-of-pipe treatment, which subsequently becomes the incentive for more cost-effective pollution-prevention solutions.
4. Definition and implementation of pollution-prevention engineering practices and technologies that improve the business' bottom line.

The "path to pollution prevention" shown in Figure 2-1 brings together the essential ingredients for a successful pollution-prevention program, whether large or small. The core pollution-prevention program or methodology is shown in the *center column*, and consists of three phases: the chartering phase, assessment phase, and implementation phase. The other boxes in Figure 2-1 (shown with dashed lines) outline supporting information, tools, and activities that are essential to the success of the program. In many ways, they help to expedite the completion of the program and increase the likelihood of choosing the best options to improve the process and reduce waste generation.

The dashed boxes on the *righthand* side of Figure 2-1 show the information and tools available to help jump-start, maintain, and increase the effectiveness of the pollution-prevention program. These include:

1. How to quickly estimate the incentive for pollution prevention.
2. Generalized pollution-prevention technologies and

practices that apply across different industries.

3. A shortcut economic evaluation method to quickly screen the better options.

The *lefthand* side of Figure 2-1 describes two techniques to divide the waste-generation problem into smaller, comprehensible parts: a waste-stream analysis and a process analysis. These two analysis techniques are used to help better define the problem as well as to focus energy on the true source of the waste-generation problem. The first technique—waste-stream analysis—is based on the premise that most waste streams contain (1) a carrier, such as water or air, that drives end-of-pipe treatment costs, and (2) compound(s) or contaminants of concern that drive the need to treat the stream.

Meanwhile, the second technique—process analysis—is based on the assumption that most processes contain (1) valuable compounds and molecules that result in a salable product (i.e., products, intermediates to make the products, and raw materials to make the intermediates/products), and (2) other compounds that add to the cost of manufacturing, which includes waste treatment costs.

2.3 Program Elements

The path to pollution prevention shown in Figure 2-1 is applicable at all phases of a project. In most cases, the methodology has been applied at the plant level. However, the same methodology can be used when a process is first conceived in the laboratory and at periodic intervals through start-up and normal plant operation.[1]

2.3.1 Chartering Phase

This initial phase of the pollution-prevention program consists of four steps: securing business leadership support, establishing the program, selecting the waste streams, and creating a core assessment team.

Business Leadership Decision to Start. The decision to begin a pollution-prevention program can be triggered by one or more of the drivers listed below:

- Legal requirement, that is, state or federal regulations.
- Public image and societal expectations. This may be fueled by an adversarial attitude in the community toward the facility or process or the desire to lead the environmental movement instead of being pushed.
- Large incentive for reducing new capital investment in end-of-pipe treatment.
- Significant return by reducing manufacturing costs.
- Need to increase revenues from existing equipment.
- Corporate goal.

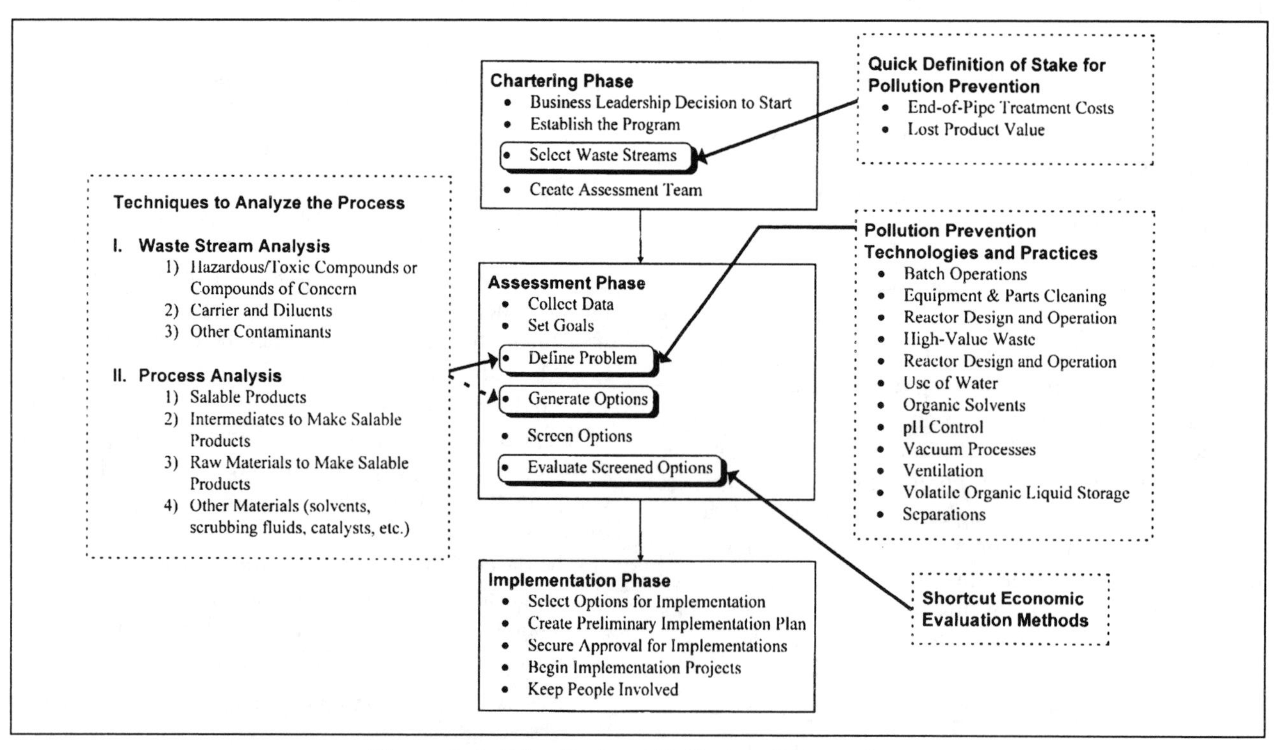

Figure 2-1. The Path to Pollution Prevention

Establishing the Program. This task helps prepare the plant or manufacturing area for a successful pollution-prevention effort. A key aspect of this task is to appoint a team leader for the program.

Selecting the Waste Streams. A typical process generates several major waste streams and many minor ones. The goal should be to select one or more of the major streams for the first round of waste assessments. If successful with these major streams, additional waste streams can be targeted, including minor ones, in a second round of assessments.

Creating an Assessment Team. In this step, a core team is selected that consists of 4–6 people who are best able to lead the program, perform the waste assessments, and implement the recommended process improvements.

2.3.2 Assessment Phase

The assessment phase in many ways represents the heart of the pollution-prevention program. It also tends to be where many engineers and scientists find the most enjoyment and personal satisfaction. For this reason, there is always a tendency to bypass the "softer" chartering phase and jump right into the assessment phase. This is a mistake. We consistently find that programs that bypass the chartering phase fail. This is because they fail to incorporate the first two major success factors listed in Section 2.2: obtaining commitment from business leadership to support change and provide resources, and seeking the early involvement of all stakeholders in the process. These two major success factors arise from the chartering phase itself.

The assessment phase consists of tasks that help the team to understand how the target waste streams are generated and how they can be reduced at the source or eliminated.

Collect Data. The amount of information to collect will depend on the complexity of the waste stream and the process that generates it. Material balances and process flow diagrams are a minimum requirement for most pollution-prevention assessments.

Set Goals. Setting goals helps the team to analyze the drivers for pollution prevention and to develop the criteria necessary to screen the options generated during the brainstorming session.

Define Problem. By defining the problem, the team begins to understand the targeted waste streams and the processes that generate them. The waste stream and process analysis techniques are used in this step to facilitate understanding of the problem.

Generate Options. When the team has developed a good understanding of the manufacturing process and the source and cause of each waste stream, they should convene to brainstorm for ideas.

Screen Options. In a separate meeting, the core assessment team will revisit the options generated during the brainstorming process to reduce the number of credible ideas carried forward.

Evaluate the Screened Options. More detailed engineering and economic evaluations are performed on the screened options to select the best option(s) to implement.

2.3.3 Implementation Phase

The goal of this phase is to turn the preferred options identified by the team into actual projects that reduce waste generation and emissions. Options are first selected for implementation. This should be a natural follow-up to the screening and evaluation stages described earlier. Next, the team needs to develop an implementation plan that includes resource requirements (both people and money) and a project time line. This is one of the reasons that having a project engineer on the core assessment team is valuable. Third, the team must secure approval and begin project implementation. Often, this step will be according to customary local practice. Finally, keep people involved throughout the entire pollution-prevention program. The team leader should always be working to build and maintain momentum.

2.4 The Incentive for Pollution Prevention

There are several ways to determine the incentive for pollution prevention. The choice will depend on your particular circumstances; that is, does a waste treatment or abatement system already exist or is a new treatment or abatement system required? Three approaches to determine the incentive for pollution prevention are described below. They are the incentive based on

- New end-of-pipe treatment;
- Raw material costs;
- Cost of manufacture.

Each of these approaches is discussed in more detail below.

2.4.1 New End-of-Pipe Treatment

Gaseous and aqueous waste streams often require capital investment for new facilities or an upgrade of existing equipment, for example, replacing an in-ground wastewater treatment basin with an above-ground treatment system in tanks. Solid wastes (both hazardous and nonhazardous) are normally handled with existing investment (e.g., site hazardous-waste incinerator) or shipped off-site for disposal. In the latter case, commercial disposal costs (including the cost of transportation) serve as the incentive for pollution prevention. A more complete discussion is given in Chapter 4, "Economics of Pollution Prevention."

Gas Streams. The major opportunity for savings is to reduce the flow of diluent or carrier gas (often air or nitrogen) at the source. For a gas stream containing both particulates and halogenated volatile organic compounds (VOCs), the minimum capital investment to abate this stream is $75 per standard cubic foot per minute (scfm) of waste gas flow.[2]

Wastewater Streams. Simply speaking, wastewater streams fall into one of two general categories: those that are biologically treatable and those requiring pretreatment or stand-alone nonbiological treatment (such as chemical oxidation, stripping, and adsorption). When treating dilute aqueous organic waste streams at the end of the pipe, consideration must be given to source reduction of both water flow and organic loading. Substantial reductions in capital investment can result by reducing water flow and contaminant loading at the source. The magnitude of these reductions will vary with technology type, hydraulic flow, and concentration; however, the *minimum incremental* capital investments for new treatment facilities are:

Biodegradable Aqueous Waste[3]

Incentive based on hydraulic flow:	$3000 per each additional gallon per minute (gpm).
Incentive based on organic loading:	$6000 per each additional pound organic per hour (lb/h).

Nonbiodegradable Aqueous Waste

Incentive based on hydraulic flow:	$1000 per each additional gpm.
Incentive based on organic loading:	Some technologies are sensitive to organic loading and some are not.

2.4.2 Raw-Materials Cost

Waste-stream composition and flow rate can be used to estimate the amount of raw materials lost as waste. The product of the amount lost to waste and the purchase price sets the incentive for pollution prevention in terms of raw-material cost alone.

2.4.3 Cost of Manufacture

The cost of manufacture includes all fixed and variable operating costs for the facility, including the cost for raw materials. The cost of manufacture should be cast in the form of dollars per pound ($/lb) of a key raw material. Another number that is readily available is the product selling price in $/lb product. Depending on the state of the business—excess capacity or sold out—one of these two numbers can be used to determine the incentive for pollution prevention.

- For a business operating with excess capacity, the product of the cost of manufacture ($/lb raw material) and the amount of raw material that goes to waste (either directly or as a byproduct of reaction) sets the incentive for pollution prevention.
- For a sold-out business, every additional pound of product can be sold; therefore, the product selling price multiplied by the additional amount of product that can be sold determines the incentive for pollution prevention.

2.5 Pollution-Prevention Engineering Technologies and Practices

There is a wealth of information available on pollution-prevention efforts and successes across many industries. However, this information has not been organized in a sufficiently generalized way so as to allow the transfer of knowledge from one type of industry to another. To help the practitioners of pollution prevention, this process- or industry-specific information has been transformed into generalized knowledge that can be more easily implemented by practicing engineers. The information has been organized in a unit operations or engineering and operating practices format. Chapters 6 through 18 present a number of these technologies and practices. When combined with a structured methodology and other chemical engineering tools, the use of these technologies and practices results in faster implementation of preferred pollution-prevention solutions. Remember, the goal is to change what is going on *inside* the pipes and vessels.

2.6 Engineering Evaluation of the Preferred Options

Engineering evaluation is the application of a full range of engineering skills to business decision making. It aids decision making by translating technical options into their economic impact—guidance that is fundamental to business decisions. The evaluation quickly focuses on only those data and analyses that are essential to quantify technical and economic feasibility. For each preferred option, the evaluation involves:

- Defining the commercial process;
- Flowsheeting;
- Analyzing the process;
- Defining manufacturing facilities;
- Estimating investment and manufacturing cost;
- Analyzing economics;
- Assessing risk.

The evaluation provides an objective view for decision making that is grounded in both engineering science and economics.

2.7 Waste Stream and Process Analyses

Section 1.5 introduced these two analysis techniques as the wisdom of pollution prevention: that is, properly defining and parsing the problem ultimately leads to the best pollution-prevention solutions. The goal is to frame the problem such that the pertinent questions arise. When the right questions are asked, the more feasible and practical solutions for pollution prevention will become obvious. These analysis techniques are presented again in detail in Chapter 3.

2.8 Case Studies

Four case studies are presented below that exemplify the role of the structured pollution-prevention program methodology, the value of quickly defining the incentive for pollution prevention using the cost of end-of-pipe treatment, and the benefits of using the waste stream and process analyses to parse the problem at hand.

2.8.1 Program Elements: U.S. EPA and DuPont Chambers Works Waste-Minimization Project

In May 1993, the U.S. EPA and DuPont completed a joint two-year project to identify waste-reduction options at the DuPont Chambers Works site in Deepwater, New Jersey.[1] As conceived, the project had three primary goals:

1. Identify methods for the actual reduction or prevention of pollution for specific chemical processes at the Chambers Works site.
2. Generate useful technical information about methodologies and technologies for reducing pollution that could help the U.S. EPA assist other companies implementing pollution-prevention/waste-minimization programs.
3. Evaluate and identify potentially useful refinements to the U.S. EPA and DuPont methodologies for analyzing and reducing pollution- and/or waste-generating activities.

The business leadership was initially reluctant to undertake the program, and was skeptical of the return to be gained when compared against the resources required. After completing a few of the projects, however, the business leadership realized that the methodology identified revenue-producing improvements with a minimum use of people resources and time, both of which were in short supply.

The pollution-prevention program assessed 15 manufacturing processes and attained the following results:

- A 52% reduction in waste generation.
- Total capital investment of $6,335,000.
- Savings and earnings amounting to $14,900,000 per year.

Clearly, this is a very attractive return on investment, while also cutting waste generation in half. No matter which methodology was used—the EPA's or DuPont's—the results were the same. The key to the site's success was following a structured methodology throughout the project and allowing their creative talents to shine through in a disciplined way.

2.8.2 Incentive for Pollution Prevention—Gas-Flow-Rate Reduction

A printing facility in Richmond, VA uses rotogravure printing presses to produce consumer-products packaging materials. Typical solvents used are toluene, isopropyl acetate, acetone, and methyl ethyl ketone. Driven by the U.S. EPA's new source performance standards for the surface-coating industry, the site installed a permanent total enclosure (PTE) around a new press so as to attain a 100% VOC capture efficiency. Leaks from the hot-air convection dryers and other fugitive emissions from the coating operation are captured in the press enclosure and routed, along with the dryer exhaust, to a carbon adsorber for recovery. Overall VOC removal efficiency for the enclosure and recovery system is greater than 95%. While many rotogravure press installations use the total pressroom as the enclosure, this facility was one of the first to install a separate, smaller enclosure around the new press. Notable features of the enclosure include

- Quick-opening access doors;
- A dryer that serves as part of the enclosure to minimize the enclosure size;
- VOC concentration monitors that control air flow to each dryer stage to maintain the dryers at 25% to 40% of the LEL;
- Damper controls that maintain a constant exhaust rate from the enclosure to ensure a slight vacuum within the enclosure.

If the pressroom had been used as the enclosure, the amount of ventilation air requiring treatment would have been close to 200,000 scfm. Instead, the use of the enclosure and the LEL monitors reduced the air flow to the adsorber to 48,000 scfm. This resulted in an investment savings for the carbon adsorber of approximately $5,000,000. The installed cost of the 1700 ft^2 enclosure was only $80,000 or $47/$ft^2$. Knowing the investment required to treat the entire 200,000 scfm provided a clear incentive for the business to reduce air flow at the source through segregation.

2.8.3 Waste Stream Analysis: Nonaqueous Cleaning

In a sold-out market, a DuPont intermediates process was operating at 56% of its peak capacity. The major cause of the rate limitation was identified as poor decanter operation. The decanter recovered a valuable catalyst, and its poor operation was caused by fouling from catalyst solids. Returning the process to high utility required a 20-day shutdown. During the shutdown, the vessel was pumped out and cleaned by water washing. The solids and hydrolyzed catalyst were then drummed and incinerated. A waste-stream analysis identified three cost factors: the volume of wastewater that had to be treated; the cost of the lost catalyst; and the incineration cost.

An analysis of the process and its ingredients indicated that the decanter could instead be bypassed and the process run at a reduced rate, while the decanter was cleaned. A process ingredient was used to clean the decanter, enabling recovery of the catalyst ($200,000 value). The use of the process ingredient in place of water cut the cleaning time in half, and that, along with continued running of the process, eliminated the need to buy the intermediate on the open market. The results were a 100% elimination of a hazardous waste (125,000 gallons per year) and an improved cash flow of $3,800,000 per year.

2.8.4 Process Analysis: Replace Solvent with a Process Intermediate, Product, or Feed

At a DuPont site, organic solvents used in the manufacture of an intermediate monomer were incinerated as a hazardous waste. These organic solvents were used to dissolve and add a polymerization inhibitor to the process. Alternative nonhazardous solvents were considered and rejected because they would not work in the existing equipment. With the help of process analysis techniques, however, the intermediate monomer itself was found to have the same dissolution capacity as the original organic solvents. As a result, the site replaced the organic solvents with the intermediate monomer. By utilizing existing equipment, realizing savings in solvent recovery, and reducing operating and incineration costs, the project achieved a 33% internal rate of return (IRR) and a 100% reduction in the use of the original solvents.

Literature Cited

1. U.S. Environmental Protection Agency. November 1993. *DuPont Chambers Works Waste Minimization Project.* EPA/600/R-93/203. Washington, DC: U.S. EPA, Office of Research and Development.
2. Dyer, J. A., and K. L. Mulholland. February 1994. "Toxic Air Emissions: What is the Full Cost to Your Business?" *Environmental Engineering: A Special Supplement to February 1994 Chemical Engineering,* 101 (Suppl.): 4–8.
3. Dyer, J. A., and W. C. Taylor. June 1994. "Waste Management: A Balanced Approach." *Proceedings* of the *Meeting and Exhibition of the Air and Waste Management Association*, Cincinnati, OH.

CHAPTER 3

Pollution Prevention Program Development

3.1 Introduction

This chapter addresses the detailed work process, hereafter referred to as the *methodology*, for implementing an effective pollution prevention program at a manufacturing site or within a business. The methodology is based on the learnings of a joint project team comprising people from the U.S. EPA and the DuPont Company's Chambers Works site;[1] however, it has been modified and expanded based on new knowledge and experiences from within the DuPont Company during the last 5 years.[2] In some ways, the path to pollution prevention presented in this book is similar to those methodologies developed by both the Chemical Manufacturers Association (CMA)[3] and U.S. EPA,[4] and can easily be modified to meet any specific requirements for your business.

Figure 2-1 introduced the path to pollution prevention. This methodology has been used on more than 35 programs within the DuPont Company. The resulting pollution-prevention projects have included not only administrative changes, such as better inventory control and improvements to housekeeping, but also basic process engineering and chemistry modifications, such as changes inside the pipes and vessels. We continue to find that it is only when one looks inside the pipes and vessels of a process that the most dramatic increases in productivity and largest reductions in waste generation and energy usage can be realized.

The heart of the pollution-prevention methodology is the brainstorming session. The process improvement ideas that are generated during this typically 4- to 6-hour session range from simple procedural changes to those requiring some level of engineering effort or, in some cases, research and development. In effect, this list of process improvement ideas describes the technology needs for the business over the next 5 to 10 years. It also identifies the technological limitations of the existing process.

In the past, environmental issues did not significantly influence technology selection, process and plant design, or the cost of new manufacturing facilities. The traditional approach to process design was to first engineer the process and then to engineer the treatment and disposal of waste streams. Designs of the past were based on minimum investment, low cost of manufacturing, and maximum product throughput. At the same time, engineers worked under the assumption that air was an infinite sink, water and solid wastes were someone else's problem, and energy costs were too low to warrant further consideration.

In stark contrast, today's new product and process designs are being increasingly influenced by ever-tightening environmental regulations and other societal pressures, such as green products and clean processes. For example, in anticipation of this trend, the DuPont Company developed its *Safety, Health, and the Environment Commitment* to drive toward zero waste generation and emissions (see Figure 1-7). This commitment presented DuPont businesses with a unique opportunity to simultaneously improve environmental performance, minimize nonreturn investment, increase production from existing facilities, and reduce variable costs, for example, by improving first-pass, first-quality yield.

3.2 Regulations

Toward the end of the 1980s, an increasing number of industries was using pollution prevention as a way to avoid installing expensive end-of-the-pipe treatment systems. While doing this, they found that they not only reduced waste generation, but they also made money! As a result of these experiences, governmental agencies began to incorporate pollution-prevention requirements into new environmental laws. Congress recognized that "source reduction is fundamentally different and more desirable than waste management and pollution control;" therefore, it ultimately passed the Pollution Prevention Act in 1990. Today, continued tightening of controls on emissions to all media in the environment (e.g., Water Sediment Quality Criteria for metals; limits on fine-particulate emissions from boilers, industrial furnaces and hazardous-waste incinerators; and maximum achievable control technology (MACT) standards for hazardous air pollutants) will continue to drive all industries toward solutions that involve source reduction, rather than end-of-the-pipe treatment, if they wish to remain competitive and profitable.

3.3 A Successful Pollution-Prevention Program

A good number of pollution-prevention methodologies have been published in the open literature.[2–8] However, many of them assume that a company or corporation is beginning at ground zero, that there is neither an established corporate culture nor visible upper-management support for pollution prevention. As a result, many methodologies prescribe activities that focus heavily on the development of an overall corporate program, and fail to place enough emphasis on where the waste reductions will actually occur—at the manufacturing-process level. The shortcomings of many pollution-prevention methodologies are:

1. Failure to consider the many activities that compete for people's attention in the process area, such as safety, maintenance, routine production, troubleshooting, start-ups, and shutdowns.
2. Failure to consider the limited time, capital, and people resources available to do pollution prevention.
3. The prescription of time-consuming and unnecessarily rigorous analytical methods for performing waste-assessment activities.

This chapter builds on the path to pollution prevention introduced in Chapter 2. It describes in more detail the methodology or work process for developing an effective, self-sustaining pollution-prevention program at the manufacturing-process level. The methodology is designed to be efficient, flexible, and conservative in its use of people resources.

3.4 Program Elements

The path to pollution prevention is shown again in Figure 3-1. It is important to understand that a pollution-prevention program can be instituted at any phase of a project; however, experience shows that the most successful programs are developed and implemented during the early stages of process development and conceptualization. Once the process chemistry is fixed and the process flow-

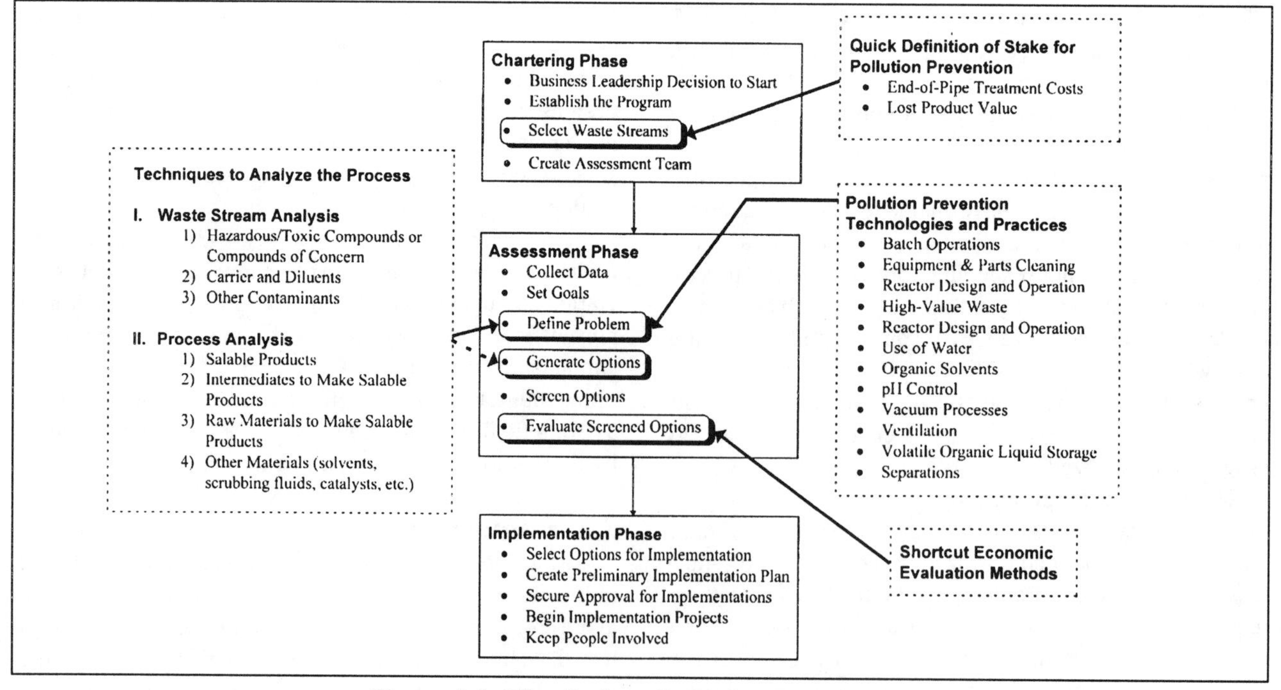

Figure 3.1. The Path to Pollution Prevention

sheet becomes cast in stone during the production design phase of a project, it becomes very difficult to significantly impact waste generation at the source. The engineers charged with minimizing emissions, then, are often left with nowhere to turn but end-of-pipe treatment.

For these reasons, we advocate that this same methodology be used in the beginning, when the process is first conceived in the laboratory; during the design phase of a project; and at periodic intervals through start-up and normal plant operation.[9] Each element or step of the pollution-prevention methodology shown in Figure 3-1 is described in detail below.

3.5 Chartering Phase

The goal of the chartering phase is to "set the table" for the pollution-prevention program. This phase consists of the following four steps:

1. Business leadership's decision to start.
2. Establishing the program.
3. Selecting the waste streams on which to focus the pollution-prevention efforts.
4. Creating the core assessment team.

The order of these four steps used to initiate the program may be different for each business or plant site. This is reflected in Figure 3-2, which shows the interactive and two-way relationship of these four steps. For example, in some situations business leadership may be committed to pollution prevention from the start. The chartering phase can then follow each step in the order just shown. At other times, it may be necessary to persuade business leadership that the advantages of pollution prevention are attractive enough to warrant people's time and other business resources. In these situations, the waste streams will need to be identified first (step 3), including an initial estimate of the incentive for waste reduction. With this information in hand, business leadership will then need to be sold on the true value of pollution prevention to business success (step 1). Once this is accomplished, the other two steps in the chartering phase—establishing the program and creating the core assessment team—can be completed.

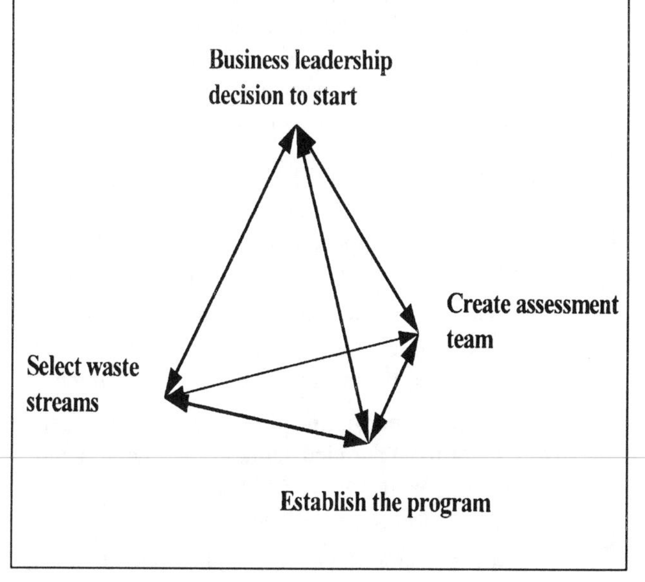

Figure 3-2.
Interactive Steps in the Chartering Phase

However, it is important to understand that the success of the entire pollution-prevention program hinges on the commitment of business leadership to support the program and provide resources. It is for this reason that "business leadership decision to start" is shown at the top of the pyramid in Figure 3-2.

3.5.1 Business Leadership Decision to Start

To reiterate what was stated in Chapter 2, the decision to start is often triggered by one or more of the following:

- A legal requirement, such as existing or impending state and federal environmental regulations.
- Corporate goals and societal expectations, such as the desire to lead the environmental movement instead of being pushed.
- A large incentive for reducing new capital investment in end-of-pipe treatment.
- A significant incentive in reducing manufacturing costs, which include the cost of waste treatment.
- The need for increased productivity and, hence, more revenue from existing equipment.
- Demands for greener processes and products from customers.

3.5.2 Establishing the Program

The goal of this step is to set the stage for a successful pollution-prevention program, and involves the recommended activities outlined below.

Convening a Meeting to Roll Out the Program. If pollution-prevention efforts in your plant or manufacturing area have become subordinated to other workday concerns, consider some sort of program rollout to highlight its upgraded importance. Convene a meeting of everyone in the area and present an overview of the pollution-prevention-program. Give the participants an idea of how they can play a role in the success of the program.

Appoint a Team Leader. Appoint someone to lead a multidisciplinary core assessment team. The team leader should be someone who is thoroughly familiar with the manufacturing process, has good leadership ability, and has cross-disciplinary experience and knowledge.

Establish Metrics. Establish measures to show progress. The simplest metric is to highlight the amount of waste *generated* per standard unit of production, for example, pounds of waste per pound of product. Another measure could be the cost of waste generation per pound of product. If air emissions are your focus, highlight gaseous wastes generated at the source. You may want to post more than one metric, but avoid having too many that distract people's attention from the main goal—to reduce total waste generated at the source. Track these metrics over time and communicate results to everyone in the manufacturing area.

Quickly Define the Initial Economic Incentive for Pollution Prevention. The operative word here is quickly. The task is to calculate and highlight for business leadership one or two high-spot incentives for instituting the pollution-prevention program. There are several ways to look at the incentive for pollution prevention. The choice will depend on your particular circumstances; that is, is an existing waste treatment or abatement system available or is a new treatment system required? The three most common incentives for pollution prevention are the investment savings realized by avoiding installation of new end-of-pipe treatment, the value of raw material losses alone, and the potential reduction in the cost of manufacture. Chapter 4 presents the techniques and tools needed to calculate these incentives.

Create Incentives for Pollution Prevention. Recognition awards are the simplest incentives. These can include cash awards, nights out on the town, and verbal recognition in front of peers. An even more effective approach is to base personnel evaluations on environmental performance in the same way that safety and financial performance affect compensation. In DuPont, corporate Environmental Excellence Award winners designate outside environmental groups or organizations to receive their cash awards.

Involve a Diverse Group of People. Incorporate a pollution-prevention mindset into the manufacturing area so that the program will survive the turnover of key players.

- Post metrics and the results of environmental audits on bulletin boards and in other visible places.
- Post relevant information from regulatory reports and note changes from report to report.
- Make environmental updates a regular agenda item for safety, state-of-the-business, and other areawide meetings.
- Keep people apprised of progress. Consider establishing a simple newsletter for the business or production area.

3.5.3 Selecting the Waste Streams

A typical process generates several major waste streams and many minor ones. Major can mean not only high volume or mass emissions, but also high toxicity, significant economic impact, and so on. You should select one or more of these major streams for the first round of waste assessments. If you are successful with these major streams, then you can target additional waste streams, including minor ones, in a second round of assessments. On the other hand, many minor streams offer opportunities for "quick hits" (i.e., projects that require little or no capital investment, offer a high probability of success, and are easy to implement). If you are having trouble building enthusiasm for the program, you may want to target some of these minor streams first to build confidence. This step in the chartering phase consists of the suggested activities described below.

Identify Area Waste Streams. Remember that minor waste streams, such as leaks from pumps, fugitive emissions, and maintenance wastes, often provide quick-hit opportunities. Do not spend time attempting to discover and document every possible fugitive emission or other nonsignificant waste stream from the process. Figure 3-3 is an example of a waste-stream selection form. This form or one similar to it can be used to document key information about each of the waste streams from the process, information that will be very valuable in subsequent steps and phases of the program. The waste-stream selection form contains key information about the source of the waste, how it is currently treated (if at all), composition and flow, and "cost" to the business in terms of dollars and cents and other measures, such as toxicity, odor, and color.

Prioritize Waste Streams. The core assessment team must ultimately decide on which waste streams to focus. This task ranks the waste streams based on one or more criteria. Some typical criteria or considerations include:

- If this waste stream is targeted, what is the likelihood of process improvements, such as increased yield, reduced cycle time, or improved product quality?
- What is the economic incentive for addressing this waste in terms of avoided capital investment in end-of-pipe treatment and lower cost of manufacture (see Chapter 4)?
- Is the waste disposed outside the plant boundaries in a landfill or in a commercial incinerator? Would the business prefer to increase control of its future by

FORM: Waste Stream Selection

Date: ____________________________

Process: ____________________________

Waste Stream ID: ____________________________

Waste Description: ____________________________

DISPOSITION
Is this stream a "quick hit"?
___Yes ___ No
If not check one:
___ Do assessment now
___ Do assessment later
___ Do nothing at this time
___ Info required

WASTE TREATMENT
Off-site Disposal? ___ Yes ___ No
Disposal Medium ___ Air Emission ___ Incineration
___ Landfill ___ Wastewater
___ Other____________________
Regulatory Exposure? ___ Yes ___ No

WASTE GENERATION
High Toxicity? ___ Yes ___ No
Special Safety Hazard? ___ Yes ___ No

Waste Compounds and Composition in Order of Importance

Stream	Compound	Wt., Vol., or Mol%	Waste Origin (Present Knowledge)
________	________	Carrier (> 80%)	________
(Annual Rate)	________	________	________
________	________	________	________
(Rate per Unit of Production	________	________	________
	________	________	________
	________	________	________

WASTE COST AND OTHER INFORMATION
Yield loss: ____________________
Treatment: ____________________
Packaging: ____________________
Transportation: ____________________
Additional Materials: ____________________
Taxes: ____________________
Cost of Manufacture: ____________________

Total Estimated Cost: ____________________

List nonquantifiable "soft-costs":
____
____
____
____
____
____

NOTICE: This form is intended for use as a preassessment worksheet. It is not a permanent record of waste generation or treatment. The information it contains may be speculative.

Figure 3-3. Waste Stream Selection Form

reducing its dependence on and the costs associated with outside resources to dispose of waste?

- What is the present or expected regulatory exposure to this waste stream?
- Is this a large-volume stream in terms of flow?
- Does the waste contain compounds that are highly toxic? Are carcinogens or suspected carcinogens present?
- Does the waste stream present a special safety hazard, such as flammability, reactivity, or volatility?
- Does the waste contribute additional "soft-costs" to the business, such as lack of greenness, poor public image, and future liabilities?

Select the Targeted Waste Streams. The number of waste streams selected will depend on the availability of resources. Choose a mix of major (at least one) and minor streams (such as quick hits). With the increased emphasis that, over the last decade, has been placed on the environmental performance of a process, the number of targeted waste streams will typically be 10 or less. The emphasis in this book is on wastes that are made by the chemical process itself, such as unwanted byproducts, contaminated water and solvent streams, cleaning wastes, and volatile organic compounds (VOCs).

3.5.4 Creating a Core Assessment Team

In creating a core assessment team, select the people who are best able to perform an assessment on a given waste stream. Each person should represent a different discipline or group in your area. If possible, try to include at least one member from outside of your area or plant site to provide a fresh, unbiased perspective.

The core assessment team's responsibilities will include (1) assessing which waste streams are to be considered for pollution prevention, (2) developing the incentive for pollution prevention, and (3) completing the assessment and implementation phases of the program. The team should include a total of four to six people. The team can be larger, but then it becomes difficult to build consensus and gather the team together at one time on an as-needed basis. Team members should represent a cross section with the following types of expertise:

- Process engineering
- Project engineering
- Research and development
- Technology specialization, such as in reaction engineering, separations, particle technology
- Operations
- Maintenance
- Marketing
- Material and logistics
- Production supervision
- Quality
- Business leadership
- Engineering evaluations
- Environmental regulations and control
- Supplier or customer representation
- Participation from outside the business and plant (central engineering function, another business, etc.)

The actual number on the core team will depend on the size of your process. Also, more than one area of expertise can be represented by a single person. Keep in mind that all areas of expertise may not be represented on the core team, but that is all right. A larger team will expand and contract around the core assessment team throughout the different stages of the pollution-prevention program. For example, during the assessment phase, the team will likely grow for the brainstorming session, then shrink again during the evaluation of screened ideas.

3.6 Assessment Phase

The assessment phase is the heart of the pollution-prevention program and consists of tasks that will help the core team to understand how the targeted waste streams are generated and how they can be reduced at the source or eliminated.

3.6.1 Collect Data

The amount of information to collect will depend on the complexity of the waste stream and the process that generates it. Material balances and process-flow diagrams are a minimum requirement for most pollution-prevention assessments. The three types of desirable information to collect are described below.

Process Flowsheets with Mass and Energy Balances. This is a minimum requirement and consists of two or more of the following:

- Process flow diagrams
- Spreadsheet material balances
- Chemical-process simulator output, for example, Aspen Technology, Inc.'s ASPEN PLUS®, or Simulation Sciences, Inc.'s PRO/II
- Operating manuals with process descriptions (particularly valuable for batch processes)
- Equipment specifications and vendor data sheets

- Piping and instrument diagrams
- Equipment layouts and logistics
- Plot and elevation plans

Process Chemistry. Process chemistry includes key physical and chemical properties for the raw materials, catalysts, byproducts, and products; definition of the chemical reactions (sequential or parallel), particularly noting if the desired product is an intermediate in the reaction sequence; acceptable reactor operating conditions; how reactants and products are added to and removed from the reactor; trace impurities; and equilibrium relationships.

Regulatory Background. The core team should address both current and anticipated future regulations and seek the input of an environmental professional if one is not on the core team. Types of basic information to collect include SARA Form R; permits and permit applications; waste audit reports; hazardous-waste reports; emission inventories; and waste shipment manifests.

Raw Materials/Production Information and Forecasts. This information and these forecasts can consist of product composition and batch sheets; material safety data sheets; product and raw material inventory records; operator data logs; operating procedures; and production schedules.

Accounting Information. Gather data on the fixed and variable costs for manufacturing the product. Be certain that they include and delineate costs for raw materials, energy, waste handling and treatment, off-site waste disposal, freshwater treatment, sewer, operations and maintenance. The team should also collect data on the selling price of the product.

Other Information. Examples of other information include a site vision, the business plan (e.g., current and future sales, projected process and product lives), financial requirements for the business (e.g., minimum internal rate of return for new capital, and capital budget), and organization charts. If available, it is also valuable to review competitive intelligence information, such as a competitor's process configuration, reported waste emissions, and environmental permits.

3.6.2 Set Goals

Setting goals helps the team to analyze the drivers for pollution prevention and to develop the criteria needed down the road to screen the options generated during the brainstorming session. Some examples of screening criteria include:

- Technical feasibility; that is, will the option work?
- Little or no capital investment is required.
- Minimal plant downtime is needed to implement the option; that is, fast turnaround time.
- Energy usage is lowered.
- Plant operability is improved.
- Waste generation and emissions are reduced.
- Operating costs are lowered.
- Regulatory objectives are met or exceeded.

3.6.3 Define the Problem

Defining the problem helps the core team to understand the targeted waste streams as well as the process steps that generate them. The following two activities are strongly recommended.

Perform an Area Inspection. Experience has shown that a plant tour is a good way to improve the core team's understanding of the processes and the pieces of equipment that generate waste. Area inspections are especially helpful to those team members who are from outside the process area. However, even insiders will benefit from area inspections. Conduct the inspection when the process is up and running, especially the process areas of concern. If there is more waste generated on a particular shift, consider performing several inspections on different shifts. Talk to the operations and maintenance people, check housekeeping, look for spills and leaks, and browse through operating procedures, reports, material purchases, and waste-collection procedures.

Perform Waste-Stream and Process Analyses. Section 1.5 introduced two techniques—waste-stream analysis and process analysis—to help divide the overall process and individual waste streams into their important parts. The goal of these two analysis techniques is to frame the problem, such that pertinent questions arise. Experience shows time and time again that when the right questions are asked, the more feasible, practical, and cost-effective solutions for pollution prevention become obvious. This is the wisdom of pollution prevention, that *properly defining and segmenting the problem ultimately leads to the best pollution-prevention solutions.* The approaches to these two analysis techniques are shown in detail in Figures 3-4 and 3-5. Analyzing the manufacturing process in this manner before and during the brainstorming session often results in an improved process that approaches zero waste generation and emissions.

3.6.4 Show Stoppers

If you determine that incoming materials are a significant cause of waste generation in your process, stop and consider whether to expand the scope of the pollution-prevention program to include an upstream process. If you are unable to determine the cause of a significant fraction of the waste, stop and consider whether any additional studies should be performed before proceeding any further with the assessment phase.

For example, a reaction/distillation process on a plant site generated a large waste stream consisting mostly of unrecovered product.[1] Extreme pH variability in one of the incoming raw materials to the process frustrated efforts to reduce the waste stream, by masking the true causes of waste generation. The only alternative available to the process engineers was to investigate changes in distillation control parameters such as temperature, pressure, and flow rate. These changes were never implemented, however, because their waste-reduction potential was small.

There was, however, another alternative. The process area decided to involve the upstream process area in their waste-reduction program. As a result of this cooperative approach, the upstream process area upgraded their pH control system and greatly reduced raw-material variability. It was not long after this change that the downstream reaction/distillation process area identified and implemented a number of worthwhile waste-reduction options in the reaction step, the main source of the waste. As a result, the downstream process area achieved a 60% reduction in waste generation resulting directly from looking at the upstream process. In other cases, you may find that you need to move upstream to the supplier of a key raw material to successfully attack the cause of waste generation.

3.6.5 Generate Options

When the team gains a good understanding of the process and the source and cause of each waste stream, the core team should convene a meeting to brainstorm for ideas. In order to generate *all* possible ideas for process improvement and waste reduction, the core assessment team will need to be supplemented with additional talents and diverse points of view.

In addition to generating new ideas, brainstorming sessions often prove successful in building enthusiasm for older ideas that were not pursued in the past (for whatever reason) by people close to the process and problem. Because ideas are not accepted or rejected during the brainstorming session, the environment encourages a creative and nonconfrontational exchange of ideas.

The goal of the brainstorming group is to identify technological changes to (1) reduce process waste generation and energy requirements, and (2) reduce the cost of manufacture. To conduct an effective technology brainstorming session, careful preparation is required by the leader, facilitator, and all group members.

There are four main steps in the brainstorming process:

1. Data collection and preparation of an information package for all participants.
2. Selection of team members
3. Information review
4. Brainstorming session

Data Collection and Information Package. The information required for the brainstorming session should be a subset of the total data collected in the "Collect Data" step above. The information package will be prepared by the core team for distribution to all brainstorming participants at least two weeks before the session. This package should include

- A sketch of the process
- Mass and energy balances of all the major process streams
- Identification of all waste streams, including available information on flow rate and composition
- Identification of the major unit operations
- Identification of the major energy consumers in the process
- Identification of any special hazards
- A general description of the chemistry involved in any of the reactors

Do *not* overwhelm the group with a large volume of information. However, the package needs to contain enough information that group members, who are not part of the core team, can come up to speed quickly.

Selection of Team Members. The composition of the brainstorming group will differ depending on what type of process is being analyzed. In general, however, the group ideally will consist of the core assessment team (four to six members), plus additional members possessing the skills required to have an effective brainstorming session. The size of the total group will typically range from a minimum of 8 to a maximum of 20 team members. Experience shows that 8 to 15 is the best size.

The brainstorming group should include people with the following skills or expertise:

- Business leadership and knowledge—Ideally, a business representative conveys upper-management

Waste Stream Analysis

1. List all components in the waste stream, along with any key parameters. For instance, for a wastewater stream these could be water, organic compounds, inorganic compounds (both dissolved and suspended), pH, etc. Use Figure 3-3 to help with this step.

2. Identify the compounds triggering the concern, e.g., compounds regulated under the Resource Conservation and Recovery Act (RCRA), hazardous air pollutants (HAPs), carcinogenic compounds, etc. Determine the sources of these compounds within the process. Then develop pollution prevention options to minimize or eliminate the generation of these compounds.

3. Identify the highest volume materials (often these are diluents, such as water, air, a carrier gas, or a solvent), because these materials or diluents often control the investment and operating costs associated with end-of-pipe treatment of the waste streams. Determine the sources of these diluents within the process. Then develop pollution prevention options to reduce their volume or eliminate them.

4. If the compounds identified in Step 2 are successfully minimized or eliminated, identify the next set of compounds that has a large impact on investment and operating cost (or both) in end-of-pipe treatment. For example, if the aqueous waste stream was originally a hazardous waste and was incinerated, eliminating the hazardous compound(s) may permit the stream to be sent to the wastewater treatment facility. However, this may overload the biochemical oxygen demand (BOD) capacity of the existing wastewater treatment facility. If so, it may be necessary to identify options to reduce organic load in the aqueous waste stream.

Figure 3-4. Waste Stream Analysis Technique

support for the program and emphasizes the importance of the group coming together. In many cases, the business representative will learn a great deal about the technology and manufacturing process, including its limitations.

- Process engineering—At least one process engineer from the business is required to play the role of process expert during the brainstorming session. Normally, this is a member of the core team. This expert will walk the group through the process flow diagrams explaining the *whys* and *hows* of the existing process. His/her main duty is to be the source of process knowledge, not to participate in the idea generation portion of the brainstorming session. One or more additional process engineers from the business should participate to share new ideas as well as other ideas that had been considered in the past, but were not implemented.
- Process chemistry—Ideally, someone from the business R&D group should participate to share both new and old approaches to the process chemistry. A significant cause of process waste is the reaction chemistry. This includes byproduct formation as well as the use of catalysts and solvents (as carriers) in the reaction steps.

Process Analysis

1. List all raw materials reacting to salable products, any intermediates, and all salable products. Use Figure 3-6 for this step. This is "List 1."
2. List all other materials in the process, such as nonsalable byproducts, solvents, water, air, nitrogen, acids, bases, and so on. Also use Figure 3-6 for this step. This is "List 2."
3. For each compound in List 2, ask "How can I use a material from List 1 to do the same function of the compound in List 2?" or "How can I modify the process to eliminate the need for the material in List 2?"
4. For those materials in List 2 that are the result of producing nonsalable products (i.e., waste byproducts), ask "How can the chemistry or process be modified to minimize or eliminate the wastes (for example, 100% reaction selectivity to a desired product)?"

Figure 3-5. Process Analysis Technique

- Environmental engineering—An environmental specialist is able to apprise the group of any regulatory impacts and to provide background on impacts on existing waste treatment operations.
- Chemical separations—Most waste streams consist of a high-volume carrier (such as water, organic solvents, air, or nitrogen) containing low levels of contaminants. Techniques for reducing the volume of the carrier or the contaminant level will depend on the type of separation technology being used.
- Energy conservation—If the process is a large energy consumer, then an energy specialist can suggest ways of improving energy usage within the process.
- Engineering evaluations—An engineering evaluator will serve as the primary coordinator of the process and economic information for each of the waste-reduction alternatives. The evaluator's duties are to facilitate the screening and evaluation of the ideas generated during the brainstorming session. This person may or may not be the lead process engineer from the business or the leader of the core team; however, the evaluator should be a member of the core team.
- Process hazards—If there are any special hazards associated with the process, then process hazards expertise will be required.
- Operations and maintenance—A lead operator and maintenance person are familiar with the day-to-day sources of waste, how operating procedures affect the amount of waste, and which waste streams are generated during start-up, shutdown, and maintenance activities.
- Wildcard or outside experts—For effective brainstorming, at least two outside experts or wildcards are needed (usually an engineer and a chemist). The duties of these experts are to provide a broader experience base for the brainstorming group, question the technology and practices of the present process, educate the group and core team on other ways of making the product, and act as a catalyst to extract the best ideas residing within the minds of the group members.

What do the outside experts or wildcards bring to the table? They help everyone in the brainstorming group to think outside of their own "box." One often finds that for a particular product all competitors have essentially the same basic approach and technology for manufacturing that product. The outside experts can bring a different perspective on how to approach any particular task; therefore,

FORM: Process Constituents and Sources

Process: ____________________ Date: ____________________

List 1

Constituent: Source (Feed, Reactor, Unit Operation):

Salable Products

Intermediates (result in salable products)

Essential Raw Materials(only those constituents used to produce the intermediates and salable products)

List 2

Constituent: Source (Feed, Reactor, Unit Operation):

Other Materials (Nonsalable byproducts, solvents, water, air, nitrogen, acids, etc.)

Figure 3-6. Process Constituents and Sources Form for Use with the Process Analysis Technique

they offer the opportunity for better and newer ways to be considered. For example, assume that you are in the business of making specialty chemicals using batch reactors, and your main source of waste is off-spec product from product changeovers. In this situation, outside expertise in batch operations from the food, pharmaceutical, or agricultural products industries could bring a different perspective on ways to improve the present operation of your process.

Not everyone you invite will be comfortable with and effective at brainstorming. During the sessions, anywhere from 100 to 200 ideas will be generated over a 2- to 6-hour period. This is a lot of ideas! To maximize the contribution of the group, each individual group member needs to follow three ground rules: participate, be concise, and be additive.

First, each individual invited to the session is expected to participate—silence is unacceptable. All participants need to understand that their ideas, no matter how off-the-wall they may sound at first, will not be judged during the brainstorming portion of the meeting. The judgment and critique will be reserved for the screening portion of the session. The concept behind brainstorming is that one idea should lead to a new idea or build on the previous one. If a person does not speak up, then that individual's ingenuity is not being fully exercised.

Second, participants need to be concise. Ideas must be conveyed clearly and completely. Answer any questions on the meaning of your idea, but do not engineer the idea. There just is not enough time during the brainstorming portion of the meeting to engineer every idea. In addition, it will restrict the flow of new ideas.

Third, participants should be additive and avoid critiquing other people's ideas. Sometimes an idea that was tried in the past and failed for either technological or political reasons will work in the current climate. Also keep in mind that all ideas will be reexamined at a later date. The goal of the brainstorming session is to get all possible ideas on the table, so that the best idea can be evaluated and chosen during the screening and evaluation stages of the assessment phase.

Information Analysis. To be effective in a brainstorming session, the participants must study the information package before coming to the session. Because 100 to 200 ideas will be generated in a short time period, an unprepared participant will contribute less than a well-prepared participant. One proven approach to aid in your preparation is to use the waste-stream and process analysis techniques described in Figures 3-4 and 3-5. If you are able to ask the right questions, then the best pollution-prevention solutions will almost always become obvious.

In preparing for the session, participants should also consider the list of questions shown in Figure 3-7. These questions are only intended to trigger the creative thought process of each participant as they review the information. Each participant is expected to bring a different perspective to all ideas that are generated. This is why it is so important to pick the right mix of people for the session. The concept is to have everyone be aware of the interdependency of any one idea on the whole, and how that idea can impact other ideas.

The Brainstorming Session. Brainstorming uses the synergy of group dynamics, coupled with the ingenuity of each participant, to generate a set of ideas superior to what could have been achieved by each person working alone.[10] To be effective, therefore, a brainstorming session must be well structured and "friendly" at the same time. According to the playwright Eugène Ionesco, "it is not the answer that illuminates, it is the question." With the proper atmosphere, questioning will generate a positive rather than a defensive response.

Logistics. In the information package sent to the participants, a cover letter should be included describing the drivers, purpose, expected products, participant responsibilities, and agenda.

Drivers. Normally, both external and internal business drivers exist for doing pollution prevention. This was discussed in Section 1.4. The premier external driver is usually regulatory, which means that there will likely be a significant incentive to avoid investment in new or upgraded end-of-pipe treatment. Using the techniques discussed in Chapter 4, the incentive for pollution prevention can quickly be determined. Other external drivers include public opinion, desired greenness of the product or process, proximity of the facility to the residential community, and market share.

Internal business drivers can include corporate or business goals and vision, realization of the cost of waste generation to the business, and the need to increase capacity in existing facilities while avoiding permitting barriers to growth.

Purpose and products. The purpose and products statement outlines the "why, how, and what" of the meeting, together with the expected products of the brainstorming session itself. An example purpose and products statement is shown in Figure 3-8.

Participant responsibilities. A short discussion of participant responsibilities should be included in the cover letter that accompanies the information package. These respon-

For the process engineer:

- What is the life of the present process? The present product?
- What is the competition doing that the group should know about?

For the chemist:

- What are the principal factors affecting yield, conversion, and selectivity?
- If the reaction is reversible, can byproducts be back-reacted to the incoming raw materials or converted to other useable products?
- If nonsalable products are homologues of the reactants or intermediates, how can they be converted and recycled?
- What other catalysts are possible?
- If excess reactants or inerts are being used, ask why?
- If air, water, or a solvent are being used, ask why?

For the separations specialist:

- If an exit gas or water stream is being generated, what other separation techniques could be used to eliminate the stream?
- For trace levels of contaminants, how can the separation unit operations be improved?
- If large amounts of energy are required, what other separation technologies are applicable?
- If significant heating, followed by cooling, and then reheating takes place, what other combinations of unit operations can be used to minimize energy usage?

For the environmental specialist:

- What are the hazardous, carcinogenic, or toxic materials in the waste and product streams that require or could require further treatment?
- What are the present and future (5-10 years out) environmental laws that impact the waste from this process?
- What end-of-pipe technologies are appropriate?

For the engineering evaluator:

- For the current waste streams, what are the end-of-pipe treatment costs?
- What is the cost of waste generation for the current process?

For the energy specialist:

- What are the opportunities to save energy in the process?
- What are the process-to-process energy exchange opportunities?
- What are the corporate energy goals?

For the lead operator and maintenance representative:

- What operating procedures are outdated or not followed?
- How does misoperation affect waste generation?
- How can startup, shutdown, and maintenance wastes be reduced?

Figure 3-7. Questions to Consider in Preparation for the Brainstorming Session

sibilities should also be reviewed at the beginning of the brainstorming meeting. The key responsibilities are to participate, be concise, be additive, and to participate. These were discussed in detail earlier in the subsections on selection of team members and information analysis.

Agenda. A typical brainstorming session will last 1 1/2 days beginning around noon the first day. During the afternoon of the first day, the purpose, products, drivers, and ground rules are reviewed, business leadership provides a business overview, and the lead process engineer introduces the manufacturing process using process flow diagrams and, when appropriate, a plant tour. The process overview should emphasize unusual process conditions that affect waste generation and highlight each waste stream of interest, including how it is generated. The group must finish the first day with a good understanding of how the process currently operates. The goal is to lay the groundwork for the generation of ideas the next morning.

The second day is a full day. It begins bright and early with a quick review of the process, followed immediately by the brainstorming session. This period will usually last 4 to 6 hours. After all ideas have been generated and recorded, the group will develop or review (preferred) two or three screening criteria that they will use to perform a first-cut ranking of the ideas. Ideally, the idea-generation and ranking periods should be roughly the same length of time; however, experience has shown that the latter can usually be accomplished in 1 to 2 hours. Quite frankly, we find that because everyone is so exhausted from the idea-generation stage, they are anxious to go home when it comes time to rank the options. For this reason, it is important to prepare the screening criteria ahead of time and to simply seek the group's buy-in before the ranking. A typical agenda is shown in Figure 3-9.

PURPOSE

To identify ideas that reduce waste generation.

WE WOULD LIKE TO:

- Take advantage of your perspective and expertise.
- Identify all ideas that reduce the amount of waste generated.
- Use the synergy of a brainstorming session to identify cost-effective ideas.
- Address any barriers and concerns.

BENEFITS:

The business can realize the desired emissions and energy reductions and improve the operation and utility of the process with maximum return to the stockholders.

PRODUCTS

- Identification of the changes (technical and operational) required to improve operation of the process.
- A prioritized list of opportunities and recommendations to be considered by the business and manufacturing site to reduce waste generation.
- A path forward with responsibilities assigned.

Figure 3-8. Typical Purpose and Products Statement for a Brainstorming Session

Ground rules. The ground rules are designed to help control the atmosphere in the room during the session. The ground rules are an extension of the participant responsibilities and will be agreed upon by the participants at the beginning of the meeting. They should be posted in the meeting room and used to facilitate an effective and productive session. Some example ground rules are shown in Figure 3-10.

Recording of ideas. During the generation of ideas, someone will need to record or scribe each idea in 1–2 lines on a flip chart. Some ideas, however, will require more text to completely document the basis and ramifications of the idea. To facilitate later understanding, a sample option generation form is shown in Figure 3-11. One person (preferably not a key participant) should be assigned to keep notes on this form during the brainstorming session to help augment the limited notes on the flip charts.

Meeting dynamics. Three or four people are needed to lead and guide the brainstorming session so that it is successful. A key person is the facilitator—the person who will lead the meeting and whose tasks are to (1) adhere to the agenda, (2) be sure the ground rules are followed, (3) keep the session open and friendly, and (4) support the scribes, process expert(s), and core assessment team leader. For groups with little experience in brainstorming, the facilitator may want to introduce the group to the concept of open-ended thinking and the synergy between group dynamics and idea generation. Figure 3-12 relates how the main catalysts for generating ideas are the questions. For example, *how* does that particular process step function? *Why* does the process step not function perfectly? *How* can the process step or flow streams be modified to improve its operation and minimize waste generation?

The process expert(s) provide the process knowledge that enables the group to understand each aspect of the process, help interpret ideas as they are being generated,

Afternoon of Day 1

Introductions

Review Agenda, Purpose and Products, Ground Rules, Drivers

Business and Environmental Overview

Process Overview and Plant Tour

Day 2

Idea Generation

Develop/Review Ranking Criteria

Rank Ideas

Develop Path Forward

Figure 3-9.
Typical Agenda for a Brainstorming Session

and make sure that no part of the process is overlooked. They can use one of two techniques to help focus the idea generation. One approach is to divide the process flow diagram into smaller blocks during the brainstorming. This allows the participants to focus their idea generation on a particular area or unit operation of the process. A second approach is to focus on a particular waste stream, and then work back through the process tracing the generation of that waste. The goal is to focus on those process steps or chemistries that lead to the generation of unwanted materials in that stream. For example, you can trace the nitrogen in a contaminated air stream from a scrubber back to the partial oxidation reactor where air was used as the source of oxygen.

Lastly, you will need to appoint one, and maybe two, scribes. The scribes are the unsung heros of the meeting. Their duties are to (1) record each idea in a concise, yet complete, way; (2) check with the generator of the idea to make certain that the text description is accurate; (3) avoid editing the idea; and (4) number each idea and page and record the initials of the person who provided the idea. The scribes will also want to group the ideas (if possible) by process area or equipment piece.

3.6.6 Screening the Options

Once the idea-generation phase is complete, the ideas need to be reexamined and ranked based on a set of criteria. The goal in this step is to perform a first-cut screening of the options. Ideally, the number of ideas that are considered to be worthy of further evaluation will be cut in half. The first-cut screening is completed during the brainstorming meeting because not all the participants are members of the core assessment team. The benefits of this are that (1) the ideas are fresh in everyone's mind, (2) those participants who generated the ideas are in the room and can add clarifications, if needed, (3) the discussions that occurred during the idea generation permit a "gut feeling" of how each idea should be ranked relative to the others, and (4) you may never assemble this much expertise in one room at the same time again. Remember that no idea is ever permanently discarded, and the core assessment team has the responsibility of recording and considering all ideas. However, the first-cut screening helps the core team to focus on the more attractive ideas (at least according to the group).

Normally, 50 to 200 ideas will result from the brainstorming. This is far too many for limited business resources to digest and evaluate. During the first-cut screening, a number of ideas will be combined, duplicates eliminated, and a fair number screened out. To accomplish this task, the group needs to agree on the criteria;

Participate

All ideas are good ideas.

Stay Focused

Keep the business needs and purpose of the brainstorming session in mind.

Build on Ideas

Use other people's ideas to create synergy.

Be Polite

Listen to understand; the person talking has the floor.

Be Positive

Work to sharpen the ideas being generated.

Figure 3-10.
Typical Ground Rules for a Brainstorming Session

FORM: OPTION DESCRIPTION **Date:** ____________________

Part 1: Option Generation

Option No.: ____________

Option Description: ______________________________________

____

Category: ______________
(new tech., equipment, oper. procedures, etc.)

Reduction Type: ______________
(source reduction, improved recovery, recycling, improved waste handling, etc.)

TSCA Status: ___No new chemicals or intermediates ___New chemicals, on TSCA inventory ___ New chemicals, not on TSCA inventory or status unknown

Notes: ______________________________________

____

____

____

____

____

____

PART 2: OPTION SCREENING

Status: ______________ Comment: ______________

Finalist? ___ Yes ___ No
(Perform technical & economic evaluation?)

More study required? ___ Yes ___ No

Figure 3-11. Option Description Form

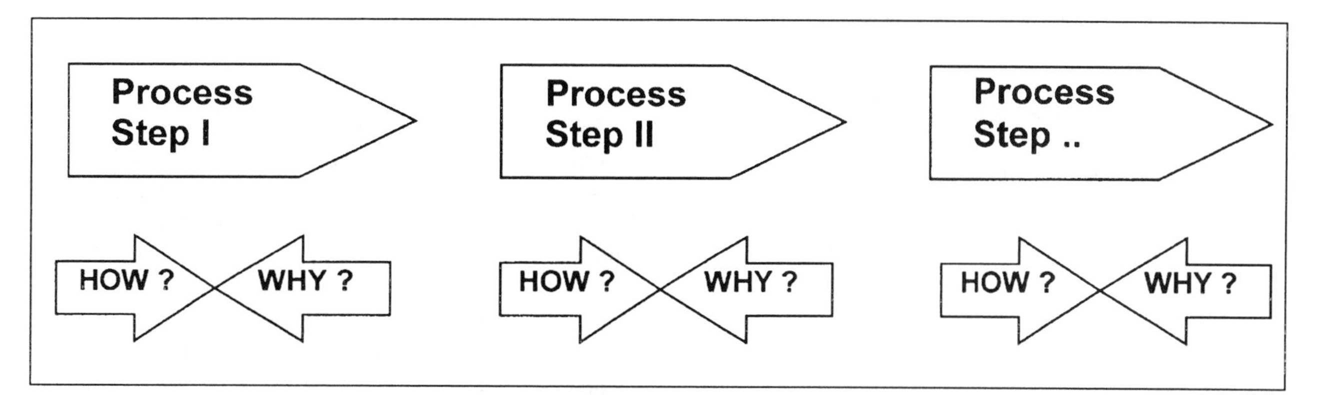

Figure 3-12. "How-Why" Questions

typically no more than three criteria are used. The criteria will be developed by the core team during the "Set Goals" step of the Assessment Phase, which was discussed in Section 3.6.2.

Avoid using more than three criteria. Because everyone will be tired by the end of the second day, they will be unable to process more than three criteria anyway. Keep it simple. Some typical methods that have been used with success are to mark each idea with

- Yes or no
- Keep or kill
- High, medium, or low

Another approach is to divide the number of ideas by 10: 120 ideas ÷ 10 = 12. Each participant would then vote on their top 12 ideas using self-adhesive dots. For smaller problems where the total number of ideas is less than about 30, the high/medium/low approach can be applied using Figure 3-13. Each idea would be ranked against two criteria (for example, technical and economic feasibility) and placed in one of the six blocks on the matrix. Ideas clustered toward the upper-right of the matrix would be carried forward into the evaluation step.

For large numbers of ideas (50 or more), it will make sense for the core team to have a second-cut screening after the brainstorming session. In this case, use the simple ranking methods listed earlier; then use Figure 3-13 for a second-cut screening of options by the core team after the brainstorming session. The goal is to further reduce the number of "feasible" options that will be carried forward and evaluated in more detail during the next step of the assessment phase—evaluate screened options. Experience shows that about 10–20% of the original ideas make it through the second-cut screening.

The path forward from the brainstorming meeting normally lists the next steps for the core team and assigns responsibilities. Everyone who participated in the meeting is kept informed of any progress on a formal or informal basis. In addition, as the best ideas are developed further by the core team during the evaluation step, brainstorming participants not on the core team can be used as technical resources.

3.6.7 Evaluate the Screened Options

In this step of the pollution-prevention program, a more detailed and thorough technical and economic evaluation of the screened ideas will be undertaken. The first two levels of screening described in the previous section minimize the number of feasible alternatives that need to be evaluated in detail. In this step, we are no longer talking about "gut feeling" or qualitative assessments. In general, revised mass and energy balances, process flow diagrams, and operating requirements will be generated for each screened idea. Based on these inputs, the capital investment, cash operating cost, and net present value (or similar economic measure of merit) for each option will then be determined and compared against the existing process. Because of the level of effort that is required to evaluate each of the screened options in more detail, the first- and second-cut screening steps become very important in minimizing the resources needed to complete the evaluation step.

At a minimum, the screened ideas will need to pass two criteria to be considered for implementation: technical and economic feasibility. First, will they work? Second, do they provide a cost savings or positive return (i.e., positive net present value). A typical checklist for technical feasibility follows:

- Will the alternative actually function as predicted?
- Does a similar system currently exist on-site? Within the company? In the industry?
- Is R&D or on-line testing required?
- How much will the option reduce waste generation or toxicity?
- Is space available in the process area for new equipment?

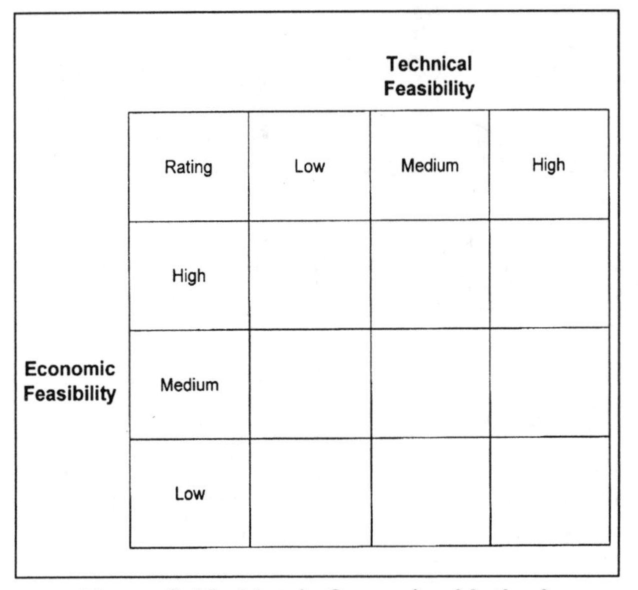

Figure 3-13. Matrix Screening Method

- Will the option affect production rates? Plant utility? Product quality?
- How rapidly can the option be implemented?
- Does implementation require a shutdown? For how long?
- Is additional operations labor and skill required? Is training needed?
- Is new design required? Can the design be done in-house?
- If the equipment is vendor-supplied, does the vendor have sufficient support capability?
- Does site engineering, production, and maintenance agree with the option?
- Are new permits required? How about regulatory requirements? Is a TSCA review needed? Do waste releases change?
- How about safety? Occupational health? Community welfare? Public relations?

The tools and techniques to complete the economic evaluation are discussed in detail in Chapter 4. Assuming the screened option is technically feasible, it will then be ranked based on an economic measure of merit, such as net present value or net present cost. Key inputs to the calculation of net present value or cost are the new investment and all cash operating costs. In reality, we find that the business will look at both initial capital investment and net present value in making its decision. This is because of the artificially high value placed on capital dollars in companies with limited capital budgets. For example, in many corporations, the hurdle rate for new capital is 25–50% internal rate of return (IRR) versus a cost of capital of only 10–12%.

As the core team evaluates and ranks the top ideas, further categorization often occurs. Either as part of the second-cut screening or as part of the evaluation step, the ideas can be further subdivided into and ranked within the following categories:

Do Now. The benefits of options in this category are obvious, little or no capital investment is required, and/or the change can be quickly implemented. Examples include procedural changes, such as the number and duration of equipment washes, setpoint changes on equipment controls, and additional sampling.

Do Later. For alternatives in this category, capital investment is required. The economic evaluation will determine the relative economic merit of each alternative in this category.

More Knowledge Needed. Ideas in this category require a small amount of laboratory work or an engineering assessment to determine both the technical and economic viability of the idea.

R&D. The alternative is a valuable concept; however, more in-depth research or engineering development work is needed. In some instances, this could lead to an entirely new process or chemistry.

Impractical. The idea does not appear technically feasible, or the business resources are insufficient to evaluate the idea further.

What happens if the evaluated options are considered technically feasible and meet the economic criteria for the business, but a consensus still cannot be achieved within the core team? Then the core team can turn to a weighted-sum rating and ranking technique, as shown in Figure 3-14. As a group, the core team would:

- Determine which criteria are important to the waste reduction effort (e.g., reduction in waste quantity, low capital cost, impact upon safety, probability of success).
- Indicate the relative importance of each criterion by assigning to it a *weight* ranging from 1 (low importance) to 10 (high importance).
- For each option, assign a *rating* from 0 to 10 for each of the criteria.

- For each option, multiply the *rating* by the *weighting* (RxW) for each criterion.
- For each option, sum all the RxWs to yield that option's overall score.
- Rank all the options according to their overall scores.

In the example shown in Figure 3-14, the high capital investment required for Options 1 and 2 (together with the large weighting placed on capital cost) led to their low scores, despite their high waste-reduction potential. Conversely, the low capital cost and ease of implementation of Options 3 and 4 made them the most promising alternatives, despite the more modest waste reductions they offer.

The value of this method is that it forces a careful consideration of each option. If, however, the options are few and well understood, then the use of the weighted-sum rating and ranking is unnecessary. In addition, the numbers will not overcome biases within the room. Despite all this effort, some options will still require further study, whether that be additional testing, engineering evaluations, or even decision and risk analysis.

3.7 Implementation Phase

The tasks listed in Figure 3-1 will turn the options identified by the core team into the projects that will actually accomplish the waste reductions. This is a very important phase of the pollution-prevention program, and in some ways, is also the most plant-, business-, or company-specific. This is one good reason to have a project engineer on the core assessment team.

3.7.1 Select Options for Implementation

The team must determine which of the preferred options will be selected for implementation. This task consists of the following recommended activities.

Achieve Team Consensus. Selection of preferred options for implementation will, of course, be based upon the results of the economic and technical evaluations or upon some compelling soft cost. It is nevertheless worthwhile to convene a final meeting of the core assessment team to review the economic and technical evaluations and to validate their underlying assumptions. Achieving consensus on the selected options will be the final act of this core assessment team.

Create the Report. If a proper paper trail is maintained throughout the entire program, then preparing the final report should be easy. This task entails gathering the recorded results of each of the steps in the chartering and assessment phases and developing an executive summary that discusses the ideas recommended for implementation. The final report should include the

- Waste-stream selection forms
- Initial development of the incentive for pollution prevention
- Discussion of the drivers for the program
- Information package sent to the brainstorming participants
- Ideas generated during the brainstorming session, ranked from high to low
- All technical and economic evaluations for ideas considered further by the core team
- Results of any sampling or other studies that were conducted as part of the program
- Implementation plan for those ideas that the core team recommends for implementation

Once the final report has been written and reviewed, it should be issued to all the team members, key plant personnel, the business team, and all appropriate liaisons. The report becomes not only a record of the effort, but is also the starting point for the next pollution-prevention exercise. With the rapid changes that can occur in technology and the business climate, pollution-prevention brainstorming sessions should usually be conducted every 3 to 5 years.

3.7.2 Create Preliminary Implementation Plan

The implementation plan should contain the following key components:

- R&D expenditures
- Capital expenditures for the projects
- Resources to complete the projects
- An outline of the required tasks
- A project time line

When identifying resources for implementing the projects, try to choose resources that are as close to the process as possible. Do not ask an engineer to do what an empowered operator can do. Do not solicit extradepartmental resources for a job that the area technical professional can handle.

Experience shows that a well-motivated production area can be remarkably self-reliant. For example, one DuPont process completely eliminated a waste stream of solvent used to wash out the process equipment during product changeovers.[1] The preferred approach for eliminating the stream—installing flanges at low points on the equipment to allow manual draining—was suggested by an operator on the core assessment team. The decision was made to allow the operator to lead a group of other operators and mechanics to plan and implement the change. The group located the equipment low points, installed the flanges, and designed and built a hand-operated cart for collecting the drainage. Not only did their efforts eliminate the waste, they significantly reduced the cycle time between product changeovers.

3.7.3 Secure Approval for Implementation and Begin Implementation Projects

The required authorizations to begin implementation need to be sought. In many respects, the approval and initiation of a project are very site-specific. Our recommendation is to work with the plant project group to secure the necessary approvals and to initiate the projects. This will vary from site to site and company to company. The key, though, is to build the good business case for each of the projects using the steps outlined earlier and to proceed according to customary local practice.

3.7.4 Keep People Involved

Successful pollution-prevention programs also keep people involved beyond the generation and implementation of ideas. Throughout the various program phases, the core team leader should be working to build and maintain momentum for the pollution-prevention program. This task consists of keeping people apprised of the project status, evaluating performance against the original goals, institutionalizing the changes, involving the same people in future pollution-prevention efforts, revisiting attractive options that did not make the final cut, and celebrating success.

3.8 Resources

In many respects, the best set of resources for generating waste-reduction ideas consists of your own people. However, you will sometimes need to provide other expertise to supplement your own resources. Some examples of other resources you will need to supply include a brainstorming facilitator, technical specialists, outsiders or wildcards, and sources of pollution-prevention ideas found in the literature.

If you cannot find a person in your business who can facilitate a brainstorming session, then you will need to hire a consultant or check with your local university or college.

Criteria	Weight (W)	Option #1		Option #2		Option #3		Option #4		Option #5	
		Replace Reactor		Install DCS		Optimize Product Transitions		Loosen Product Specs		Increase Column Pressure	
		R	RXW	R	RXW	R	RXW	R	RXW	R	RXW
Safety	10	10	100	10	100	8	80	8	80	4	40
Capital Cost	10	0	0	4	40	10	100	10	100	8	80
Ease of Implementation	8	2	16	5	40	9	72	10	80	7	56
Probability of Success	10	6	60	8	80	8	80	8	80	4	40
Effect on Quality	6	10	60	10	60	6	36	6	36	6	36
Waste Reduction Potential	6	8	48	8	48	4	24	5	30	2	12
Final Evaluation	Sum of RXW		284		368		392		406		264
	Option Ranking		4		3		2		1		5

Figure 3-14. Weighted-Sum Rating and Ranking Technique

The outsiders or wildcards should be good chemical engineering and process chemistry generalists, and not directly associated with the process. The technology specialists should be skilled in the engineering unit operations or technology areas that are most critical to waste generation in the manufacturing process, such as drying, particle technology, reaction engineering, and pumps. Most midsize to large companies can identify the outsiders, wildcards, and technology specialists internally. For smaller firms, sources of wildcards and technology specialists include academe, research institutes and engineering consultants.

There is a large body of literature and case histories on pollution prevention in the public domain. In fact, the second half of this book represents the synthesis of a good number of these case histories and other resources into a collection of generalized pollution-prevention engineering technologies and practices. Yet, you should be aware of other available sources for ideas. A number of these can now be found on the Internet. We only mention a few to give you a perspective on the wide range of resources that are available.

- The CMA's "Designing Pollution Prevention into the Process: Research, Development & Engineering," Appendices A and B.[3]
- The *Industrial Pollution Prevention Handbook* by Harry M. Freeman.[11]
- The U.S. EPA's Pollution Prevention Directory (published annually).
- The U.S. EPA's Pollution Prevention Information Clearinghouse (PPIC).
- The U.S. EPA's Office of Pollution Prevention and Toxics (OPPT).
- The U.S. EPA's Pesticide Environmental Stewardship Program.
- The U.S. EPA's Environene (environsense) database.
- Case histories in journals such as *Chemical Engineering Progress, Chemical Engineering, Environmental Progress,* and *Pollution Prevention Review.*
- State pollution-prevention offices or centers. You would be surprised how many states offer services to small- and medium-sized businesses.
- Private consultants or consulting firms.
- Private consortia and organizations, such as AIChE's Center for Waste Reduction Technology (CWRT), the Center for Clean Industrial Treatment Technology (CenCITT), and the National Center for Manufacturing Sciences (NCMS).
- Pollution-prevention or waste minimization centers at universities, for example, the UCLA Center for Clean Technology and the Emission Reduction Research Center at the New Jersey Institute of Technology (NJIT),
- Numerous other Internet sites, such as the Great Lakes Pollution Prevention Centre in Canada and the Pacific Northwest Pollution Prevention Resource Center.

A recent review on using the Internet for pollution prevention was published by Scott Butner at the Battelle Seattle Research Center.[12] In addition, Appendix A is a compilation of some ideas that have been used within DuPont with some success. All of these resources can be used to help prepare you and other members of the brainstorming team for the generation of ideas.

3.9 When Should You Do Pollution Prevention?

The continuum depicted in Figure 3-15 shows the relative points at which a pollution-prevention program should be implemented. The decision of how far to move toward a zero waste and emissions design will depend on a number of factors, including corporate and business environmental goals, regulatory pressures, economics, the maturity of the process, and product life. It is safe to say, "the earlier, the better." If you can make changes during the R&D stage of the process or product life cycle, then you have the best opportunity to make significant reductions in waste generation at the source. However, as you move down the continuum from R&D through process design and engineering and post-start-up operation, your dependence on end-of-pipe treatment increases. At the bottom of the continuum is a total reliance on end-of-pipe treatment. Here, pollution prevention may manifest itself in the form of energy savings or a reduction in air flow to the abatement device, and so on.

3.9.1 Pollution Prevention during Research and Development

Research and development programs typically progress through three distinct phases: process conception, laboratory studies, and pilot-plant testing. The level of effort and detail required in pollution-prevention assessments depends on the particular R&D phase. Generally speaking, studies are qualitative during process conception, semiquantitative in laboratory studies, and quantitative in pilot-plant testing. The basic steps in a pollution-prevention study, however, are the same in each phase.

During process conception, reaction pathways, inherent process safety, general environmental impacts of prod-

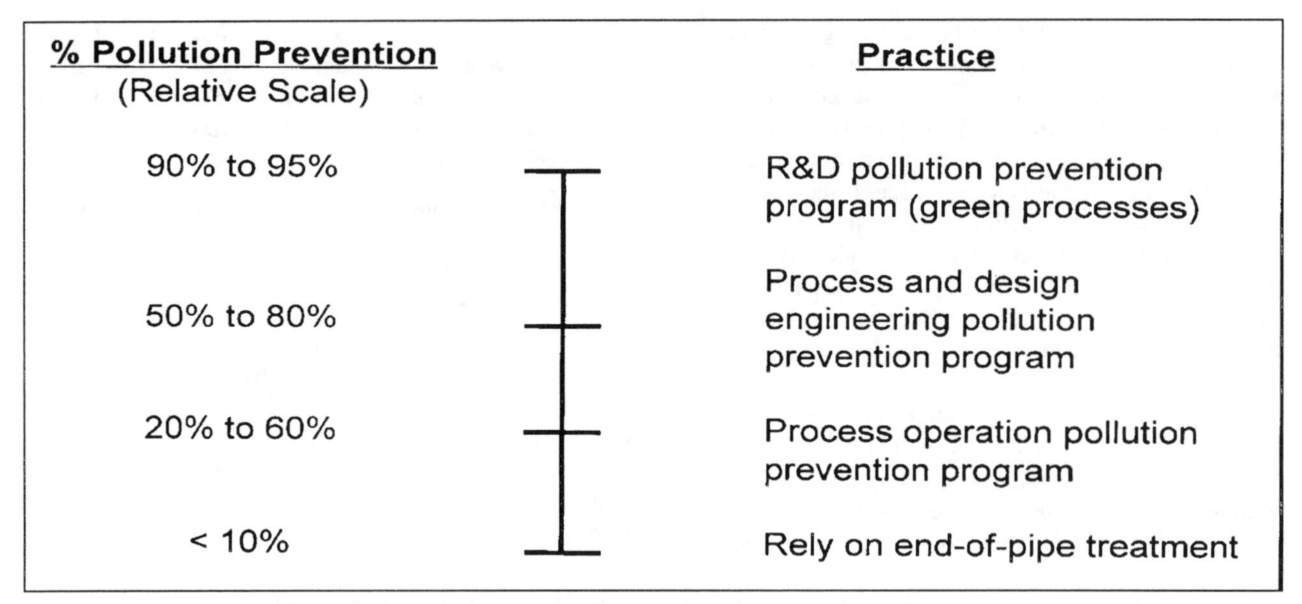

Figure 3-15. The Pollution Prevention Methodology Continuum

ucts, and waste streams are studied, and pollution-prevention concepts are formulated.

During laboratory studies, reaction chemistry is confirmed, waste streams are characterized, process variables are tested, pollution-prevention options are identified, data are collected for the pilot plant and process design, and the potential impact of environmental regulations is determined.

During pilot-plant studies, laboratory results are confirmed, process chemistry is finalized, key process variables are tested, equipment design is evaluated, and waste characteristics are defined. It is especially important at this stage of R&D that all major environmental cost areas are understood as they relate to the overall viability of a commercial project.

3.9.2 Pollution Prevention during Process and Design Engineering

While the greatest opportunity for cost-effective waste reduction at the source exists at the R&D stage, additional opportunities may exist during process engineering, and should be explored. The potential to reduce waste and pollutant releases in this stage is impacted by the selection of process configuration (batch versus continuous, for example), process conditions (such as temperature and pressure), manufacturing procedures, design and selection of processing equipment, and process control schemes.

As a project moves into the detailed design stage (sometimes referred to as the "mechanical design stage" or "production design"), source-reduction opportunities typically diminish. The main reason is that the process and preliminary plant design become fixed and the project becomes schedule-driven. The focus at this stage shifts from the chemical process to equipment and facility design. The emphasis at this point should be to protect groundwater from spills and to minimize or eliminate fugitive emissions.

3.9.3 Pollution Prevention during Process Operation

If the pollution-prevention program began during the research stage, then a pollution-prevention analysis is not necessary until 3 years after start-up of the process. Ideally, a pollution-prevention program should be completed every 3 to 5 years.

For a process that does not have a history of doing pollution prevention, a pollution-prevention program should realize a greater than 30% reduction in waste generation and a greater than 20% reduction in energy usage.

3.10 Case Studies

3.10.1 R&D Phase

Waste Reduction Through Control of the Reaction Pathway[3]. In hydrocarbon oxidation processes to produce alcohol, there is always a degree of overoxidation. The alcohol is often further oxidized to waste carboxylic acids and carbon oxides. If boric acid is introduced to the reactor, the alcohol reacts to form a borate ester that protects the alcohol from further oxidation. The introduction of boric acid terminates the byproduct-formation pathway and greatly increases the product yield. The borate ester of alcohol is then hydrolyzed, releasing boric acid for recycle back to the process. This kind of reaction pathway control has been applied to a commercial process, and resulted in about a 50% reduction in waste generation once the process was optimized.

Waste Reduction Through Catalyst Selection[3]. For chemical processes involving catalysis, proper selection of catalysts can have a major impact on product formation. One example is the ammoxidation of propylene to form acrylonitrile. Different catalysts result in a wide range of product and byproduct yields. Byproduct yields of 50 to 80% (based on carbon) have been reported in the literature. Use of a different catalyst provided a 50% reduction in waste generation by increasing product yield from 60% to 80%.

3.10.2 Process and Design Engineering Phase

Reuse Reaction Water in Wash Step[3]. A dehydration reaction generates a continuous stream of water that requires disposal. A separate product wash step used deionized water that was also disposed. Testing verified that the dehydration water could replace the deionized water in the wash step without impacting product quality. Initial concerns about product quality were unfounded. Total waste generation is reduced by the quantity of dehydration water that is reused.

Groundwater Protection.[3] At a grassroots facility, one company utilized a groundwater protection strategy that included several construction tactics not required by current environmental regulations. Chemical storage tanks were designed with double bottoms to allow leak detection before environmental damage. Similarly, one nonhazardous process-water pond was constructed with synthetic liners to eliminate the possibility of groundwater impact from any pollutants. Nonhazardous process-water ditches, traditionally used in chemical plants, were replaced with hard-piped sewer lines to eliminate the leak potential inherent with concrete.

3.10.3 Existing Process Operation

Solvent Elimination. In a DuPont plant, organic solvents used in the manufacture of intermediate monomers were incinerated on-site as a hazardous waste. Alternative nonhazardous organic solvents were considered and rejected. However, the intermediate monomers themselves were found to have the dissolution capacity of the original organic solvents, and could replace them. By utilizing existing equipment, realizing savings in ingredients (solvent) recovery, and reducing operating and incineration costs, the project achieved a 33% internal rate of return (IRR) and a 100% reduction in the use of the original solvents.

Literature Cited

1. U.S. Environmental Protection Agency. November 1993. *DuPont Chambers Works Waste Minimization Project.* EPA/600/R-93/203. Washington, DC: U.S. EPA, Office of Research and Development.
2. Dyer, J. A. and K. L. Mulholland. January 1998. "Follow This Path to Pollution Prevention." *Chemical Engineering Progress,* 94: 34–42.
3. Chemical Manufacturers Association. May 1993. *Designing Pollution Prevention into the Process: Research, Development and Engineering.* Chemical Manufacturers Association.
4. U.S. Environmental Protection Agency. May 1992. *Facility Pollution Prevention Guide.* EPA/600/R-92/088. Washington, DC: U.S. EPA, Office of Research and Development.
5. Doerr, W. W. August 1996. "Using Guidewords to Identify Pollution Prevention Opportunities." *Chemical Engineering Progress.* 92: 74–80.
6. Gouchoe, S., M. James, K. M. Lynch, M. F. Rose, and S. M. Usher. November 1996. "Integrate Your Plant's Pollution Prevention Plans." *Chemical Engineering Progress.* 92: 30–43.
7. Rossiter, A. P., H. D. Spriggs, and H. Klee. January 1993. "Apply Process Integration to Waste Minimization." *Chemical Engineering Progress.* 89: 30–36.
8. Mizsey, P. 1994. "Waste Reduction in the Chemical Industry: A Two Level Problem." *Journal of Hazardous Materials.* 37: 1–13.
9. Fromm, C. H. Autumn 1992. "Pollution Prevention in Process Design." *Pollution Prevention Review.* 2(4): 389–401.
10. Pojasek, R. B. Autumn 1996. "Identifying P2 Alternatives with Brainstorming and Brainwriting." *Pollution Prevention Review.* 6(4): 93–97.
11. Freeman, H. M. 1995. *Industrial Pollution Prevention Handbook.* New York: McGraw-Hill.
12. Butner, S. Autumn 1997. "Using the Internet for Pollution Prevention." *Pollution Prevention Review.* 7(4): 67–74.

CHAPTER 4

Economics of Pollution Prevention

4.1 Introduction

The failure of many pollution-prevention programs can be traced to the inability of the engineers and scientists involved in the programs to convince business leadership to change the manufacturing process. Often, this reluctance to change is not because the recommended process improvements were not technically sound, but because of the team's failure to speak the language of the business person—that is, dollars and cents. The role of economics in pollution prevention is very important, even as important as the ability to identify technology changes to the process.

In DuPont, we teach an internal course on pollution prevention. When it comes time for course evaluations, the subject area that is always the lowest rated is the material on economics and engineering evaluations. The feedback is interesting. "We don't see the connection between economics and pollution prevention." "This is a business person's job." "I've seen all this before." The reality is that some, but not all, engineers fail to see their role in convincing business leadership to change. These engineers are often uncomfortable in a sales and dollars-and-cents kind of environment. They do not like selling their ideas to upper management. They like technology. The reality, though, is that the core team will need to use economics to help them sell the program, secure resources, and identify the best options.

As is shown in the path to pollution prevention in Figure 4-1, economics comes into play in two key areas:

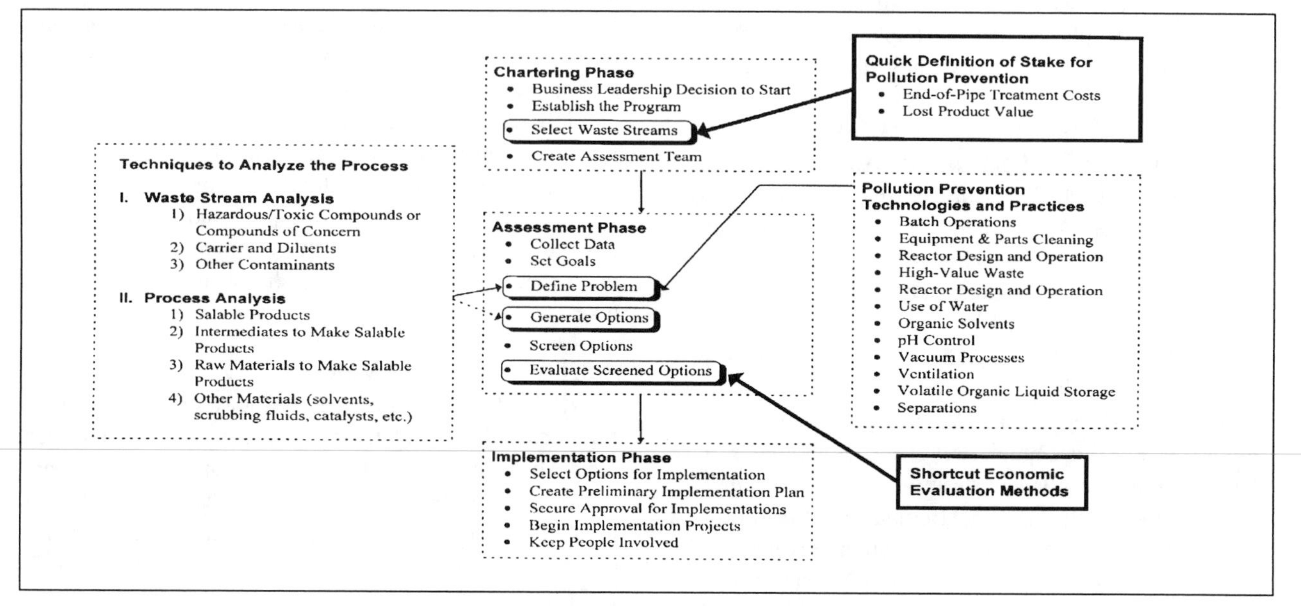

Figure 4-1. The Path to Pollution Prevention

the chartering phase and the assessment phase.

In the chartering phase, we are interested in quickly defining the incentives for pollution prevention. These incentives are largely financial (for example, avoided investment in end-of-pipe treatment, lower manufacturing costs, avoided environmental fines and penalties, increased yield, and higher plant uptime), but they can also be less quantifiable in financial terms, such as improved public image, an environmental friendly product, and fewer permitting barriers to business growth. In particular, quantifying the financial incentives will help to rank the process waste streams for source reduction during the waste-stream selection step and to secure support for the program from business leadership. As seen in Figure 4-1, there are two ways to define the incentive for pollution prevention: (1) the cost associated with new end-of-pipe treatment for the selected waste streams; and (2) the lost product value associated with the waste leaving the process, either in terms of loss of raw materials alone or the cost of manufacture. In this chapter, we show how to quantify the incentive for pollution prevention using these two approaches.

In the assessment phase, economics plays a key role in evaluating and ranking the screened options from the brainstorming session (Figure 4-1). In other words, what is the economic viability of the best ideas? How will these changes make the business more profitable? After all, this is what business leadership is most concerned about—making money. In this chapter, we show how to use engineering evaluations techniques, which include shortcut net present-value calculations, to determine the economic value of the technically feasible options to the business. These techniques will help you to speak the language of the business people, so that the implementation phase is seen by the business people as a business opportunity.

4.2 End-of-Pipe Treatment Cost as the Incentive for Pollution Prevention

To meet the 1990 Clean Air Act Amendments in a cost-effective manner, DuPont conducted an extensive survey of air-emissions control technologies for both nonhalogenated volatile organic compounds (VOCs)[1] and halogenated VOCs (HVOCs).[2] The control technologies were evaluated not only on a technical basis, but also in economic terms. Similar surveys were done for gas-entrained particulates and aerosols and for biological and other nonbiological aqueous organic treatment technologies other than incineration. A side benefit of this work was that the costs for end-of-pipe treatment could be used to quickly assess the incentive or value for pollution prevention. For solid-waste treatment, commercial disposal costs for incineration and landfilling have been assembled. The large economic penalty associated with treatment of gaseous, liquid- and solid-waste streams provides an important focus on the value of pollution prevention in the form of source reduction.

Keep in mind, though, that the total cost to control air emissions includes not only the end-of-pipe VOC treatment cost, but also permitting, stream characterization, particulate removal, and record-keeping costs.[1]

The biological wastewater treatment survey focused on understanding and developing economical biotreatment technologies, which include the more familiar conventional aerobic-activated sludge technology as well as other aerobic, anaerobic, and anoxic technologies.

The nonbiological water treatment survey focused on aqueous organic treatment technologies, including mature technologies, such as air and steam stripping, reverse osmosis, and granular-activated-carbon adsorption, as well as new-to-DuPont technologies, such as advanced oxidation processes and wet oxidation. Potential applications for these technologies include

- Pretreatment of toxic and recalcitrant wastes to make them amenable to biological treatment (e.g., halogenated organic compounds);
- Stand-alone treatment of aqueous organic wastes when biotreatment is not feasible or cost-effective (e.g., *ex situ* groundwater treatment, solvent recovery).

4.3 Economic Criteria for Technology Comparisons

Technology evaluation encompasses not only the technical feasibility of a particular technology but also the economics of its implementation. While there are many measures of economic merit, two measures—net present value (NPV) and investment—provide the business person with a complete set of information to make an informed, economic decision. Investment refers to how much money must be spent initially to design and build the new facilities; net present value is the total value or cost (if negative) of the alternative to the business, including all contributions to the cash flow of the alternative, that is, new investment, working capital, cash operating costs, income taxes, and revenues.

4.3.1 Net Present Value

NPV is the after-tax worth in today's dollars of all the future cash that an alternative will either consume or generate. It includes the effects of the four primary economic elements: investment, cash costs, revenues, and taxes. NPV is the most popular single measure of the economic merit of an alternative, and many spreadsheet programs can easily calculate it.

The concept of NPV is derived from the time value of money. Money can be invested to yield a profit; therefore, a dollar in hand today is worth more than a dollar received in the future. The converse is also true—foregoing the receipt of a dollar today can be justified by the prospect of receiving a larger sum in the future. The relationship of the future value of money to the present value of money is shown by the following equation:

$$F = P(1+i)^n$$

where

F = future value
P = present value
i = annual earnings rate
n = years in the future

This equation is used to determine how many future dollars will be realized by investing present dollars at a given interest rate, i, for n years. Discounting is using this equation to determine the present value of a future value of money. Thus, if $i = 12\%$, $n = 5$ years, and if $F = \$1.50$, then $P = \$0.85$.

A complete financial analysis requires the inclusion of cash flow. On a year-by-year basis, the cash flow from a manufacturing investment is determined through a series of calculations.

Cash Costs (Operating and Overhead). Cash costs include raw materials (chemicals, catalysts, etc.), utilities (steam, electricity, natural gas, etc.), maintenance materials and labor, operations labor, technical support, startup costs, taxes and insurance, and administrative costs.

Total Costs. Total costs include the sum of cash costs from the previous subsection and depreciation. Note that depreciation is not a cash cost and will only be used to calculate income taxes.

The investment may generate revenues or savings from which the total costs are subtracted. This resultant is known as the "pretax earnings." Income taxes (calculated on pretax earnings) are next subtracted to yield the after-tax earnings. A year-by-year *cash flow* can then be determined by summing the four real cash flows: investment, cash costs, revenues, and income taxes.

The NPV is the sum of the present values of a series of cash flows:

$$NPV = CF0 + CF1*D1 + CF2*D2 + CF3*D3 + \cdots$$

where

CF0, CF1, CF2, etc. = Cash flow in year 0, 1, 2, and so on

$D1$, $D2$, etc. = Discount factor for year 1, 2, and so on

The NPV calculation embodies all parts of the effect of an original investment—costs, income, taxes, and the original investment outlay. For waste-management or pollution-abatement investments, the NPV is normally negative. A negative NPV is really a cost and will be identified in this book as a positive net present cost (NPC). The lowest NPC is the best option, but normally the two values—investment and NPC—are considered together. Because waste management is often a cost, investment is important. Thus, the business might choose to minimize current investment while pursuing a future pollution-prevention goal. The NPC is important because it includes not only investment, but also operating costs and how they impact the business for a given period of time (normally 10 to 15 years).

4.3.2 Investment

Investment includes all the monies required for plant, property, and equipment. For the end-of-pipe treatment technology surveys, investment is based on venture guidance appraisals for installed inside-battery-limits, skid-mounted systems. The estimates do not include site-specific items, such as extensive dismantlement and rearrangement, long tie-ins, custom engineering, roads, and pipe bridges. To provide a common ground for comparing the different technologies, the estimates are based on the following:

- Equipment as provided by the vendor
- All allowances for utilities, motor starters, and so on, and any necessary investment for power, general, and service facilities (PG&S) are included
- Appropriate platform or concrete pad for the equipment is included
- Current cost index for escalation (ENR cost index = 5,700 for all estimates)
- Appropriate contingency included for uncertainties in the process scope of work
- A greenfield project is assumed
- Short tie-ins to the process are included

In other words, the equipment is ready to operate except for other site-specific, outside-battery-limits requirements.

The overall ratio of the project-level investment to the vendor quote depends on the scope of the budget quote from the vendor. For a turn-key vendor quote that includes installation, the ratio is on the order of 1.4–1.6, whereas for an uninstalled, skid-mounted system, the ratio is about 2–3, and for bare equipment that must be field erected, the ratio can be as high as 6–10.

4.4 End-of-Pipe VOC and HVOC Treatment Technology Selection

The technologies were compared in both engineering and economic terms over a wide range of gas flow rates and VOC and HVOC concentrations. In general, the studies found that, for VOC-containing streams, end-of-pipe treatment will

- Require a capital investment of $40–$1000 per standard cubic foot per minute (scfm) of waste gas treated
- Incur annual cash operating costs of 10%–30% of the installed investment

For HVOC-containing streams, the study found that end-of-pipe treatment will

- Require a capital investment of $60–$3000 per scfm of waste gas treated
- Incur annual cash operating costs of 10%–25% of the installed investment

More importantly, this knowledge allows one to quickly assess the cost of abatement, so that more effort can be devoted to reducing waste at the source. More specifically, *reducing air flow at the source* will play the most important role in minimizing capital investment in air-emissions control equipment.

The treatment technologies that were evaluated in the two studies are shown in Figures 4-2 and 4-3.

In general, these technologies were sized to provide either 98% destruction or less than 20 parts per million by volume (ppmv) discharge concentration. Not all technologies are capable of this destruction efficiency for all compounds.

Figures 4-4 and 4-5 are technology selection maps summarizing what the evaluations showed to be the most cost-effective (i.e., lowest NPC) end-of-pipe technologies for a particular gas flow rate and VOC or HVOC concentration. In areas of the map where technologies overlap, the NPCs differ by less than 25%. For these cases, an application-specific technology evaluation is required to determine which technology is the most economical.

If the VOCs or HVOCs have a high value-in-use (greater than about $0.20 per pound), recovery by carbon adsorption should be considered for gas flow rates greater than 1000 scfm and for concentrations greater than 500 ppmv.

At low waste-gas flow rates (less than 100 scfm), the

Thermal Destruction Technologies

Flaring
Thermal Oxidation Without Heat Recovery
Recuperative Thermal Oxidation
Catalytic Oxidation
Regenerative Thermal Oxidation

Concentration Technologies

Zeolite Wheel Plus Catalytic Oxidation
Zeolite Wheel Plus Condensation
Zeolite Wheel Plus Recuperative Thermal Oxidation
Zeolite Wheel Plus Carbon Adsorption

Recovery Technologies

Refrigerated Condensation
Membrane Separation
Carbon Adsorption
Polymeric Adsorption

Other Technologies

Water Scrubbing
Biofiltration

Figure 4-2. Treatment Technologies Evaluated in VOC Study

Thermal Destruction Technologies

Thermal Oxidation Without Heat Recovery
Recuperative Thermal Oxidation
Catalytic Oxidation Without Heat Recovery
Flameless Thermal Oxidation
Regenerative Thermal Oxidation
Regenerative Catalytic Oxidation

Concentration Technologies

Zeolite Wheel Plus Catalytic Oxidation
Zeolite Wheel Plus Carbon Adsorption

Recovery Technologies

Refrigerated Condensation
Membrane Separation
Carbon Adsorption
Polymeric Adsorption

Other Technologies

Biotrickling Filters

Figure 4-3. Treatment Technologies Evaluated in HVOC Study

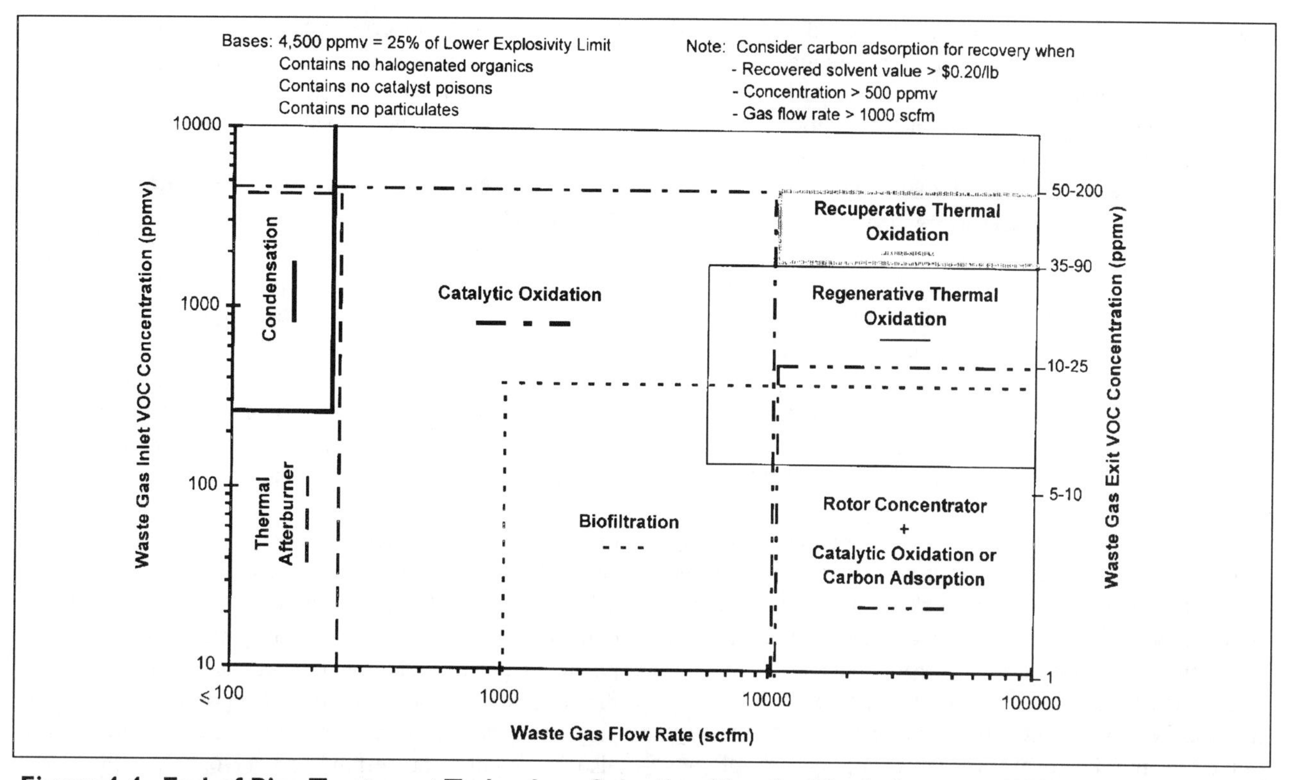

Figure 4-4. End-of-Pipe Treatment Technology Selection Map for Nonhalogenated VOCs. Identifies the Lowest Net Present Cost Technology(ies) for a Region of Waste Gas Flow and Concentration

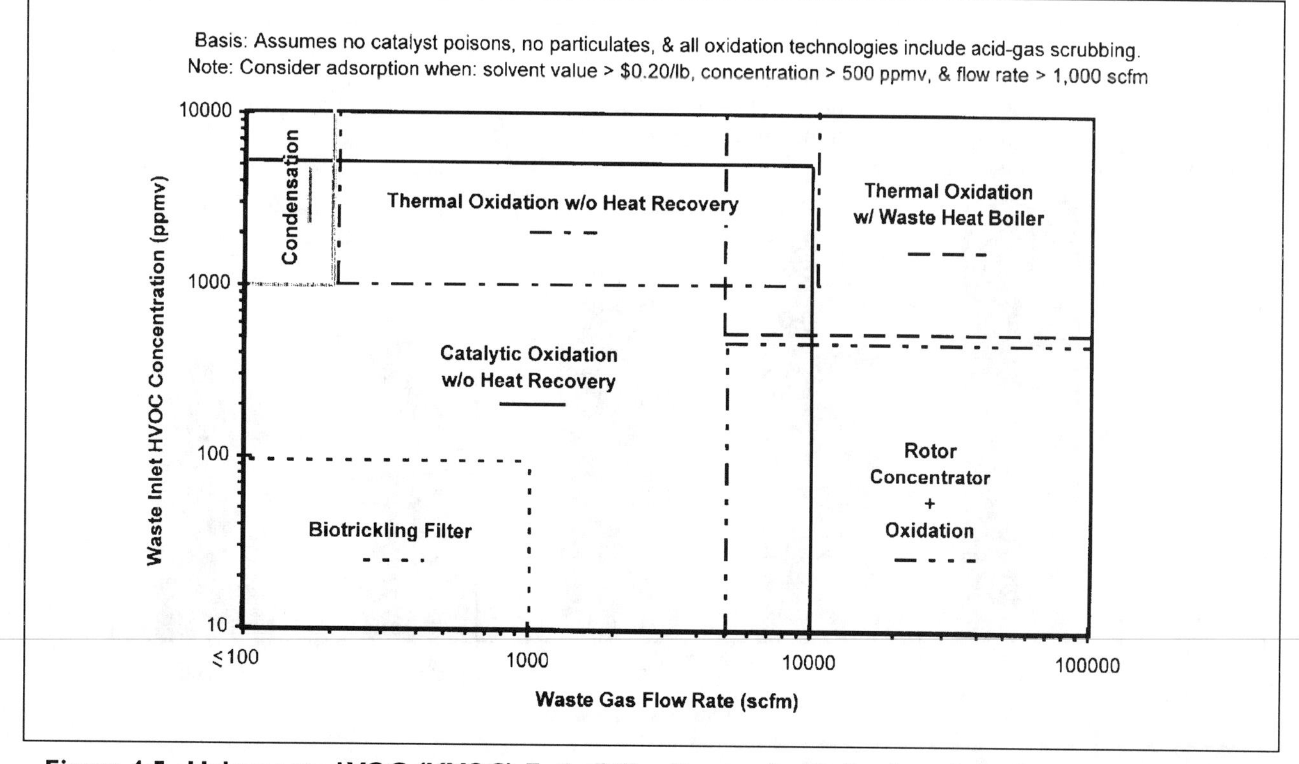

Figure 4-5. Halogenated VOC (HVOC) End-of-Pipe Treatment Technology Selection Map. Identifies the Lowest Net Present Cost Technology(ies) for a Region of Waste Gas Flow and Concentration

low-cost technologies are refrigerated condensation and combustion—either thermal oxidation without heat recovery or catalytic oxidation.

Regenerative thermal oxidation is the most frequently applied abatement technology for VOC gas streams with waste-gas flow rates greater than 10,000 scfm. For HVOC-containing gas streams, recuperative thermal oxidation (waste heat boiler for energy recovery) is the most cost-effective technology. When abating gas streams between 200 and 10,000 scfm, on the other hand, catalytic oxidation is most often the economically preferred choice.

A zeolite rotor concentrator wheel system is used to preconcentrate a very dilute (less than 100 ppmv VOC or so) high-flow-rate waste-gas stream before further processing in a downstream abatement or recovery device, such as a catalytic oxidizer or carbon adsorber. For very dilute streams, this results in an overall reduction in the cost of treatment. The zeolite technology, however, is not universally applicable. In some cases, an activated-carbon wheel may be preferred over a zeolite wheel.

In some applications, biofiltration and biotrickling filters are the lowest NPC abatement technologies for very dilute streams; however, they are emerging technologies with limited commercial experience in the United States. Because of the uncertainties involved in biological processes, pilot testing is almost always necessary before full-scale application. Pilot testing can substantially add to the cost and schedule, especially for small applications.

4.4.1 Costs Associated with Permit Applications

The cost to a business to treat a waste-gas stream includes not only the end-of-pipe VOC or HVOC treatment costs but also costs for obtaining and renewing permits, stream characterization, and record keeping. The permit and record-keeping costs consist of two elements: one-time or periodic costs (costs incurred every time the permit is renewed) and annual costs. One-time costs included in the studies were (1) stack testing to characterize the stream; (2) compliance testing of the installed treatment device; (3) a regulatory assessment to determine applicability of regulations and what would be maximum achievable control technology (MACT); (4) permit development to ensure process operating flexibility; (5) compliance report development, such as quarterly reports, material balances, and automated reporting; and (6) dispersion modeling. One-time costs (2) through (6) will also be required at permit renewal or when there is a major process modification.

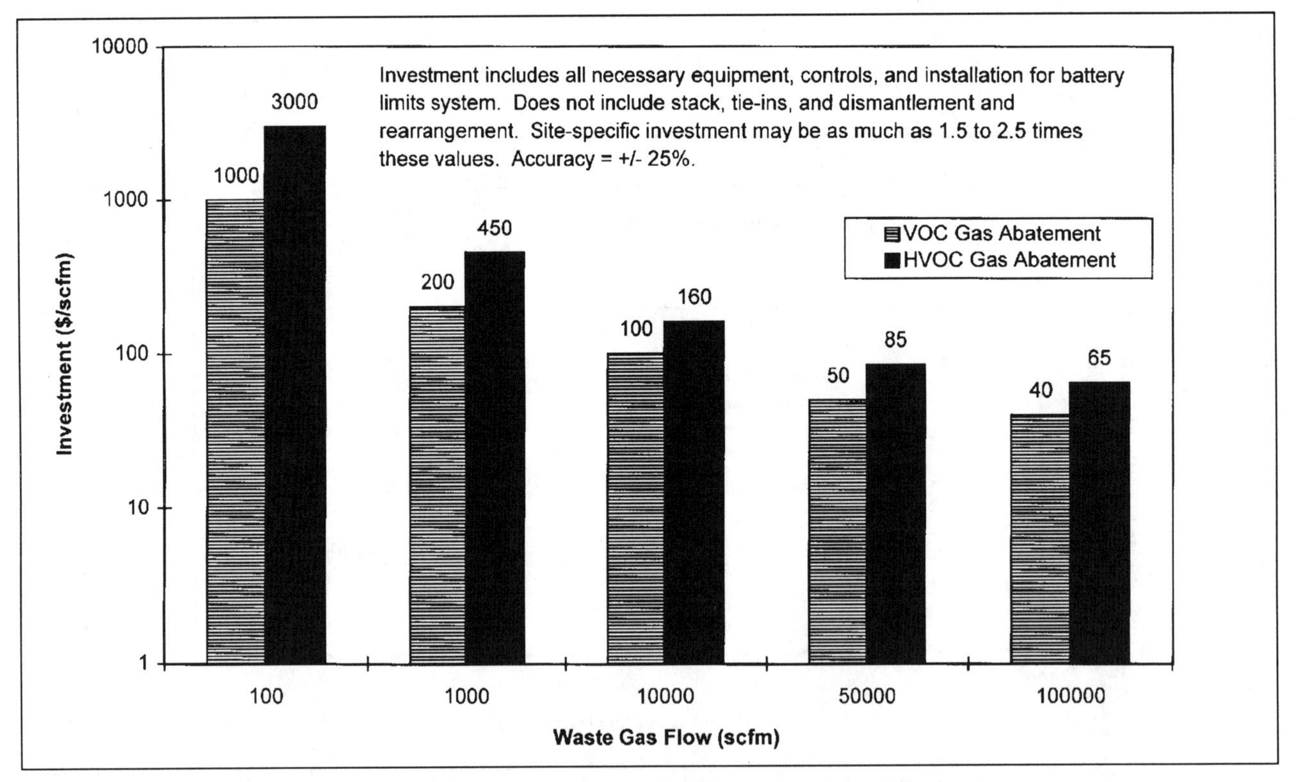

Figure 4-6. Investment For Treatment of VOC and HVOC Waste Gas Streams (U.S. dollars, ENR = 5700)

Annual permit costs include (1) permit fees of $25/ton for emissions of VOCs, air toxics, and criteria pollutants; (2) report preparation and submission; and (3) equipment leak detection and repair. Continuous monitoring costs were not included in the evaluation.

4.4.2 Investment, Cash Operating Cost, and Net Present Cost

Figures 4-6, 4-7, and 4-8 summarize investment, annual cash operating cost, and NPC, respectively, for the technologies shown in Figures 4-4 and 4-5. Figure 4-8 includes the NPC associated with permit and record keeping costs. For a waste-gas stream that requires a new permit, the total NPC is the sum of the NPCs for permitting and end-of-pipe treatment. Note that these costs represent the minimum cost for an inside-battery-limits system. Site-specific installations will probably be 1.5 to 2.5 times the capital investment shown in Figure 4-6 because of outside-battery-limits requirements.

4.5 Particulate Control Technologies

Most end-of-pipe treatment technologies for HVOC- and VOC-containing waste gases require a particulate- or aerosol-free feed stream to avoid plugging or fouling the equipment. In addition, waste-gas streams containing only particulates or aerosols (for example, the air from conveying bulk solids and ventilation air in contact with finish oils in fibers manufacture) often require treatment before the stream can be vented to the environment. This section discusses the control technologies available, along with their associated investment and costs, to remove particulates from a waste-gas stream. As with other gas cleaning technologies, there is a significant incentive for reducing air or diluent flow at the source. In the case of particulate control technologies, the incentive is on the order of $20 to $50 per scfm, depending on the technology type.

4.5.1 Investment, Cost, and Economics

Figures 4-9 to 4-11 summarize investment, operating cost, and NPC for the principal particulate cleaning technologies shown in Figure 4-12. Note that these costs represent minimum costs for an inside-battery-limits system. Site-specific investment can be as much as 1.5 to 2.5 times the values shown in Figure 4-9.

4.5.2 Technology Description

Particulates in a gas stream fall into four basic categories: dust, condensation aerosols, entrainment mists, and sticky solids.

Dusts consist of solid particles, larger than 1 micron, which result from mechanical disintegration of matter.

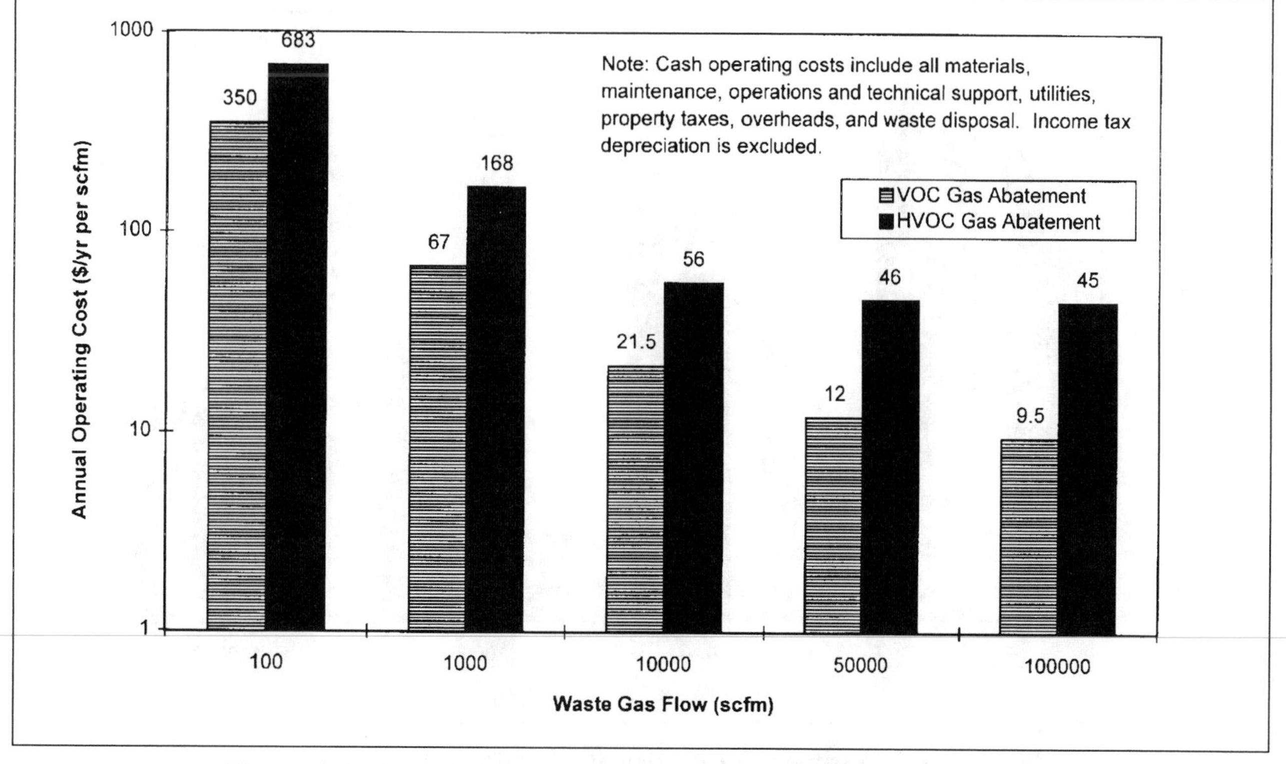

Figure 4-7. Cash Operating Costs (1999 U.S. dollars) For Treatment of VOC and HVOC Waste Gas Streams

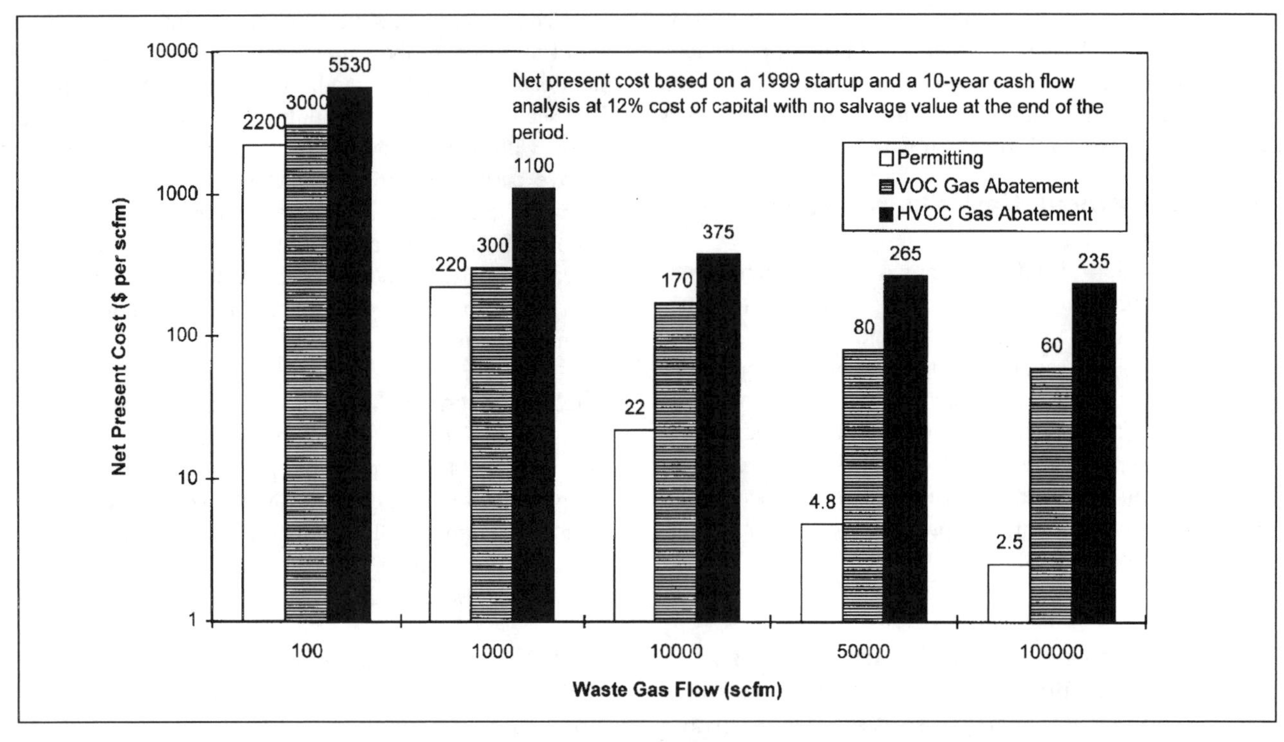

Figure 4-8. Net Present Cost For Treatment of VOC and HVOC Waste Gas Streams (1997 U.S. dollars, 10 years, 12% discount rate)

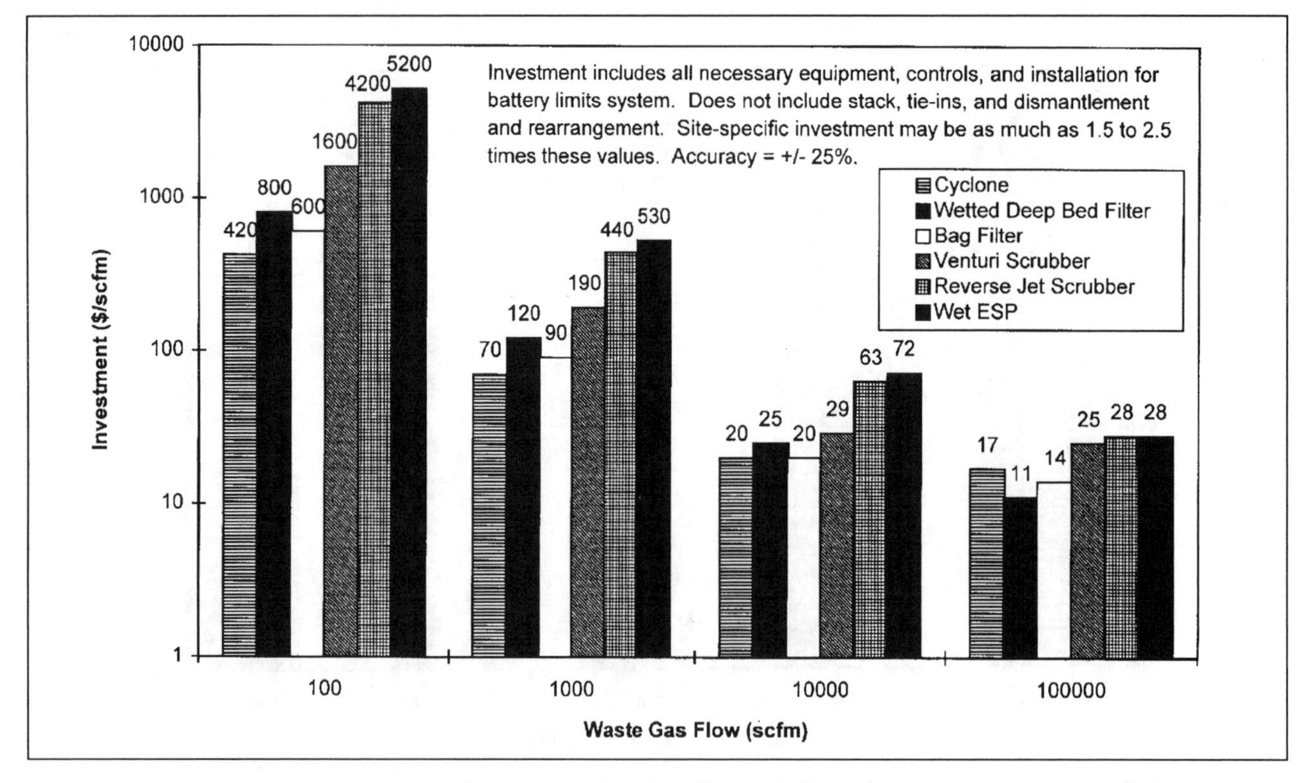

Figure 4-9. Investment For Particulate Gas Cleaning Technologies (U.S. dollars, ENR = 5700)

Examples include coal dust, cement dust, ground talc and pigment, and milled flour.

Condensation aerosols are liquid particles, smaller than 1 micron, that form due to the condensation of a vapor in a gas stream as the gas cools. Some are self-nucleating, while others require a nucleation site (such as a solid particle or seed) to condense. Examples include sulfuric and hydrochloric acid aerosols and atmospheric fog.

Mists are liquid particles, larger than 1 micron, that are dispersed in a gas stream (if larger than 10 microns, they are referred to as sprays). Mists often form because of the entrainment of small liquid droplets from the surface of a liquid by a gas stream: thus the term entrainment mist. Examples include liquid entrainment in the overhead vapors from distillation columns, gas scrubbers, vacuum systems, and sulfuric acid mist formed in contact plants.

Finally, sticky solids are solid particles that are tarlike in nature. They tend to adhere or stick to particle-collection equipment surfaces; therefore, they usually require "wet" collection systems to remove them.

Meanwhile, there are five basic gas cleaning technologies to address these four categories of particulates: demisters, cyclones, wet scrubbers, filters, and electrostatic precipitators.

Demisters. Demisters are primarily used to separate liquid aerosols from a gas stream. They are subdivided into two classes: the diffusional demister and the inertial demister. The diffusional demister or wetted deep-bed filter, is used to collect submicron aerosols. On the other hand, the inertial demister, which includes the Chevron-type demister and knitted mesh pads, is effective in collecting entrainment mists. In both classes, the aerosols coalesce to form a liquid film that drains from the vessel.

Cyclones. Cyclones are used extensively to recover dry solid products. Functioning as coarse dust and mist eliminators, cyclones operate by the increased gravitational field produced by the centrifugal spin or vortex imparted to the gas stream by a spin vane or tangential entry. Both reversing and nonreversing vortices are used to make the separation.

Wet Scrubbers. Wet scrubbers bring the particulate-laden gas into intimate contact with the scrubbing liquid. As a result, the particulates are captured by a combination of inertial impaction and interception. Wet scrubbers are used to remove both solids and liquids. The minimum particle size that can be collected is a function of the energy input as measured by the pressure drop through the ves-

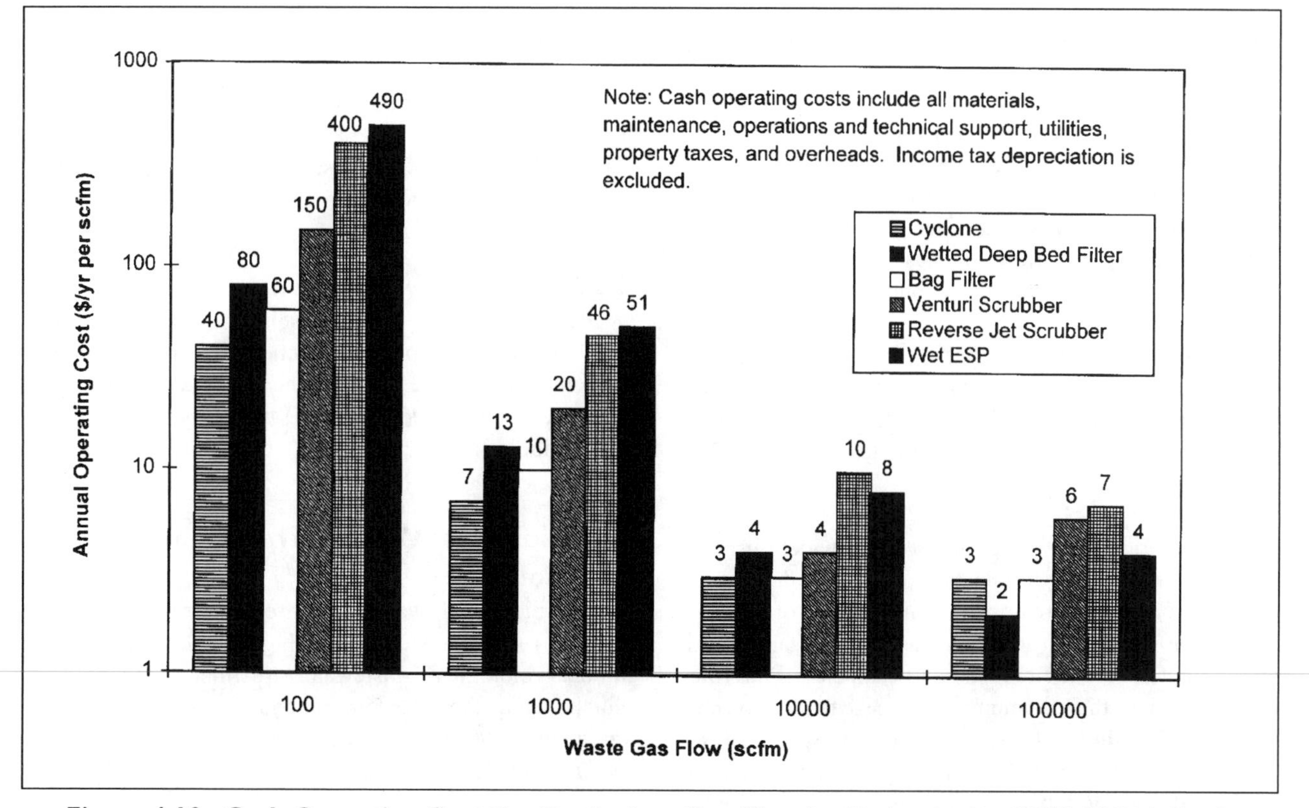

Figure 4-10. Cash Operating Cost For Particulate Gas Cleaning Technologies (1999 U.S. dollars)

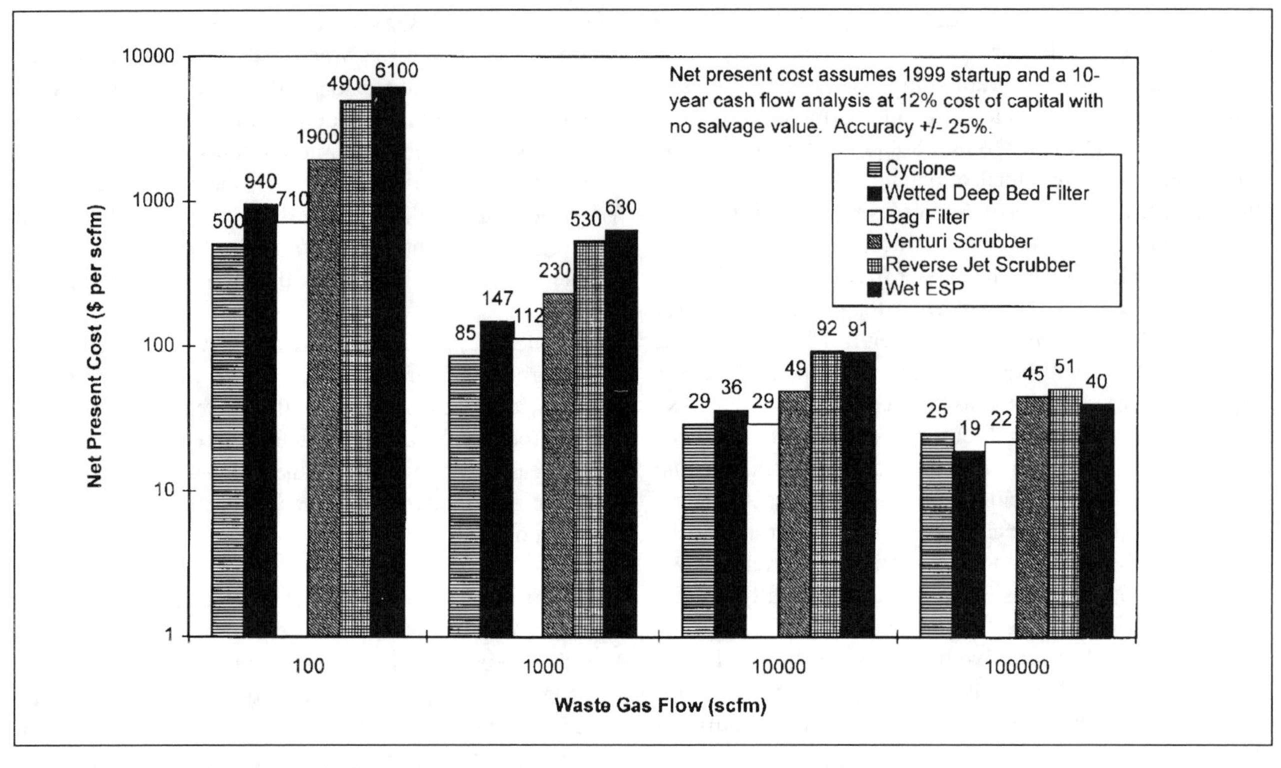

Figure 4-11. Net Present Cost For Particulate Gas Cleaning Technologies (1997 U.S. dollars, 10 years, 12% discount rate)

sel. Wet scrubbers are categorized as either low energy or high energy.

The reverse-jet scrubber is an example of a low-energy scrubber. It is a device that sprays a scrubbing liquid countercurrent to the flowing waste-gas stream, thereby increasing the relative velocity between the gas and the liquid. Gas/liquid contacting is designed to be in the froth regime of two-phase flow. Because the liquid spray is turned around, a highly turbulent zone is produced for particulate collection.

The venturi scrubber is a high-energy, cocurrent device that accelerates the gas in the throat of the venturi to atomize the scrubbing liquid. The venturi scrubber can be designed to collect submicron particles with sufficient energy input.

Filters. Fabric filters, or baghouses, are widely used in industry to recover dry products or to remove dust contaminants from an air stream with a high degree of efficiency—down to 1 micron in size. The filtration takes place on the filter cake. The porosity and character of the filter cake determines the minimum particle size that can be collected. HEPA filters rely on filter fibers to capture submicron particles down to 0.2 micron in size. They are used where extremely low particulate emissions are required.

Electrostatic Precipitators. The electrostatic precipitators (ESP) (both wet and dry types) rely on electrostatic forces, instead of inertial forces, to remove particles from a waste-gas stream. The particle is charged by gaseous ions produced by a corona discharge; the charged particles are subsequently removed by rapping in a dry ESP or by a liquid spray in a wet ESP.

4.5.3 Technology Selection

Figure 4-12 is a technology selection map for particulate emissions control as a function of particle size and particulate type. The *boldface* technologies are economically preferred (i.e., lowest NPC) where they apply, while the nonboldface technologies represent the alternate choice.

4.6 Biological Wastewater Treatment Technologies

The biological wastewater treatment survey within DuPont focused on understanding and developing economical biological wastewater treatment technologies, which include the more familiar conventional aerobic activated-sludge technology as well as other aerobic, anaerobic, and anoxic technologies.

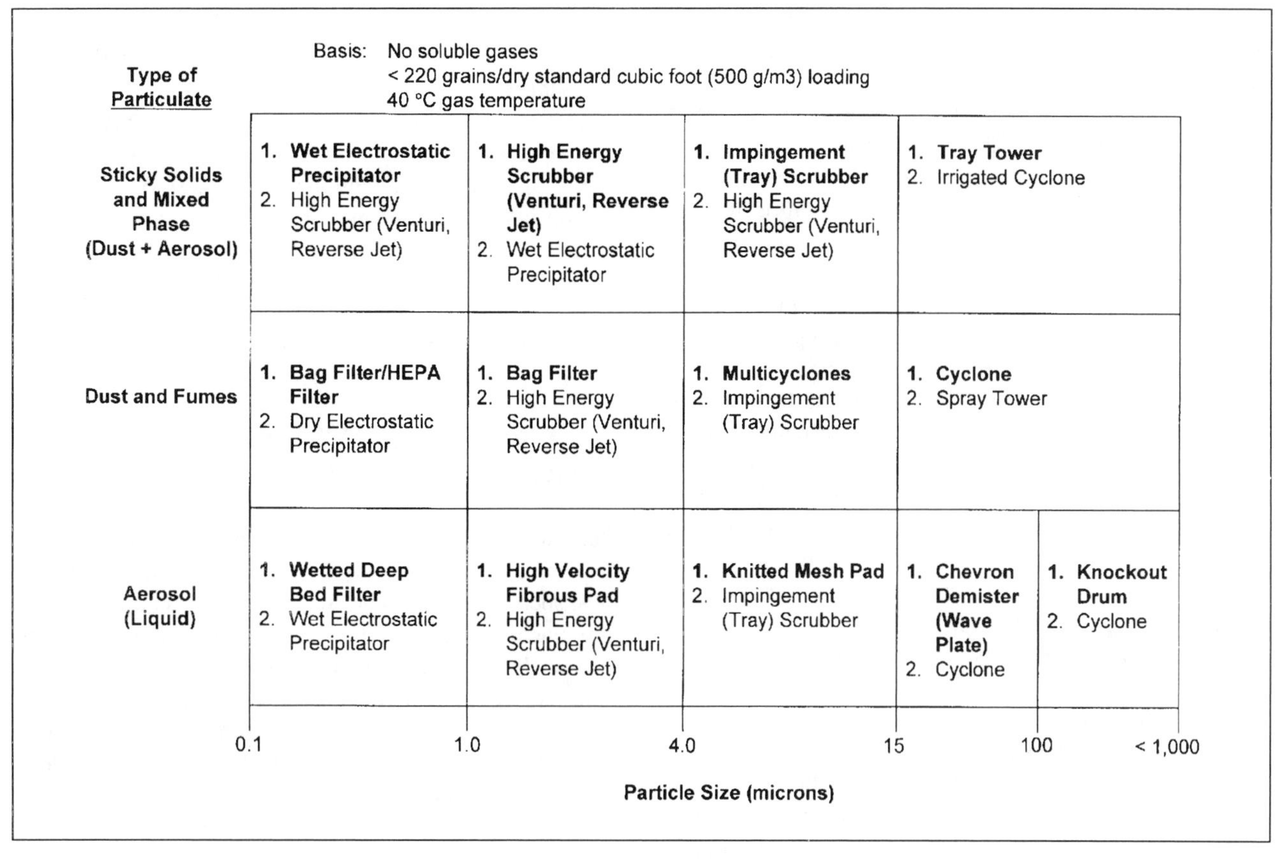

Figure 4-12. Technology Selection Map For Particulate Emissions Control

4.6.1 Applicable Technologies

Biological wastewater treatment technologies are generally categorized into one of three processes—aerobic, anoxic, and anaerobic.

Aerobic processes are those in which the biotreatment microbes use oxygen to degrade the organics and ammonia in the wastewater. The common applications are aeration lagoons and ponds, and activated sludge treatment in tanks. Both the energy required to aerate the wastewater and the amount of excess biosolids (sludge) produced are relatively high.

Anoxic processes are similar in nature to aerobic processes, except that the microbes do not require molecular oxygen (O_2). Instead, the oxidizing power comes from certain compounds already contained in some wastewaters, typically nitrate (NO_3^-). Compared with aerobic processes, energy requirements are very low and the amount of excess biosolids produced is about one-half. However, it has not been found to be economical to add nitrate ion to the wastewater in place of aeration.

Anaerobic processes are the fermentation and methane-producing microbial processes that can occur only in the total absence of oxygen. These processes can never be used for total treatment. Typically, an aerobic system must be used to "polish" the discharge from an anaerobic process. However, energy requirements are very low and very few excess biosolids are produced—only around one-fifth to one-tenth of that produced by aerobic systems. Therefore, an anaerobic process may be appropriate as a pretreatment or source treatment for more highly contaminated wastewaters, typically above 2000 mg biochemical oxygen demand per liter (BOD/L). This is roughly equal to 2000 mg organic/L for a common oxygenated organic compound like acetic acid. Less oxygenated organic compounds will have a mg BOD/mg organic ratio greater than 1.0.

4.6.2 Incentive for Source Reduction for New Facilities

A common biological wastewater treatment technology today is aerobic deep-tank activated sludge (ADTAS). Substantial reductions in the investment and operating cost required for future ADTAS facilities can result from reducing water flow and contaminant loading at the source. The magnitude of these reductions will vary with the waste stream's characteristics, hydraulic flow, contaminant concentration, and excess biosolids disposal method; however, the *minimum* incentive for *new* ADTAS facilities is shown in Table 4-1.

How should these *incremental* investment and cost numbers be used? If you are at a site that is planning to or will eventually install new investment for biological wastewater treatment, then these numbers have significance. A possible scenario would be an existing in-ground aeration pond that needs to be replaced with an above-ground, deep-tank system to comply with corporate groundwater protection policies. In this case, if the aqueous waste going to the planned new treatment facility can be reduced by 1 gallon per minute (gpm) *and* 1 lb organic/h, then the new investment and cash operating cost will be reduced by at least $9000 ($3000 per gpm + $6000 per lb/h) and $2300/yr ($300/yr per gpm + $2000/yr per lb/h), respectively. The incremental investment and cost then represent the partial derivative of the investment and operating-cost curves with respect to flow and concentration. Also, keep in mind that the numbers given in Table 4-1 are additive.

If your business or site is planning to use a technology other than ADTAS, these values still represent a good starting point for defining the minimum incentive for waste reduction at the source.

The incentives for ADTAS can be higher, too. The incentive for investment can be as much as $13,000 per gpm and $32,000 per lb organic/h. The incentive for cash operating cost will increase for more expensive biosolids disposal methods, such as incineration or land disposal as a hazardous waste. The numbers in Table 4-1 assume land disposal of the excess biosolids as a nonhazardous waste.

4.6.3 Investment and Costs for Aerobic Deep-Tank Activated-Sludge Treatment Facilities

Figures 4-13 to 4-15 summarize actual investment, cash operating cost, and NPC, respectively, to build and operate new ADTAS facilities. These estimates include appropriate pretreatment (diversion, equalization, and neutralization), ADTAS treatment, excess biosolids dewatering, and nonhazardous land disposal. Also included are general and service facilities, such as roadways, utilities, and groundwater protection facilities. As you can see, end-of-the-pipe biological treatment is not cheap.

4.7 Nonbiological Wastewater Treatment Technologies

The nonbiological wastewater treatment survey within DuPont focused on treatment technologies for dilute aqueous wastes. It included the evaluation of mature technologies, such as air and steam stripping, reverse osmosis, and granular activated carbon as well as emerging technologies, such as advanced oxidation processes and wet oxidation. As mentioned in Section 4.2, potential applications of these technologies are for pretreatment of toxic and recalcitrant wastes to make them amenable to biological treatment and for stand-alone treatment of aqueous organic wastes when biotreatment is not feasible or cost-effective.

4.7.1 Incentive for Pollution Prevention

For waste-gas streams, reducing air or diluent flow at the source is the best way to minimize the total cost of treatment. However, when treating dilute aqueous organic waste streams at the end of the pipe, consideration must

	Incremental Investment (1997 U.S. $)	Incremental Cash Operating Cost (1999 U.S. $)
Hydraulic Flow	$3,000 per gpm	$300/yr per gpm
Organic Loading*	$6,000 per lb organic/hr	$2,000/yr per lb organic/hr

* assumes 1 pound BOD per pound organic

Table 4-1. Minimum Incremental Investment and Operating Cost for New Activated, Deep-Tank Activated Sludge Treatment Facilities

be given to source reduction of *both* water flow and organic loading.

Substantial reductions in the NPC and investment of nonbiological technologies will result from reducing water flow and contaminant loading at the source. The magnitude of these reductions will vary with technology type, flow, and concentration; however, the *minimum* savings for new facilities are $2000 per gpm for hydraulic flow and $200 per lb organic/day for organic loading. These savings represent the minimum incremental net present cost of new nonbiological treatment technologies.

In some instances, the savings will be as high as $300,000 per gpm and $250,000 per lb organic/day, respectively. Minimizing organic loading at the source will be more important for technologies or treatment systems sensitive to mass loading, such as carbon adsorption and advanced oxidation.

4.7.2 Nonbiological End-of-Pipe Technology Selection Map

Figure 4-16 is a technology selection map that enables users to *narrow the field* of nonbiological technologies to be considered for a specific problem. The final choice will depend on a number of factors, including waste components, technical feasibility, wastewater flow and concentration, effluent requirements, and economics (investment, cash operating cost, and NPC). The map ranks technologies by their NPC as a function of wastewater flow rate and contaminant concentration with the assumption that the organics are not amenable to biotreatment. *The map is applicable to technologies in stand-alone applications and must be applied with caution.* There are always exceptions to generalized technology selection criteria, such as those reflected in Figure 4-16.

The map shows that at the present time, mature technologies, such as air stripping and carbon adsorption, will usually represent the minimum NPC options in stand-alone applications when they are feasible. This is because mature technologies are inherently "simple and straightforward," they have withstood the test of time, and their manufacture and delivery have been optimized by the vendors. Therefore, where mature technologies are feasible, vendors of new technology must focus on reducing investment and increasing operating efficiency to be competitive.

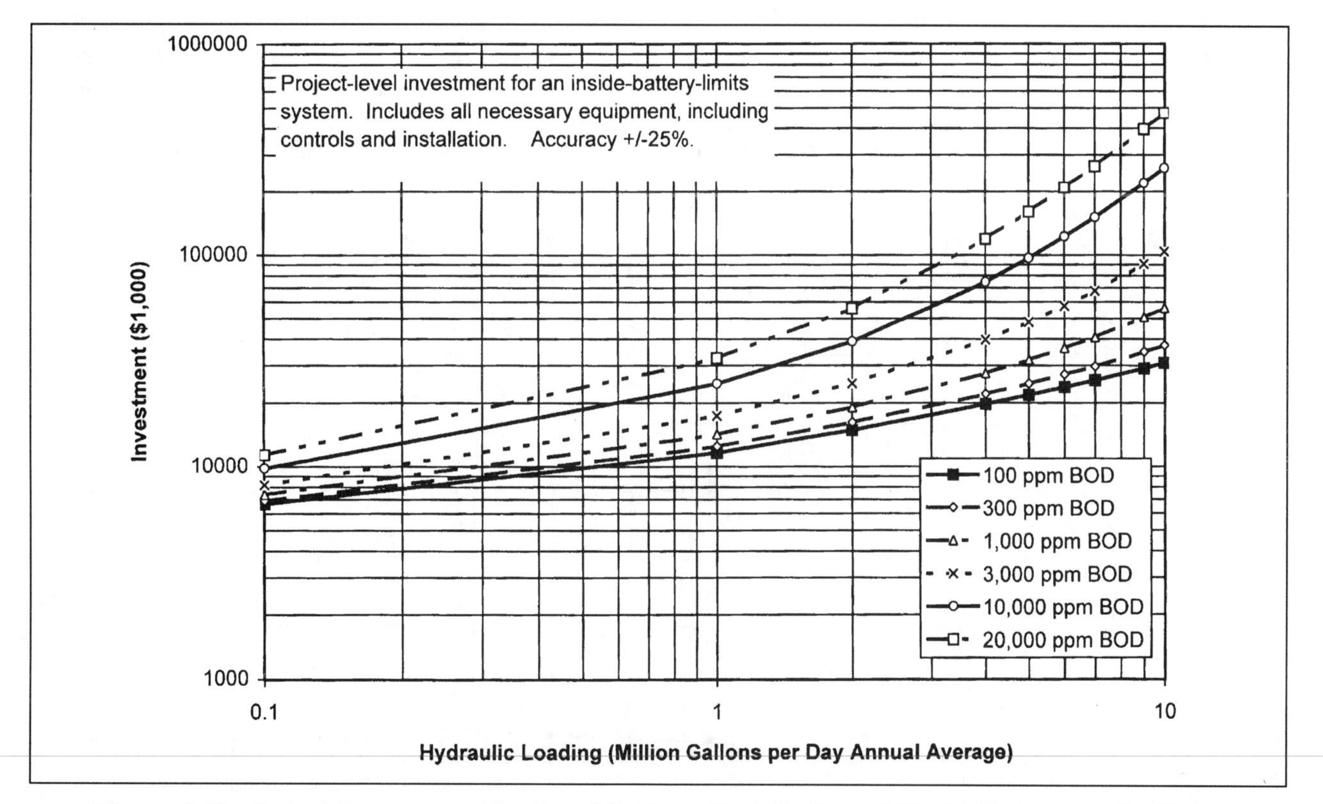

Figure 4-13. Actual Investment For Aerobic Deep-Tank Activated Sludge Treatment Facilities (U.S. dollars, ENR = 5700)

4.7.3 Investment, Cost, and Economics for Individual Technologies

The preceding discussion focused on technology selection (based on minimum NPC) for nonbiological wastewater treatment technologies. In this section we will discuss each of the available control technologies in more detail and present their individual investment, operating cost, and NPC as a function of flow and concentration.

The figures that follow summarize investment, operating cost, and net present cost for the principal dilute aqueous organic treatment technologies shown in Figure 4-16. Note that these costs represent minimum costs for an inside-battery-limits system. Site-specific investment will probably be 1.5 to 2.5 times the values shown in Figures 4-17, 4-20, and 4-23. The technologies were subdivided into three groupings.

1. Air stripping plus abatement of the off-gas and stream stripping plus incineration of the overhead condensate (Figures 4-17, 4-18, and 4-19).
2. UV/H_2O_2 advanced oxidation (AO), granulated activated carbon, and reverse osmosis plus incineration of the concentrate stream (Figures 4-20, 4-21, and 4-22.)
3. Wet oxidation and electron-beam irradiation (Figures 4-23, 4-24, and 4-25).

Air Stripping Plus Thermal Oxidation of the Off-Gas. VOCs are stripped from the wastewater with ambient or heated air in a packed or tray column. The off-gas from the stripper column is thermally oxidized in a thermal or catalytic oxidizer. If halogenated VOCs are present, a water or caustic scrubber may be required downstream of the oxidizer to scrub acid gases, such as HCl, HBr, and HF. In some cases, a carbon adsorber may be more attractive than an oxidizer if recovery of the stripped VOCs is desirable or if natural gas is not available.

Steam Stripping Plus Incineration of the Condensed Organic Phase. This technology is similar to air stripping, except that steam is used instead of air to strip the VOCs from the wastewater. A water-cooled condenser and decanter are used overhead to condense the steam/organic vapor and separate the condensate into organic and aqueous phases. The aqueous phase is recycled to the stripper, while the organic phase is incinerated. In some cases, it may be feasible to recycle recovered VOCs to the process.

Granular Activated Carbon Adsorption Plus Carbon Regeneration by a Vendor. Organics are removed by activated carbon in fixed-bed adsorbers. A typical system contains two beds in parallel (one adsorbing and one spare) and is skid-mounted. The adsorption capacity of the carbon (lb organic/lb carbon) will vary with compound type, concentration, pH, degree of polarity, and so on. When the carbon is saturated with organics, the vendor will thermally regenerate the carbon at its own facility.

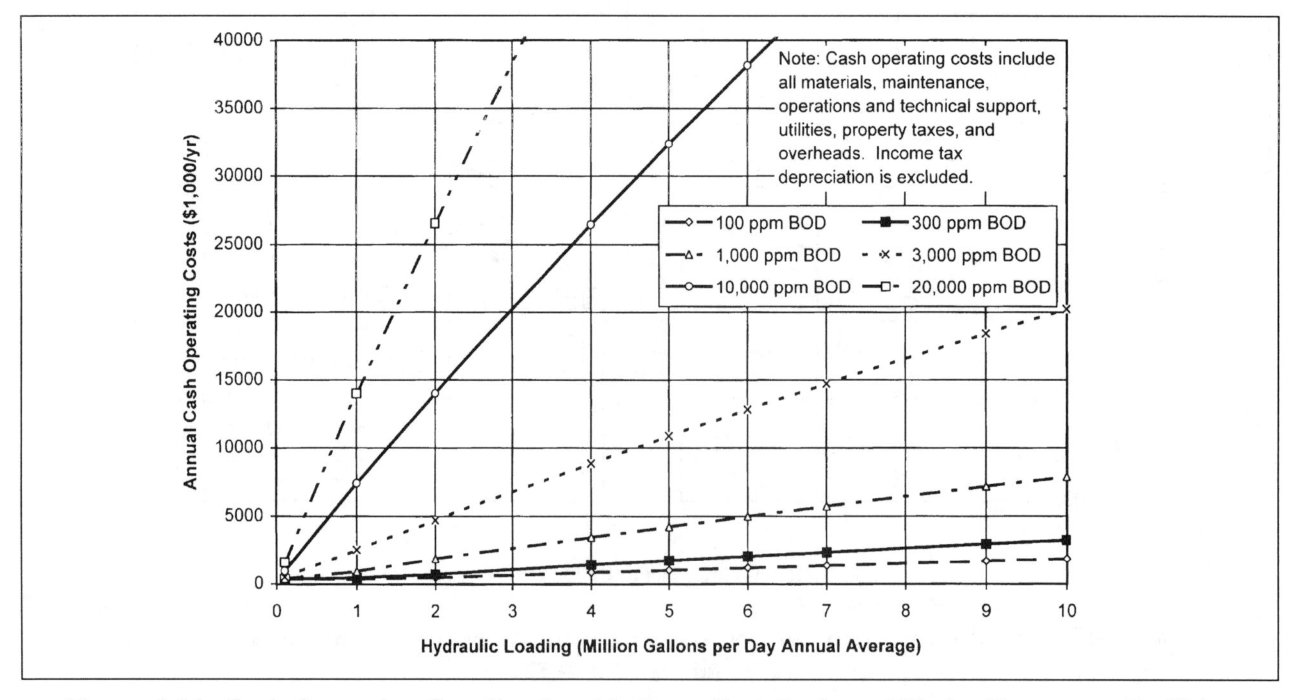

Figure 4-14. Cash Operating Cost For Aerobic Deep-Tank Activated Sludge Treatment Facilities (1999 U.S. dollars)

Reverse Osmosis Plus Incineration of the Concentrate. Reverse osmosis is a membrane process that is selectively permeable to water, but excludes ionic species (dissolved salts) and some organics. Pressures higher than the naturally occurring osmotic pressure force "pure" water through a semipermeable membrane, producing a concentrated brine waste. The concentrate is then incinerated or, in some cases, recycled to the process.

UV/Hydrogen Peroxide Advanced Oxidation. This is a chemical oxidation process that uses ultraviolet light and hydrogen peroxide to partially or fully oxidize organic contaminants to nontoxic or less hazardous compounds. For example, partial oxidation may make a recalcitrant compound biotreatable. Highly reactive hydroxyl radicals are generated *in situ*. In some cases, iron is used as a catalyst. The technology is applicable for dilute (typically < 100 parts per million by weight (ppmw) total organic carbon) aqueous organic wastes.

Fenton Reaction and Precipitation. Also a chemical oxidation process for dilute aqueous wastes, the Fenton reaction utilizes an iron catalyst (e.g., ferrous sulfate, ferric chloride) and hydrogen peroxide to oxidize the contaminants in tanks. Additional organics can be removed by controlled precipitation of the iron catalyst. The spent catalyst solids must be discarded via incineration, landfilling, and so on.

Electron-Beam Irradiation. Electron-beam irradiation is an advanced oxidation process for dilute aqueous wastes that uses an electron beam for *in situ* generation of highly reactive oxidizing species, such as hydroxyl radicals and solvated electrons. Hydrogen peroxide is sometimes used to enhance oxidation rates, but the technology is developing.

Wet Oxidation. Aqueous wastes are oxidized at high temperature (greater than 150°C) and elevated pressure (200 to 2000 psig) using compressed air or oxygen. The operating pressure is a function of the degree of oxidation required and the specific application, that is, biosolids mass reduction, carbon regeneration, or mineralization of aqueous organic waste. The technology is applicable for more concentrated aqueous wastes (greater than 10,000 ppm by weight).

4.8 Solid-Waste Treatment Costs

The classification of a "solid" waste as either nonhazardous or hazardous as defined under the Resource Conservation and Recovery Act (RCRA) has a significant impact on the cost of disposal. For example, solid, liquid, semisolid, or containerized-gas waste streams containing organic solvents will be classified as a hazardous waste under RCRA if they

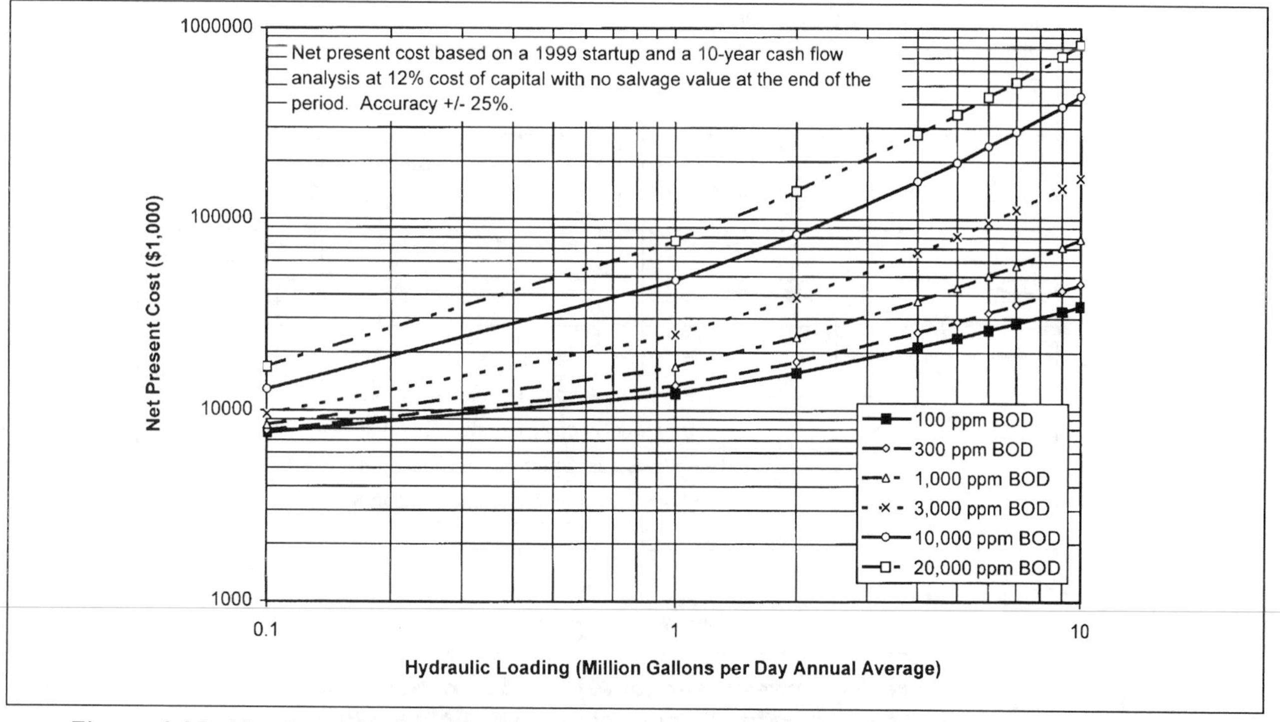

Figure 4-15. Net Present Cost For Aerobic Deep-Tank Activated Sludge Treatment Facilities (1997 U.S. dollars, 10 years, 12% discount rate)

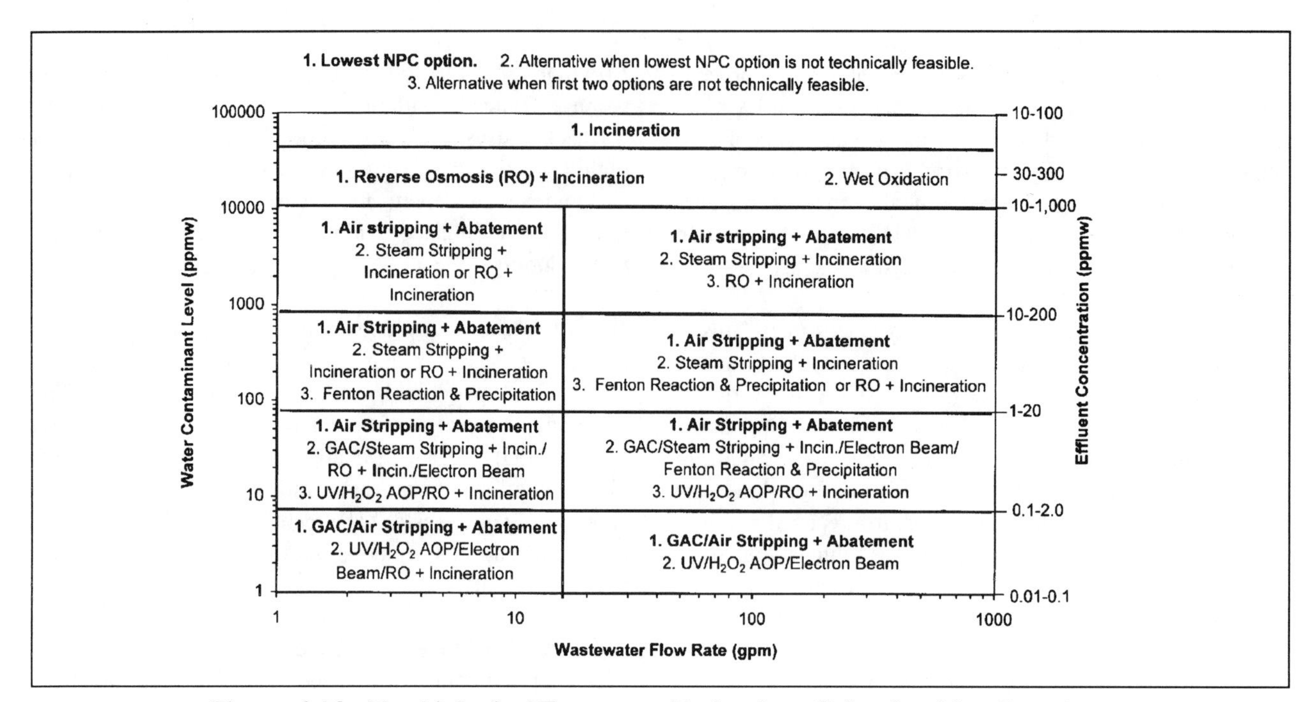

Figure 4-16. Nonbiological Treatment Technology Selection Map Based on Engineering and Economic Evaluations for Stand-alone Applications

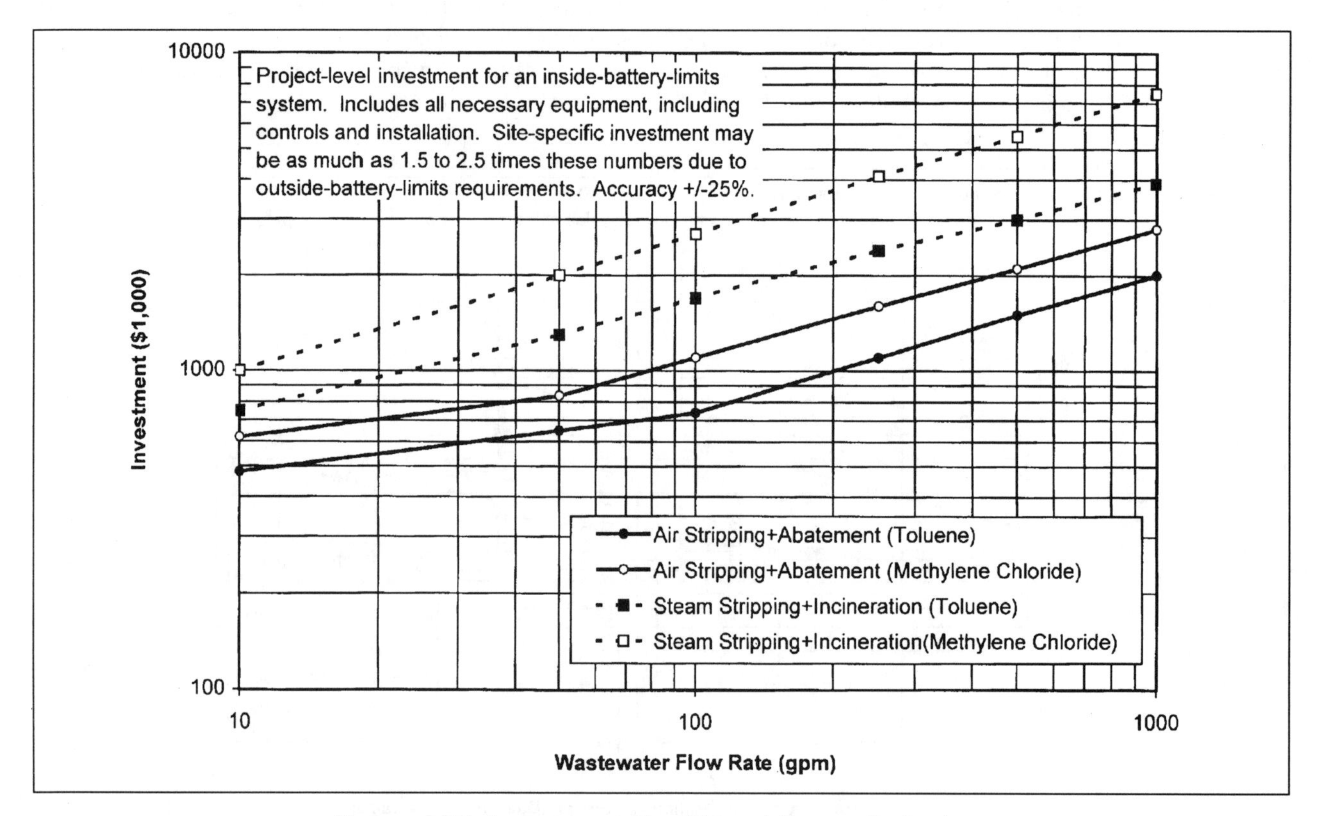

Figure 4-17. Investment For Air and Steam Stripping (Concentration Implicit in Cost Curves, ENR = 5700)

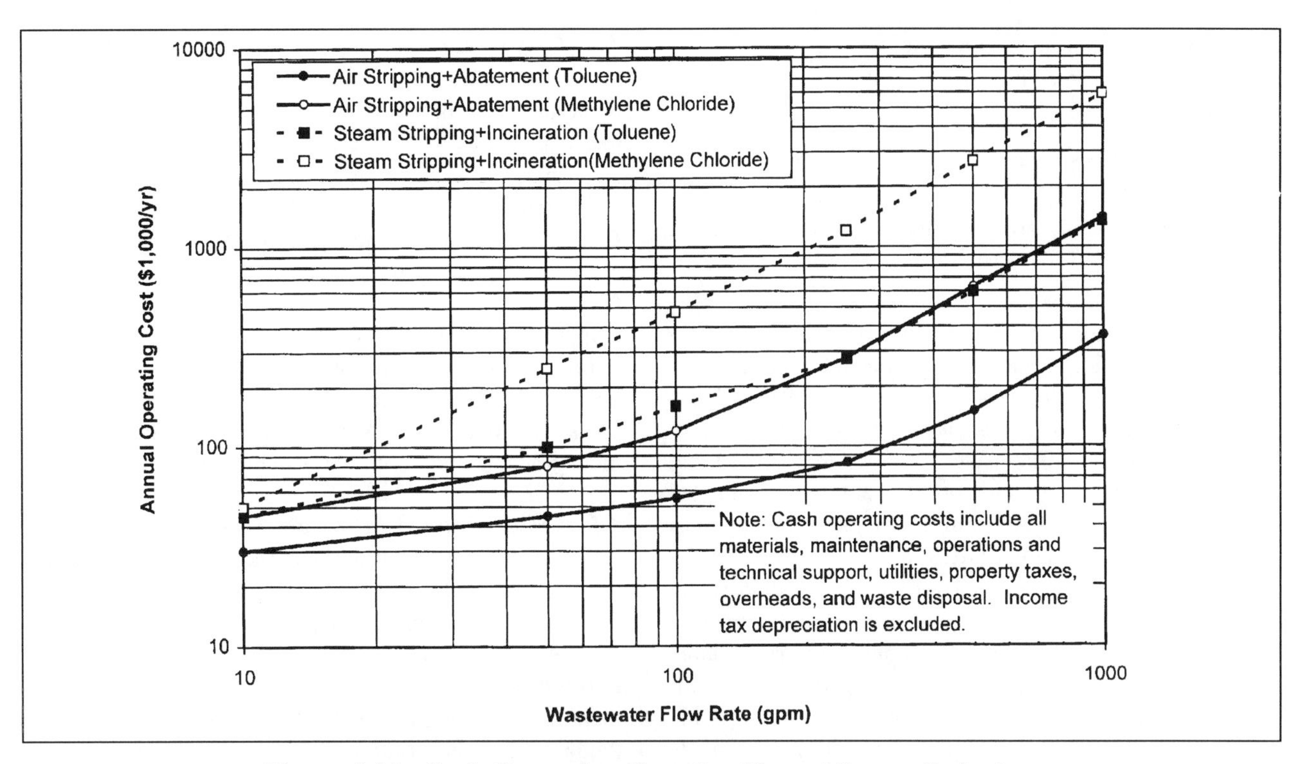

Figure 4-18. Cash Operating Cost For Air and Steam Stripping (1999 U.S. dollars)

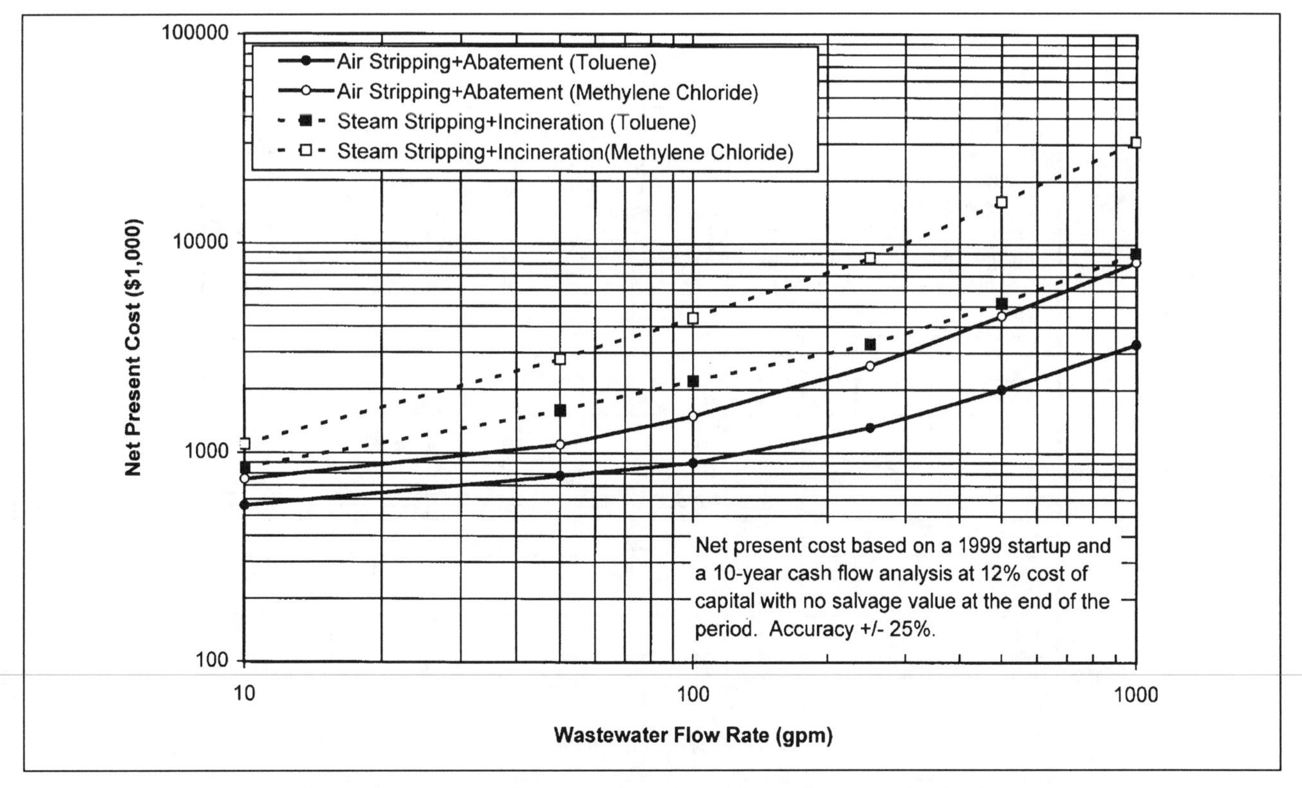

Figure 4-19. Net Present Cost For Air and Steam Stripping (1997 U.S. dollars, 10 years, 12% discount rate)

- Exhibit one or more of the four EPA-defined characteristics: corrosiveness, ignitability, reactivity, and toxicity
- Are specifically listed in the regulations, that is, F-, K-, P-, and U-listed wastes
- Are declared hazardous by the generator based on its knowledge of the waste.

A good number of organic solvent wastes will exhibit the characteristic of ignitability (flash point below 140°F) or toxicity [by failing the toxicity characteristic leaching procedure (TCLP)]. The objective of the TCLP is to determine whether the toxic constituents in a solid-waste sample will leach into groundwater if the waste is placed in a solid-waste landfill. In addition, many spent halogenated and nonhalogenated solvents from "nonspecific" sources are listed as F-wastes under RCRA.

Table 4-2 summarizes typical commercial solid-waste treatment costs. These are ballpark costs, and can vary substantially depending on waste composition, available commercial incinerator or landfill capacity, and geographic location. The intent of Table 4-2 is to provide approximate costs, so that a quick estimate can be made of the incentive for pollution prevention. Readers may want to obtain their own estimates based on their own site experience.

4.9 Examples

Four scenarios follow that demonstrate how the investment and cost curves for end-of-pipe treatment technologies can be used to define the incentive for pollution prevention. Each scenario is built around a typical situation encountered within the manufacturing industries. The four scenarios are (1) understanding the incentive for segregating VOC-contaminated air within a large building; (2) defining the cost to treat the air exhaust from a dryer containing chlorinated VOCs; (3) calculating the economic price tag for abating a waste-gas stream containing not only VOCs but also particulates; and (4) estimating the costs to treat wastewater streams containing both biodegradable and nonbiodegradable materials.

4.9.1 High-Flow, VOC-Laden Air Stream

The first scenario illustrates the investment and cost savings that are possible by isolating or segregating general building ventilation air from the contaminated air surrounding a piece of VOC-emitting equipment. The example case involves a large building housing process equipment that uses nonhalogenated solvents. Stack testing of the building exhaust air vent reveals that VOC emissions are significantly higher than the state air permit allows, and end-of-pipe treatment might be required to bring the process vent into compliance. The waste gas has a flow

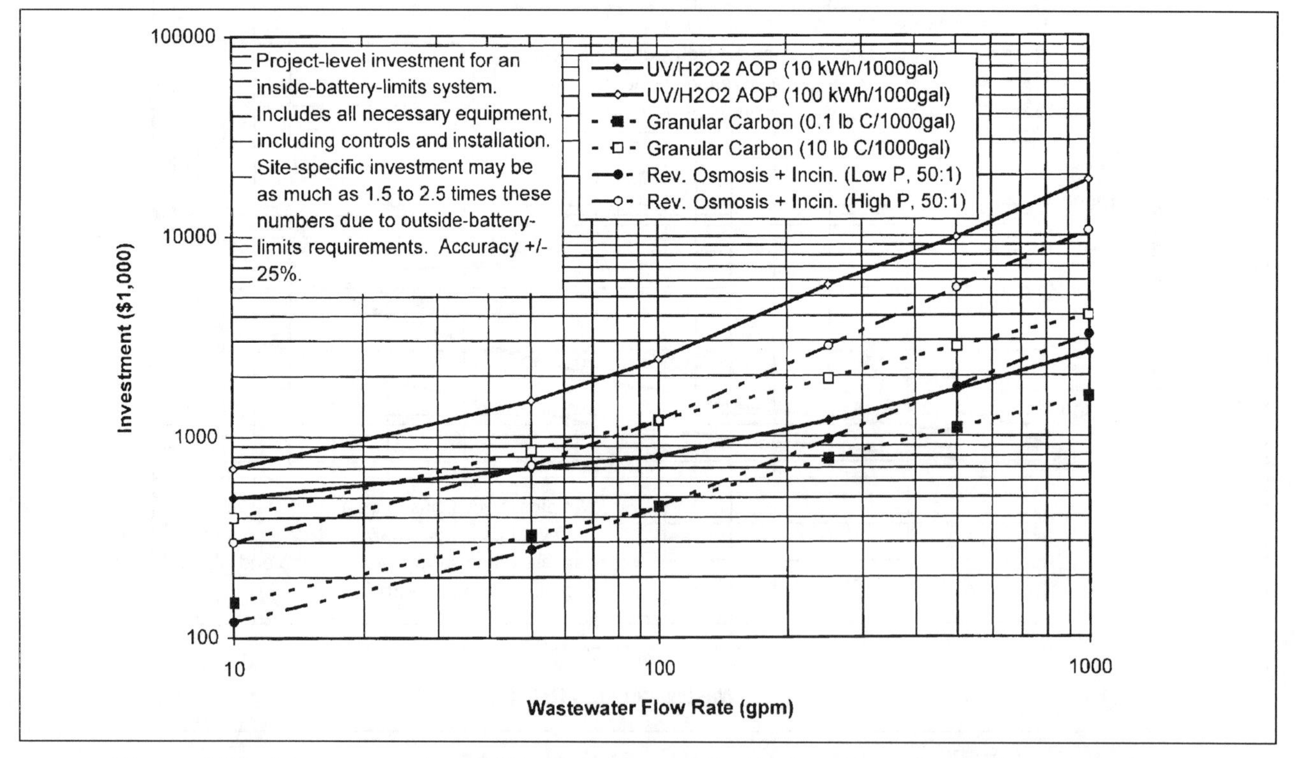

Figure 4-20. Investment For UV/H_2O_2 Advanced Oxidation, Granular Activated Carbon, and Reverse Osmosis (Concentration Implicit in Cost Curves, ENR = 5700)

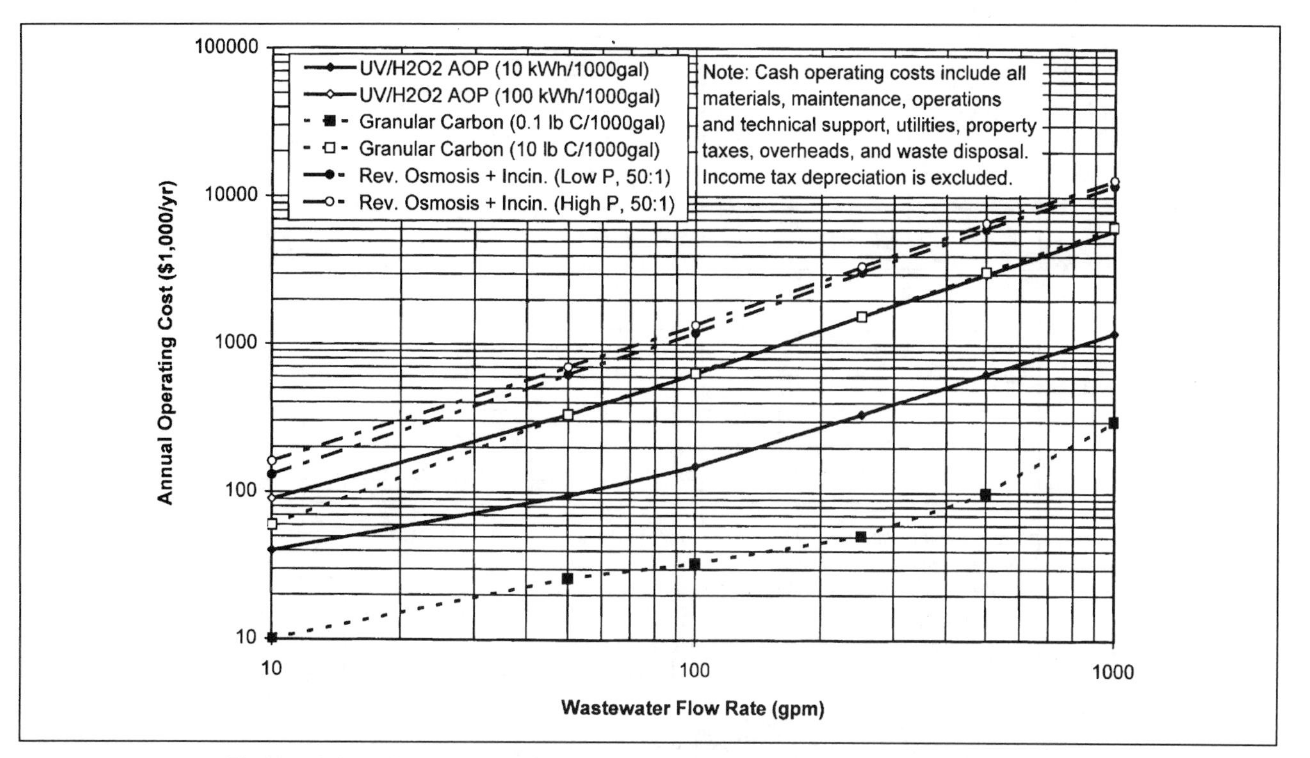

Figure 4-21. Cash Operating Cost For UV/H_2O_2 Advanced Oxidation, Granular Activated Carbon, and Reverse Osmosis (1999 U.S. dollars)

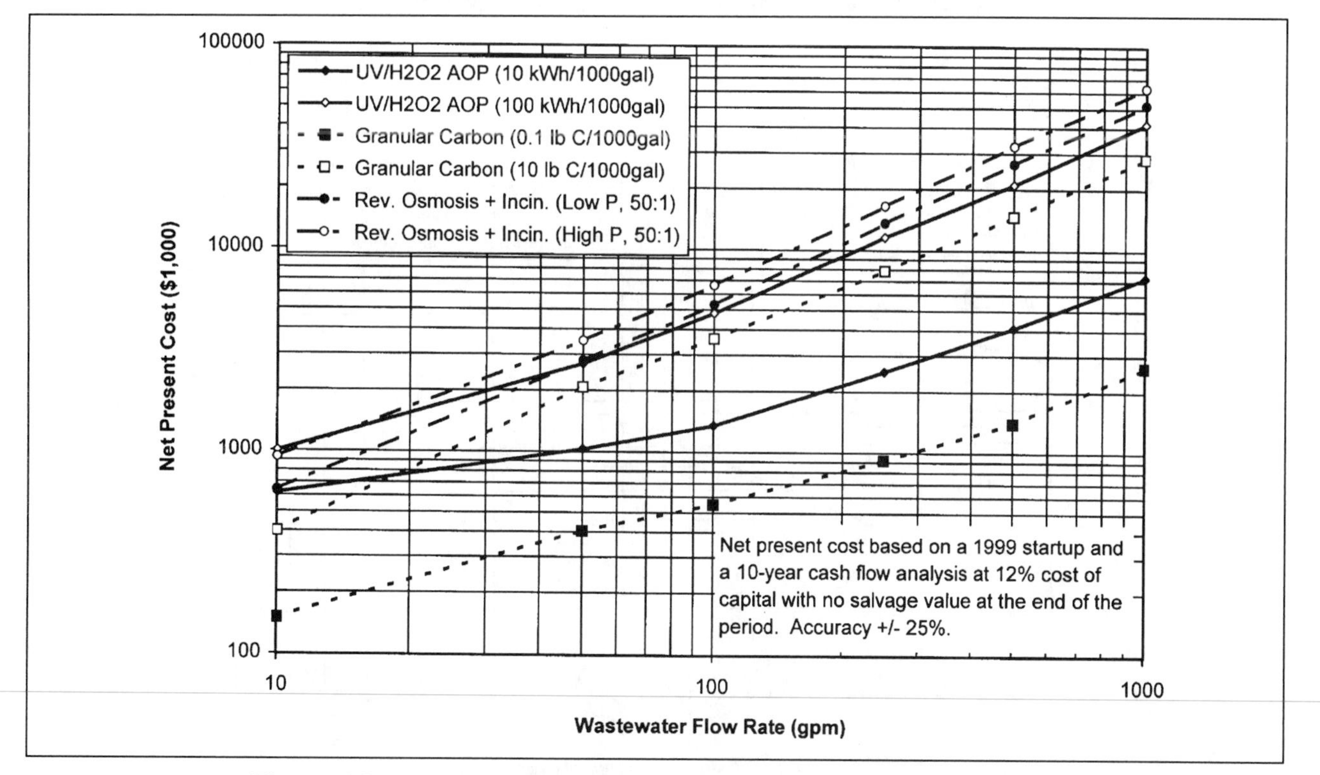

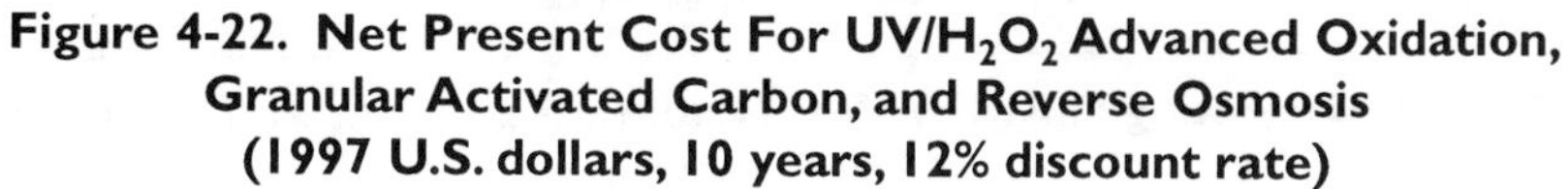
Figure 4-22. Net Present Cost For UV/H_2O_2 Advanced Oxidation, Granular Activated Carbon, and Reverse Osmosis (1997 U.S. dollars, 10 years, 12% discount rate)

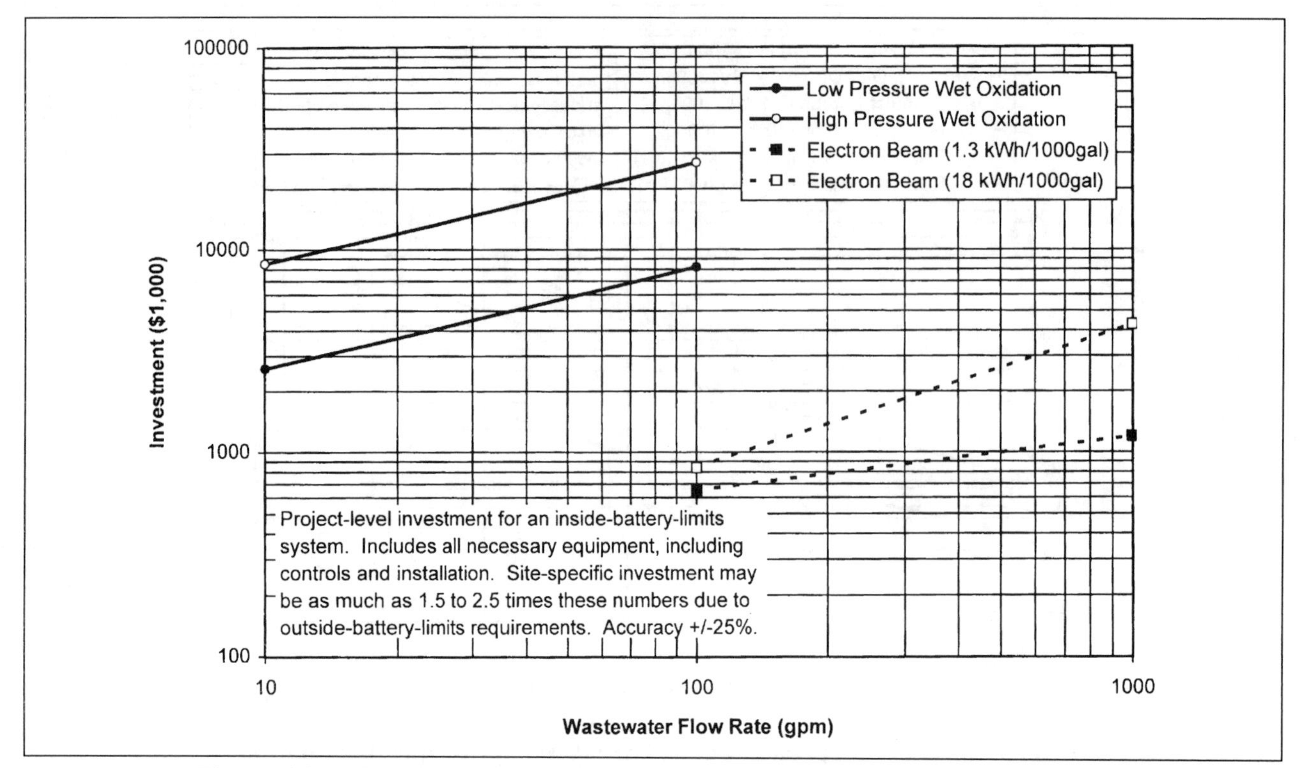

Figure 4-23. Investment For Wet Oxidation and Electron Beam Irradiation (Concentration Implicit in Cost Curves, ENR = 5700)

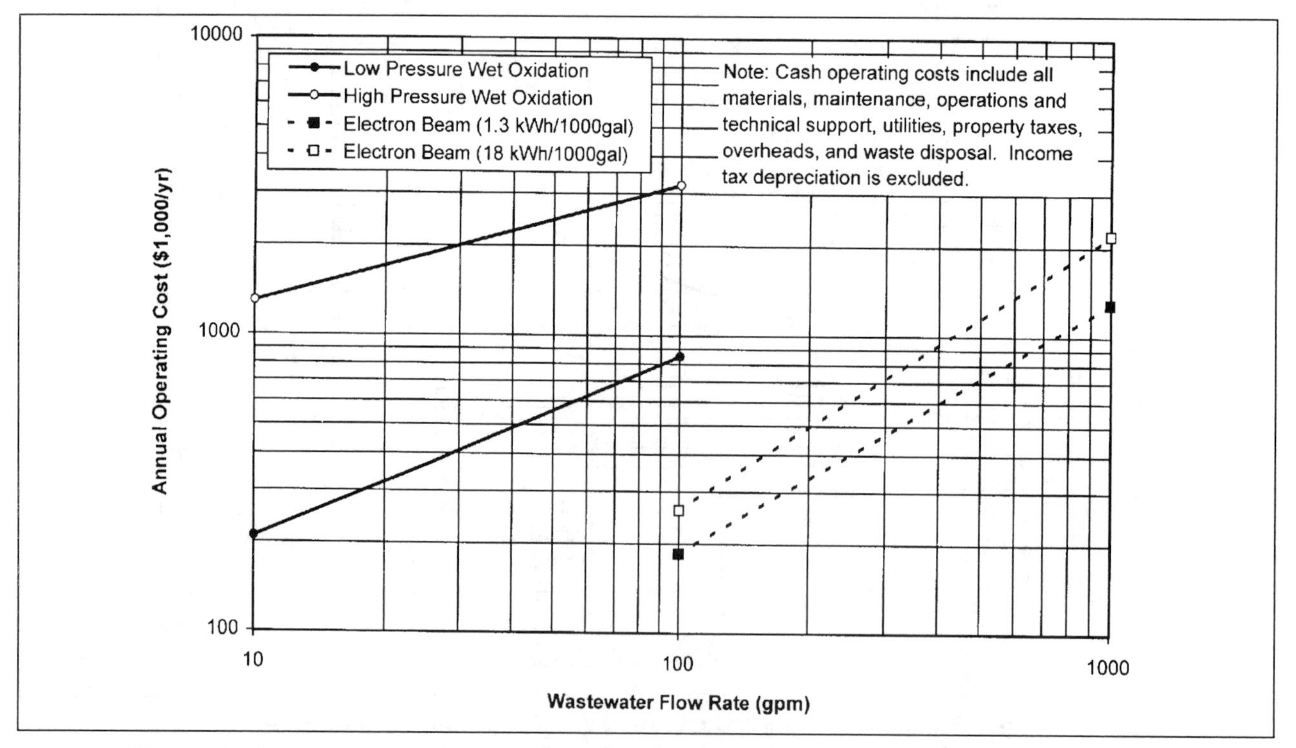

Figure 4-24. Cash Operating Cost For Wet Oxidation and Electron Beam Irradiation (1999 U.S. dollars)

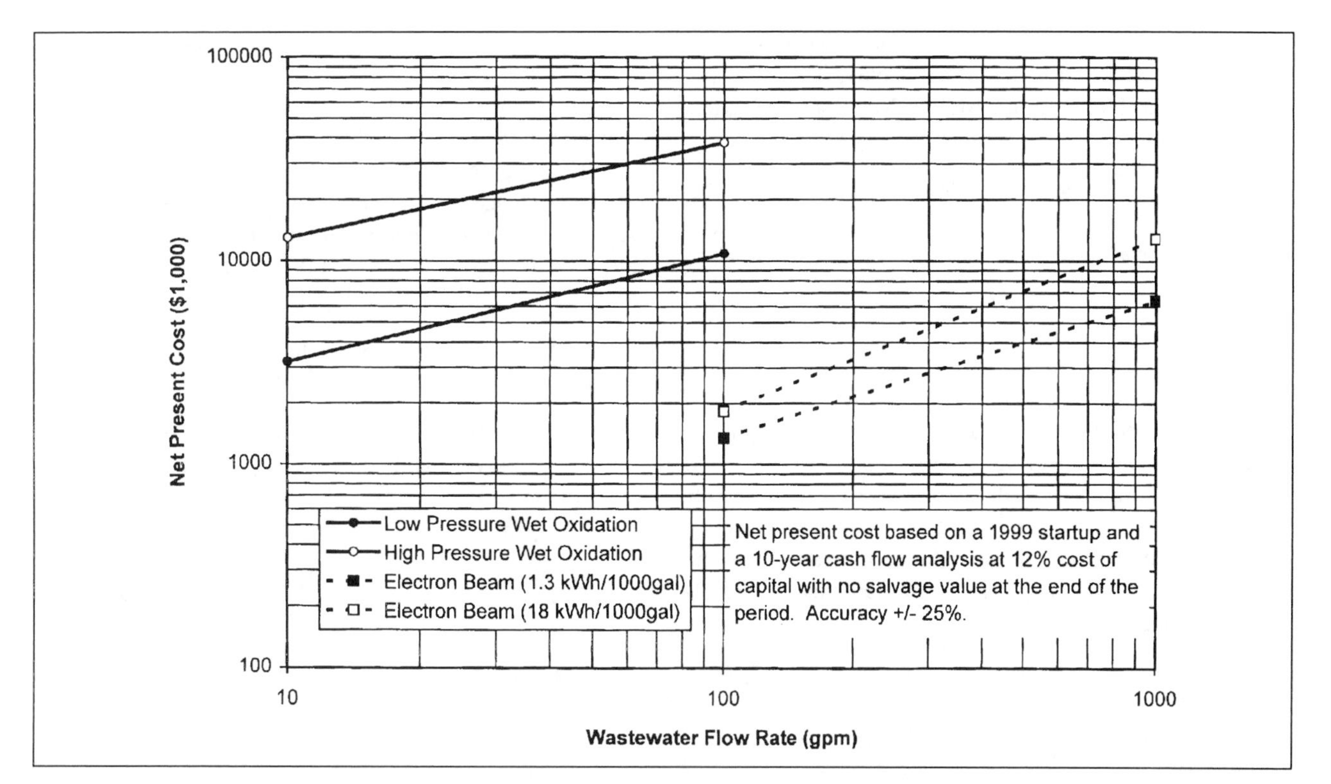

Figure 4-25. Net Present Cost For Wet Oxidation and Electron Beam Irradiation (1997 U.S. dollars, 10 years, 12% discount rate)

Solid Waste Type	U.S. $/lb*	Disposal Method
Bulk Organic Liquids	0.35	Incineration
Sludges (w/ organics)	1.15	Incineration
Sludges (w/ inorganics)	0.43	Stabilization + Secure Landfill
Solids (w/ organics)	0.85	Incineration
Solids (w/ trace organics)	0.13	Secure Landfill

* excludes freight

Table 4-2. Typical Commercial Solid Waste Disposal Costs (Excludes Freight)

rate of about 100,000 scfm and contains nonhalogenated VOCs at a concentration of 100–200 ppmv. The high-spot investment, cash operating cost, and NPC for abating this stream would be:

Investment (Figure 4-6)	= \$40/scfm x 100,000 scfm	= \$4,000,000
Cash operating cost (Figure 4-7)	= \$9.5/scfm x 100,000 scfm	= \$950,000
NPC (Figure 4-8)	= \$60/scfm x 100,000 scfm	= \$6,000,000

The VOCs in the building exhaust are solvents from process equipment in the middle of a large building enclosure. The process supervisor knows of a similar situation where a printing press in the center of a large room was enclosed and the air to be treated was reduced in flow by tenfold. Assuming a tenfold reduction in air flow, the investment, operating cost, and net present cost of end-of-pipe treatment would drop to:

Investment (Figure 4-6)	= \$100/scfm x 10,000 scfm	= \$1,000,000
Cash operating cost (Figure 4-7)	= \$21.5/scfm x 10,000 scfm	= \$215,000
NPC (Figure 4-8)	= \$170/scfm x 10,000 scfm	= \$1,700,000

This reduction in air flow results in a reduction in investment of \$3,000,000. Past experience has shown that the enclosure would cost from \$250,000 to \$700,000; therefore, the net investment savings would be from \$2,300,000 to \$2,750,000. A pollution-prevention program could potentially identify other options to reduce or even eliminate investment in end-of-pipe treatment, such as using less volatile and/or unregulated solvents, changing operating procedures for solvent usage, and better sealing the process equipment.

4.9.2 Moderate-Flow, HVOC-Laden Air Stream

The second example demonstrates the high costs associated with treatment of halogenated VOCs in a waste gas, and the need to consider treatment of a single stream versus a combined vent header comprising multiple waste gas streams.

A chemical facility generates several waste-gas streams containing different levels of VOCs and HVOCs. These gas streams are combined in a common header and vented to the atmosphere untreated. The business wishes to expand the facility; however, any expansion will result in the site exceeding the VOC mass emission limits allowed by the current state air permit. Therefore, any expansion in plant capacity will require that one or more of the existing waste-gas streams be treated to reduce HVOC/VOC emissions. One such stream is the exhaust from a drying operation with a flow rate of 35,000 scfm and containing chlorinated organics.

Using Figures 4-6 through 4-8 for a HVOC-containing stream, one needs to interpolate between a flow of 10,000 scfm and 50,000 scfm to obtain the investment, cash operating cost, and NPC for 35,000 scfm. The calculations would go as follows using Figure 4-6:

Investment at 10,000 scfm = \$160/scfm x 10,000 scfm = \$1,600,000
Investment at 50,000 scfm = \$85/scfm x 50,000 scfm = \$4,250,000

A power scaling factor, which relates the two flow rates and the two investment values, is required to calculate the investment at 35,000 scfm. This scaling factor is calculated as follows:

$$\left(\frac{50{,}000 \text{ scfm}}{10{,}000 \text{ scfm}}\right)^{z} = \frac{\$4{,}250{,}000}{\$1{,}600{,}000} = 2.66$$

Rearranging and solving for z gives z = 0.61. Therefore, at 35,000 scfm,

$$\left(\frac{35{,}000 \text{ scfm}}{10{,}000 \text{ scfm}}\right)^{0.61} = 2.15$$

and

$$(\$1{,}600{,}000)(2.15) = \$3{,}440{,}000$$

which gives the investment at 35,000 scfm.

Cash operating costs are calculated in the same way. Using Figure 4-7,

Cash operating cost at 10,000 scfm = \$56/scfm x 10,000 scfm = \$560,000
Cash operating cost at 50,000 scfm = \$46/scfm x 50,000 scfm = \$2,300,000

The scaling factor, which relates the two flow rates and two cash operating-cost values, is calculated as follows:

$$\left(\frac{50{,}000 \text{ scfm}}{10{,}000 \text{ scfm}}\right)^{z} = \frac{\$2{,}300{,}000/\text{yr}}{\$560{,}000/\text{yr}} = 4.107$$

Rearranging and solving for z gives z = 0.88. Therefore, at 35,000 scfm,

$$\left(\frac{35{,}000 \text{ scfm}}{50{,}000 \text{ scfm}}\right)^{0.88} = 0.73$$

and

$$(2{,}300{,}000/\text{yr})(0.73) = \$1{,}680{,}000/\text{yr}$$

which gives the cash operating cost at 35,000 scfm.

If the net present cost calculations are done in the same way using Figure 4-8, one obtains z = 0.79 and NPC at 35,000 scfm equal to $10,000,000 over 10 years at 12% discount rate.

Similar analyses can be done for the other waste-gas streams feeding the common vent header. The hope is that the required emissions reduction can be achieved by abating only a single stream for less investment and cash operating cost, such as a lower-flow-rate stream containing a large fraction of the VOC emissions. The best path forward for the business, then, is to use these investment and cost values as the incentive for establishing a pollution-prevention program at the facility. This program should consider segregation of waste-gas streams in the vent header to reduce the cost of treatment.

4.9.3 Waste-Gas Stream with VOCs and Particulates

The third example considers the impact of particulates and permit costs on the investment, operating cost, and NPC of treating a waste-gas stream containing sulfur-bearing VOCs and submicron-size aerosols.

Expansion of an existing textile fiber manufacturing facility will consist of a new spinning line that uses a special finish oil for the fiber. The finish oil is applied to the fiber surface to lubricate the fiber as it is pulled over and through various guides. During this process, the finish oil flashes into the surrounding ventilation air; the exit vapor stream contains VOCs and aerosols. The special finish oil has sulfur-based compounds that decompose into odoriferous disulfides.

The waste-gas stream has a flow rate of 1000 scfm and contains 0.1 to 1-micron-size aerosols and VOCs, some of which contain sulfur. Investment, cash operating cost, and NPC for this waste-gas are calculated by summing the contributions of both the VOC and the aerosol particulate. Because sulfur is present in the VOCs, the cost curves for HVOC-containing waste-gas streams are used. The assumption is that the sulfur dioxide, which is generated by oxidation of the sulfur atom in the VOCs, will need to be removed in a caustic scrubber before discharge to the environment. The cost calculations would proceed as outlined below. For VOC abatement,

Investment (Figure 4-6)	= $450/scfm x 1000 scfm	=	$450,000
Cash operating cost (Figure 4-7)	= $168/scfm x 1000 scfm	=	$168,000
NPC (Figure 4-8)	= $1100/scfm x 1000 scfm	=	$1,100,000

For aerosol control, Figure 4-12 provides guidance on which technology is most appropriate. In this particular case, we assume that a wetted deep-bed filter will be installed upstream of the VOC abatement unit to remove the aerosols from the waste-gas stream. The cost calculations for particulate control are as follows:

Investment (Figure 4-9)	= $120/scfm x 1000 scfm	=	$120,000
Cash operating cost (Figure 4-10)	= $13/scfm x 1000 scfm	=	$13,000
NPC (Figure 4-11)	= $147/scfm x 1000 scfm	=	$147,000

This is a new vent stream for this facility; therefore, permitting costs should also be considered. These costs can be obtained from Figure 4-8. In this case, the NPC for permitting at 1000 scfm would be about $220,000 (1000 scfm x $220/scfm). The total investment is the sum of the individual investments for VOC and aerosol control. The total NPC is the sum of the net present costs for VOC and aerosol control plus permitting.

Total investment = $450,000 + $120,000 = $570,000
Total NPC = $1,100,000 + $147,000 + $220,000 = $1,467,000

A pollution-prevention program developed around this problem should focus on alternative finish oils that do not contain sulfur and that do not volatilize as easily, because the sulfur atom in the VOC drives the investment and cost for treatment. Additional efforts should focus on air flow reduction at the source and techniques to minimize aerosol formation.

4.9.4 New Biotreatment Facility for an Existing Manufacturing Site

The fourth scenario looks at the costs for treating wastewater streams containing biodegradable as well as nonbiodegradable compounds. In this case, pretreatment of the wastewater streams containing biorefractory compounds will be considered before final treatment in a new biological wastewater treatment plant.

An existing manufacturing site produces five wastewater streams, three that are sent to a local municipal wastewater treatment plant (streams 1, 2, and 3) and two that are incinerated because they contain nonbiodegradable material (streams 4 and 5). Due to new regulations, the municipality can no longer treat streams 1, 2, and 3. At the same time, new incinerator regulations will increase the cost of incinerating streams 4 and 5 tenfold. For these reasons, the business has decided to (1) install pretreatment technology for streams 4 and 5 to make them biodegradable, and (2) send the effluent from the pretreatment unit, along with streams 1, 2, and 3, to a new aerobic deep-tank activated-sludge biotreatment system. Table 4-3 contains the key flow and concentration parameters for each stream. Assume that after pretreatment, the sum of streams 4 and 5 is 150 gpm with 10 ppmw BOD. Therefore, the total flow and BOD concentration to the new wastewater treatment facility will be 850 gpm [1.22 million gallons per day (MGPD)] and 1180 ppmw BOD, respectively.

For the new wastewater treatment facility, we will use Figures 4-13 through 4-15 to obtain the investment, cash operating cost, and NPC, respectively. The calculations are as follows, based on 1.22 Mgpd and 1180 ppmw BOD:

Investment (Figure 4-13) = \$16,000,000
Cash operating cost (Figure 4-14) = \$1,000,000
NPC (Figure 4-15) = \$20,000,000

For the pretreatment of streams 4 and 5, we use the investment, cost, and economics for nonbiological aqueous organic treatment technologies. Assume for stream 1 that the appropriate pretreatment technology is reverse osmosis plus incineration (Figure 4-16). The investment, cash operating cost, and NPC for stream 1 can be obtained from Figures 4-20 through 4-22 at 100 gpm:

Investment (Figure 4-20) = \$450,000 to \$1,100,000
Cash operating cost (Figure 4-21) = \$1,200,000 to \$1,300,000
NPC (Figure 4-22) = \$5,200,000 to \$6,500,000

For stream 2, the appropriate pretreatment technology is assumed to be air stripping plus abatement (see Figure 4-16). The investment, cash operating cost, and net present cost for Stream 2 can be obtained from Figures 4-17 through 4-19 at 50 gpm:

Investment (Figure 4-17) = \$650,000 to \$850,000
Cash operating cost (Figure 4-18) = \$45,000 to \$80,000
NPC (Figure 4-19) = \$800,000 to \$1,200,000

Therefore, total investment, cash operating cost, and NPC to the business to treat the existing wastewater streams (pretreatment plus new biotreatment) are

Investment = \$17,100,000 to \$17,950,000
Cash operating cost = \$2,245,000 to \$2,380,000
NPC = \$26,000,000 to \$27,700,000

Thus, any pollution prevention program would focus on both reducing the hydraulic flow and organic loading. As was noted in Section 4.6.2, the incentive for source reduction is a minimum reduction in investment of \$3000 per gpm and \$6000 per lb organic/h.

Stream Number	Flow (gpm)	BOD (ppm by weight)
Biodegradable		
1	100	100
2	300	300
3	300	3,000
Total	700	1,430
Nonbiodegradable		
4	100	2,000
5	50	8
Total (before pretreatment)	150	1,336
Total (after pretreatment)	150	10

Table 4-3. Wastewater Stream Flows and Concentrations for New Biotreatment Facility

4.10 Engineering Evaluations and Pollution Prevention

Choosing the best options requires knowledge not only of the technical feasibility of the options, but also of the economic feasibility. In the next four subsections, we introduce engineering evaluation techniques and tools that help in the selection of the best options for implementation.

4.10.1 What Is Engineering Evaluation?

Engineering evaluation is the application of a full range of engineering skills to business decision making. It aids decision making by translating technical options into their economic impact—guidance that is fundamental to business decisions. It simultaneously answers the questions: "What is technically feasible?" and "What is it worth?".

4.10.2 How Does It Work?

The evaluation should quickly focus on only those data and analyses that are essential to quantify technical and economic feasibility. The evaluation may involve identifying technical options, defining the commercial process, flowsheeting, analyzing the process, defining manufacturing facilities, estimating investment and manufacturing costs, analyzing economics, and assessing risk. It requires the integration of these elements, such that the relationship of technology to economics can be clearly seen. The evaluation should provide the missing pieces and connect them together into a clear picture, leading to a deeper understanding. The appropriate level of definition ranges from a quick assessment to an ongoing, in-depth analysis.

4.10.3 Where Is It Valuable?

Engineering evaluation (Figure 4-26) is a holistic approach that provides valuable insight to research, manufacturing, environmental programs, plant projects, and business management. The evaluation aids by envisioning, defining, and assessing the commercialization of an idea. This leads to such things as:

- Identification of the research critical to technical and economic success
- A more economic manufacturing process
- The most economic achievement of environmental goals
- The idea with the most economic blend of project investment and manufacturing costs
- More competitive technologies
- Determining the economic impact of a technology on business strategy

Through this holistic approach, the evaluation provides an objective view for decision making, grounded in both engineering science and economics, to improve competitive position.

The 10-Step Method
"A Road Map to the Best Solution"

Define Problem and Goals

1. Define the Problem
2. Set the Goals

Identify the Alternatives

3. Identify All Possible Alternative Solutions

Define the Alternatives
For Each Alternative:

4. Develop Flowsheets & Facility Scopes
5. Define Development and Operating Requirements

Evaluate the Alternatives
For Each Alternative:

6. Estimate the New Investment Required
7. Estimate the Change in Cash Costs and Revenues
8. Estimate the Net Present Value
9. Determine All "Non-Economic" Considerations

Then

10. Choose the Best Alternative

Each Step Builds Upon Previous Steps,
So Don't Skim Over The Early Steps

Iterative Evaluations
Give insight on what's important, the detail required, potential problems & risks, and how to formulate better alternatives

Figure 4-26. The 10 Steps of Engineering Evaluations

4.11 The 10-Step Method

The 10-Step Method was developed by the Engineering Evaluations group in DuPont to outline the necessary steps to properly identify, evaluate, and choose the best alternatives for a particular engineering problem. It has been successfully applied throughout DuPont as part of not only new process development programs, but also environmental programs. If you compare the 10-Step Method (Figure 4-26) to the path to pollution prevention (Figure 4-1), you will notice a lot of similarity, particularly in the assessment phase. This is not surprising, because the principles of economic and engineering evaluations are critical to any pollution-prevention program, not only in choosing the best options for waste reduction, but also for securing business buy-in to the program.

Each of the 10 steps are described in more detail below. Although the words are slightly different, each of the 10 steps can be found in one of the three phases of the path to pollution prevention. Steps 1 and 2 align most closely with the chartering phase and the initial steps of the assessment phase. Steps 3 through 9 coincide with the last three steps in the assessment phase: generate ideas, screen options, and evaluate screened options. Finally, Step 10 in the 10-Step Method represents the first step of the implementation phase. Figure 4-27 lists engineering evaluation terms and definitions that will be used throughout this section. Refer to it if you are uncertain about the meaning or significance of a particular term or phrase.

4.11.1 Define Problem and Set Goals (Steps 1 and 2)

Steps 1 and 2 define exactly what the problem is and, specifically, what goals you hope to achieve. It is hard to get someplace if you do not know where you are and where you are going.

Start Early. An early start with a sound approach often requires less overall effort and allows time to develop a wider range of solutions. For example, if you start late, such as well into the process design phase of a project, you may not have enough time to develop and implement process waste reduction ideas and may end up with expensive end-of-pipe treatment only.

Provide Background Information. Background information can include business economics, production forecasts, manufacturing process information, waste-stream characterization, applicable company policies, and environmental regulations.

Engineering Evaluation Terms and Definitions

Cash Costs (business out-of-pocket costs). Dollars spent to make and sell products. This includes R&D, ingredients, utilities, wages, salaries & benefits, all direct or indirect costs, and non-capitalized project-related costs (project front-end-loading, project liaison, and start-up costs). Basically, any dollars spent that are not an investment. Differs from "investment" due to its different impact on income taxes.

Depreciation. In economic analyses, depreciation can be viewed as a "tax deduction" that can be subtracted as a cost when calculating pretax earnings. It is equal to a portion of new facilities investment. The allowable amount (specified by the IRS) varies by investment class (Chemical and Allied Products equipment, steam generating equipment, etc.) and the number of years since the investment was made. Most of chemical industry's U. S. investments are "depreciated" over six years (20%-year of start-up, 32%-1st year after start-up, then 19%, 12%, 12%, and 5%). Although depreciation does impact the amount of income taxes paid, it is not a real cash cost and, therefore, cannot be saved.

Development and Operating Requirements. As used here, this is all the other bases needed (other than that information contained in the flowsheets and investment estimates) for estimating all "cash costs" resulting from implementing an alternative. It will include R&D (i.e., development facility costs, researcher's time, etc.), on-going operating personnel (i.e., an additional operator per shift at $25/hr, wages and benefits), annual maintenance labor & materials (i.e., 3% of project investment), technical support (a plant support engineer), and non-capitalized project-related costs (i.e., 20% of project investment).

Flowsheet. A process diagram showing major process steps/equipment, processing conditions (temperature, pressure, etc.), and material and utility (steam, electricity, etc.) flows.

Income Taxes. A portion of earnings (profit) paid to governments. Currently, combined U.S. and state income taxes equal 38% of pretax earnings.

Investment. Dollars committed to execute and support the manufacture and sale of products. Includes both working capital and permanent investment (project or physical facilities). New project investment is generally the primary new investment. Investment differs from "costs" due to its different impact on income taxes.

Net Present Value (NPV). The single best measure of economic merit; an "economic price tag." It is the value to a business, expressed in today's after-tax dollars, of the alternative. Mathematically, it is the sum of all the discounted (to the present) net cash flows for the years of the analysis. The appropriate discount rate to use is the business's opportunity cost of capital, which is higher than the company's cost of capital. Unless the discount rate is specified, it is assumed that a discount rate equal to business's cost of capital, currently 12%, was used. Net Present Cost (NPC) is the negative of NPV.

Figure 4-27. Engineering Evaluations Terms and Definitions

Be Specific Enough, But Do Not Be Too Specific. Be specific enough for focus and direction, but not so narrowly defined that you preclude some very good solutions.

Prioritize Goals. Determine which goals are necessary and which are desirable. Are all goals firm?

Recognize Constraints, Limitations, and Opportunities. Are there other regulations that affect your process facility that you wish to address at this time? What other plant areas will potentially be utilized or impacted? These areas might include existing wastewater treatment facilities, utility production, or other manufacturing areas. Are other benefits desired, such as improving groundwater protection, reducing ground-level odors, and internalizing site waste? Is the waste-management hierarchy accepted within the business? That is, (1) minimizing waste generation; (2) minimizing waste introduction; (3) segregation and reuse; (4) recycle; (5) recover energy value in waste; (6) treat for discharge; and (7) disposal. What are the economic guidelines or constraints?

Non-Economic Considerations. These are important considerations that are not stated in economic terms such as safety, attaining company environmental goals, operability, technological risk, adaptability to tightening regulations, public relations, etc. Differences in these "non-economic" considerations must be weighed against the differences in the economic price tag of the alternatives.

Real Cash Flows. Actual dollars that flow in or out of the corporation. The only significant real cash flows (remember the **"TRIC"**) that change as a result of implementing any alternative are 1) income Taxes, 2) **R**evenues, 3) new **I**nvestment, and 4) cash **C**osts. To define the economics of any alternative, you need to know the future changes in these four real cash flows caused by implementing the alternative.

Revenues. Dollars received from customers for products. A business only major source of dollars.

Scope-of-work. A description of physical facilities adequate to estimate the cost to build the facilities. This description can vary from very detailed to a less-detailed scope specifying only major equipment pieces, along with significant costing parameters such as material of construction, key size and rating parameters, new building requirements, new utility facilities (new steam boiler, etc.), and other general facilities (maintenance shops, roads, waste treatment, etc.).

Working Capital. Working capital is defined as current assets (e.g., inventories, accounts receivable, and prepaid expenses) less current liabilities (e.g., accounts payable and other accrued liabilities). For most engineering evaluations the inventory portion of working capital does not change because

- IRS tax rules state that the value of the inventories (raw material, intermediates, semi-finished product , and product) are that when the material was first put in the storage facility (Last In First Out), thus the value of the inventories do not change over time, and
- for most alternatives the inventory amounts do not change.

Unless there is a drastic change in the flow of materials along the value chain (purchases and sales) the accounts receivable and payable portion of working capital is also ignored.

Figure 4-27. Engineering Evaluations Terms and Definitions

4.11.2 Identify the Alternatives (Step 3)

The key is to identify *all* possible solutions. You cannot pick the best alternative if it is not on your list!

Provide Background. Appropriate background might cover site infrastructure (powerhouse, land available, waste treatment facilities, etc.); manufacturing process description, including processing conditions, yields, impurity levels, and cost overview; and all background information listed under "Define Problem and Set Goals" in the previous subsection.

Involve Diverse Viewpoints. Diverse viewpoints lead to diverse alternatives. Valuable viewpoints include (1) product and process research; (2) plant technical, operations, and maintenance; (3) business leadership; (4) marketing; (5) technology specialists; (6) diverse cultural and educational backgrounds; (7) organizational level; (8) age; (9) gender; and (10) viewpoints from outside the company, such as the community, suppliers, and customers.

Brainstorm, Do Not Judge. A brainstorming session will generate many ideas. This flow of creativity can be stifled by critical judgment of the ideas. Promote a positive, "can-do" atmosphere and have fun playing with the ideas.

Think Concepts, Not Details. Do not get bogged down with details, particularly during brainstorming.

Think Big. Consider integrated solutions involving other plants on the site, other sites in the company, suppliers, customers, and the rest of the world.

Think Combinations. The best alternatives may be a blend of approaches in the pollution-prevention hierarchy.

4.11.3 Define the Alternatives (Steps 4 and 5)

During this phase, each alternative will be taken from a concept to an alternative with complete process material and energy flowsheets, facility scopes-of-work, and development and ongoing operating/maintenance requirements. The details required will vary depending on the type of evaluation (i.e., rough initial screening to detailed project design). Figure 4-28 is a helpful guide that provides an overall picture of the flow of information, beginning with Step 4 and proceeding through Step 10.

In Step 4, a process flowsheet (including mass and energy balances) and facility scope-of-work are developed for each alternative. Figure 4-29 outlines the important information that must be included in a process flowsheet. Figure 4-30 defines guidelines for preparing a scope-of-work for each alternative. The key is to define all new facilities that must be built and all existing facilities that must be utilized. In Step 5, development and operating requirements are defined for each of the alternatives. These will largely be based on the process flowsheets and scopes-of-work prepared in Step 4.

Properly Define Alternative Boundaries. Make sure the "system boundaries" are broad enough to include all differences among the alternatives. For example, if one or more alternatives affect the operation of the site wastewater treatment plant, include the wastewater treatment plant as part of each alternative.

Have Equivalent Outcomes. Try to define all alternatives so that they equally achieve all "hard" goals over the same time period. For example, define each alternative so that the same abatement level is attained over the same time period. When this is not practical, acknowledge the differences between them.

Include Anything Affecting Economics. Define alternative boundaries to include anything that changes new investment, development or operating costs, company revenues, or the timing of these cash flows. Anything affecting the business must be included. Frequently overlooked items include development required, effect on site waste treatment or powerhouse facilities, project front-end-loading/liaison/startup (project-related noncapital costs), and additional maintenance costs. For example, noncapital project-related costs can range from 10% to 25% of the total project cost.

Be Consistent. Inconsistency can result from knowledge strengths and weaknesses, political pressure, idea favoritism, and from inconsistency in formulating system boundaries, flowsheets, facility scopes-of-work, or operating requirements. Having someone responsible for "consistency" can prove invaluable.

4.11.4 Evaluate the Alternatives (Steps 6 through 10)

This phase takes you from alternative definition through choosing the best alternative. It includes estimating new capital investment (Step 6), the change in manufacturing costs and revenues (Step 7), and the NPV (Step 8), listing all other "noneconomic" considerations (Step 9) and, finally, choosing the best alternative (Step 10). Figure 4-31 defines the important components that should be included as new permanent investment, while Figure 4-32 describes the information required to estimate changes in operating costs and NPV.

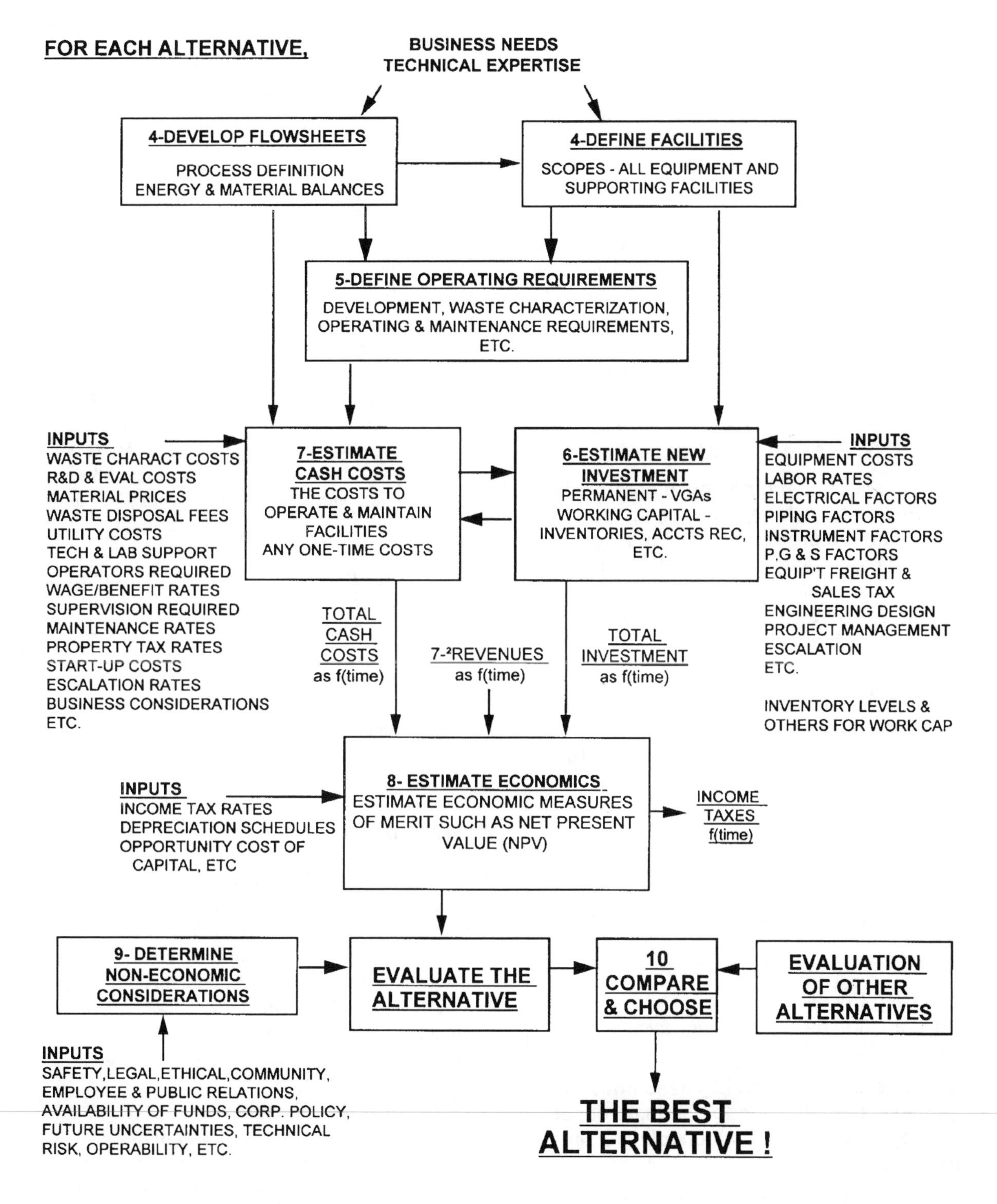

Figure 4-28. Information Flows for Steps 4 through 10 of the Engineering Evaluations 10-Step Method

Income Taxes Are Important. Because combined U.S. and state income taxes are about 38% of pretax earnings, a savings of $1 nets only $0.62 in cash. Investment dollars are always after-tax.

The Past Is Irrelevant. We often confront projects on which a lot of money has already been spent. We hear such comments as, "It's time to stop pouring money down a rat hole," or "After spending so much, it's a shame to stop when with a little more we might reach our objective." Both these sentiments are wrong! The money that has been spent is gone, and "nor all your piety nor wit" can bring it back. We are concerned only with money yet to be spent. This is the money over which we still have control. Do not let emotions get in the way of practicing this principle. Exclude the past from all your evaluations.

Timing Is Important. Money has a time value based on its ability to earn a return, if invested. You must consider the timing of cash flows. Also, any money already spent or committed should not be included in the economics, because it cannot be "unspent."

Process Flowsheet

(The first step in defining any alternative is to develop a flowsheet of the commercial-scale treatment process. It can vary from a sketched block diagram to a detailed flowsheet required for final project design. The flowsheet serves as the foundation for the facility scope-of-work)

What to Include

- Major Equipment Pieces
  - ensures all commercial process steps are included.
  - show interconnections with existing process and equipment.
- Material and Energy Balances
  - there is often a significant difference between average and design rates. Specify both.
  - make sure to track minor components.
- Processing Conditions
  - temperature, pressure, pH, physical state, presence of solids, etc.

Make sure to document all key assumptions. Use process simplification to eliminate unnecessary steps and to shorten/consolidate necessary steps.

Figure 4-29. Development of an Alternative's Process Flowsheet

Economic Merit = Business Wealth. Economics should be on a total business, out-of-pocket cash basis, not a process or plant basis. Be careful with "cost sheet" costs and allocated costs because they may not be equivalent to true company cash costs. Use a sound measure of economic merit, such as net present value; never use just investment or annual costs. When optimizing the economics of an integrated alternative, optimize the entire alternative, not just each of its individual parts. Do not confuse economic merit with accounting—accounting concerns itself with how the company pie is divided among processes or plants; economics is concerned with the total size of the pie.

The Future Is Uncertain. Factors affecting the relative economics can change over time. Consider how tightening of regulations or changes in production forecasts or specific utility costs may affect the relative rating of the different alternatives. At some point, a decision and risk analysis may be justified.

Money Is Not Everything. There are many other important considerations that are not stated in economic terms, such as safety, company environmental goals, community relations, operability, technological risk, availability of people and capital, and adaptability to tightening regulations. The differences in these other "non-economic" considerations must be weighed against the differences in the economic price tag of all alternatives.

Be Consistent...Again! Consistency in all aspects of the evaluation is critical.

4.12 Shortcut NPV Method

One criticism of detailed cash flow and NPV calculations is that they require a great deal of time and effort. Often, the amount of uncertainty in the cash flows does not warrant a detailed analysis, but an approximate NPV is still desired. The shortcut, or "back-of-the-envelope" method presented in this section provides a quick, approximate NPV that can be used for preliminary studies.

4.12.1 Data Requirements

First, in order to calculate NPV with this method, one must estimate four types of cash flow. Not all alternatives will have all four cash flows, but they are included here for completeness. Then, one selects the appropriate cost of capital to use for the discount rate and the appropriate life

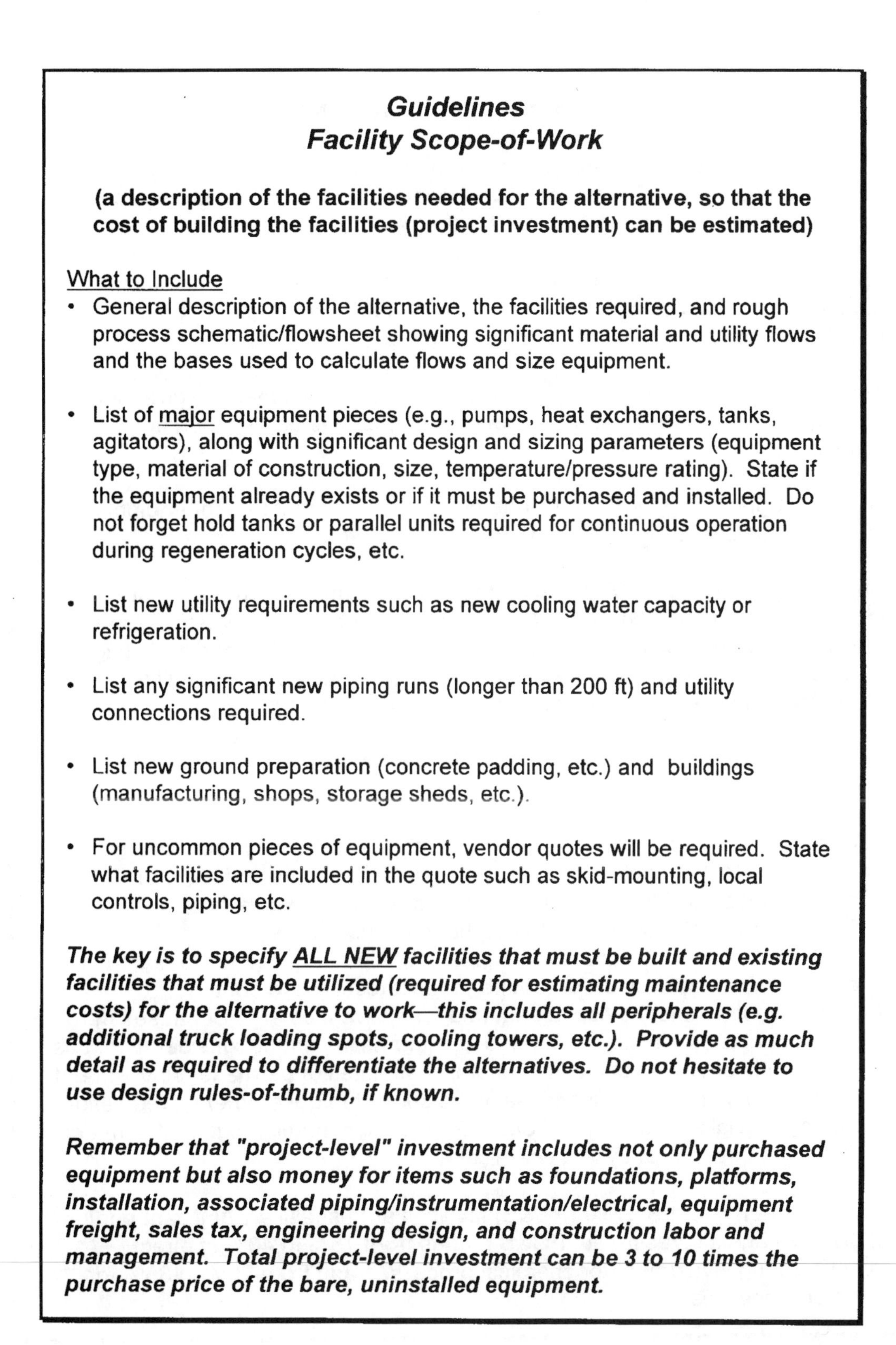

Guidelines
Facility Scope-of-Work

(a description of the facilities needed for the alternative, so that the cost of building the facilities (project investment) can be estimated)

What to Include

- General description of the alternative, the facilities required, and rough process schematic/flowsheet showing significant material and utility flows and the bases used to calculate flows and size equipment.
- List of major equipment pieces (e.g., pumps, heat exchangers, tanks, agitators), along with significant design and sizing parameters (equipment type, material of construction, size, temperature/pressure rating). State if the equipment already exists or if it must be purchased and installed. Do not forget hold tanks or parallel units required for continuous operation during regeneration cycles, etc.
- List new utility requirements such as new cooling water capacity or refrigeration.
- List any significant new piping runs (longer than 200 ft) and utility connections required.
- List new ground preparation (concrete padding, etc.) and buildings (manufacturing, shops, storage sheds, etc.).
- For uncommon pieces of equipment, vendor quotes will be required. State what facilities are included in the quote such as skid-mounting, local controls, piping, etc.

The key is to specify ALL NEW facilities that must be built and existing facilities that must be utilized (required for estimating maintenance costs) for the alternative to work—this includes all peripherals (e.g. additional truck loading spots, cooling towers, etc.). Provide as much detail as required to differentiate the alternatives. Do not hesitate to use design rules-of-thumb, if known.

Remember that "project-level" investment includes not only purchased equipment but also money for items such as foundations, platforms, installation, associated piping/instrumentation/electrical, equipment freight, sales tax, engineering design, and construction labor and management. Total project-level investment can be 3 to 10 times the purchase price of the bare, uninstalled equipment.

Figure 4-30. Guidelines for Developing a Facility Scope-of-Work for Investment Estimating

Permanent Investment

Permanent investment is often the single most important element in an economic evaluation. Permanent investment is defined as capital expenditures for plant, property, and equipment. It is the capital necessary to provide the installed equipment and all the auxiliaries that are needed for complete process operation. We must consider all the physical facilities need to manufacture and to sell our products in the marketplace, even though some of these facilities may be remote from the process under study.

Total investment is working capital plus permanent investment. Permanent investment is the sum of direct manufacturing facilities and supporting investment.

Direct manufacturing facilities include the capital expenditures for equipment and buildings directly involved in product manufacture.

Supporting investment. Manufacturing facilities cannot operate without substantial supporting permanent facilities. The supporting investments are

- Power facilities are the centralized facilities for generating or distributing electricity, steam, air, water, brine, etc. They include both the buildings and equipment necessary to provide this function.
- General facilities are the common facilities of a general nature at a plant site. They include such items as sewers, site improvements, fences, parking areas, railroads, road, walkways, and alarm, guard, and communications systems.
- Service facilities are common facilities that provide specific services necessary to the function of a plant site. Included in the category are administration buildings, cafeterias, shops, stores, sewage treatment facilities, mobile equipment, change houses, etc.
- Technical facilities includes buildings and equipment used for research and development activities or for routine product testing and control.
- Land investment is the capital required for site real estate. Because land investment is small and common to all alternatives under consideration, this category is often neglected in economic estimates of permanent investment.
- Precious materials includes those items subject to governmental regulation and/or special accounting and often are precious metals, e.g., platinum and its alloys, used to catalyze a chemical reaction.

Power, General, and Service (PG&S) facilities are usually a significant portion of the permanent investment. PG&S ranges from 15% to 40% of direct manufacturing investment for chemical processes and from 20% to 35% for polymer processes.

Figure 4-31. Factors Included in Permanent Investment

of the alternative. Factors are then looked up in a table and plugged into the NPV shortcut formula shown below. The method handles details of inflation, income taxes, and depreciation automatically.

The first type of cash flow that must be estimated is *new permanent investment.* This is for the equipment and other facilities required by the alternative. It is expressed as project-level investment at today's dollars.

The second type of cash flow to be estimated is *new annual cash costs minus revenues.* Cash costs are all those incremental costs for which a business writes a check. Revenues are incremental sales dollars, either for products or coproducts. Many alternatives do not have any incremental revenues. This cash flow is for a full year of operation after startup of the facilities. It is expressed in today's dollars; that is, before inflation, and before considering income taxes. Depending on the magnitude and sign of the costs and revenues, this type of cash flow can be either positive or negative. Be careful to keep signs straight. A cost savings is a negative cost. Maintenance, property taxes, and miscellaneous other costs are automatically included in the calculations, based on the new permanent investment. Therefore, the primary types of costs that have to be estimated are raw materials, energy (utilities), and manpower.

The third type of cash flow accounts for a *change in working capital.* While it is difficult to imagine a pollution-prevention alternative that would significantly change product and raw-material inventories, it is included here for completeness. Seasonal businesses, such as agricultural products, may need to consider significant changes in working capital among alternatives. This type of cash flow is expressed as a change in working capital for a full year of operation and in today's dollars.

The last type of cash flow is for a *one-time cash cost.* This may be required for large feasibility studies. It is also useful for remediation projects, where the "investment" is actually treated for tax purposes as a one-time cash cost. It is expressed in today's dollars. No recurring costs, such as maintenance, are factored from this one-time cost, so

Guidelines
Operating Costs and Economics
Description of Information Required for Estimating

Material Flows
The mass flowrate of raw materials required. Include replacement of adsorption resins and catalysts at the end of their life. Delivered prices.

Utility Flows
Utility flowrates and their incremental costs.

Miscellaneous Costs

Basic Data Development Requirements
The costs to test, demonstrate, and develop basic data for the final design and operation of the alternative. Include the dollars spent for testing, the cost of building a test facility, if required, and the technical/research people required.

Operator Requirements
The amount of operator attention required to operate the alternative. Measure in hours per 24 hour day of operation. Wage and benefit rates.

On-going Technical Support
The amount of technical/research required to support the on-going alternative operation. Measured as a percentage of a full-time person.

Other Miscellaneous Costs/"Non-Economic" Considerations
Anything such as extraordinary maintenance requirements, etc. Also, list any key technical uncertainties discovered or key assumptions made during the flowsheeting and scoping of the alternative. List any "non-economic" advantages or disadvantages of the alternative.

Figure 4-32. Factors Required to Estimate An Alternative's Cash Flow

exercise caution when using this term.

The life of the alternative and the cost of capital must then be decided upon. These factors have a large impact on the NPV. The appropriate life will frequently be determined by the expected life for the pollution-prevention facility, until it is no longer useful, either technically or due to regulatory pressures. It may also depend on the expected life of the rest of the plant or business, depending on whether the product is a core chemical process or a high-tech product that will become obsolete in a few years and the plant shut down.

The appropriate cost of capital should be determined by the business. If the business is aggressively pursuing pollution-prevention alternatives, it may choose to use a relatively low hurdle rate, even as low as the normal cost of capital, 12%. On the other hand, if the business's capital expenditures are constrained, even pollution-prevention alternatives may have to meet a relatively high hurdle rate, such as 25% to 30%. If a discount rate other than the cost of capital is used, it must be stated explicitly, such as NPV_{25} for a 25% discount rate.

4.12.2 Calculating NPV

The NPV can be calculated once we know all the cash flows as well as the facility life and cost of capital. Each nonzero cash flow is multiplied by a coefficient which is tabulated as a function of the life and cost of capital. These terms are then added together to get the NPV, as shown below:

$$NPV = (a*\text{investment}) + (b*(\text{cost} - \text{revenue})) + (c*\Delta \text{ working capital}) + (d * \text{one time cost})$$

Tables of the four coefficients (*a, b, c,* and *d*) as a function of facility life and cost of capital are given in Tables 4-4 through 4-6. The following abbreviations are also used:

I = Investment
C = Cash cost
R = Revenue
WC = Working capital change (Δ)
OTC = One time cost

For a typical large project (greater than $20 million investment), assuming 12% cost of capital and a 10-year life, the shortcut NPV equation reduces to

$$NPV_{12} = -0.94 * I + (-3.3) * (C - R) + (-0.5) * WC + (-0.54) * OTC$$

Examples - Other Bases Needed To Estimate Costs And Economics

Other Costing Bases

- Wage rate = $20/hr, benefits =45% of payroll, supervision = 20% of operating wages + benefits
- Non-capital project-related costs (start-up, project front-end-loading, project liaison) = 10%-25% of project
- Annual maintenance at 2%-6% of project investment
- Property taxes and insurance at 1% of project investment
- General plant overheads = 14% of wages, salaries and benefits

Cash Flow Analysis Bases

- 38% combined income tax rate (35% U. S. and 3% average state)
- Annual creep investment = 1%-2% of project
- Start-up in 1Q97
- Income tax depreciation over 6 years (20%, 32%, 19%, 12%, 12%, 5%), U. S.
- Change in working capital
- 10 years of operation
- NPC discount rate of 25% or the hurdle rate for new capital for your business

Figure 4-32. Factors Required to Estimate An Alternative's Cash Flow

This shows that a dollar of additional investment has an NPV of -$0.94, while a dollar of *annual* cost ($1/yr) has an NPV of -$3.30. The ratio of these two coefficients (-0.94/-3.3) shows that if a 12% cost of capital is acceptable, a dollar of investment can be justified by about $0.28 of annual cost savings or revenues. This is handy number to remember.

For the case of 25% cost of capital and 10-year life, the equation reduces to

$$NPV_{25} = -0.67 * I + (-1.4) * (C - R) + (-0.41) * WC + (-0.44) * OTC$$

If 25% is the appropriate cost of capital to use, a dollar of investment can be justified by about $0.48 (- 0.67/-1.4) of annual cost savings or revenues. Thus, it is substantially more difficult to justify investment based on cost savings at a 25% cost of capital than it is at a 12% cost of capital.

4.12.3 Basis of the Shortcut Method

The shortcut NPV method is based on a full, detailed cash flow calculation, using a particular set of assumptions. Under these assumptions, the shortcut NPV method gives exactly the same result as a full NPV calculation. It is mathematically equivalent and rigorous. If the assumptions are not correct, however, the shortcut method is only an approximation. The equation parameters are further refined by the type of project—large, midsize, and small improvement. The assumptions and other characteristics of each project type are shown in Table 4-7. Each type of project has its own table of coefficients for use in the shortcut NPV equation. These coefficients are listed in Tables 4-4 to 4-6. To estimate the NPV of an alternative using the shortcut method, you need only to

- Identify project type
- Choose the appropriate discount rate
- Select the desired facility life (labeled as "Lives" in the tables

4.12.4 Adjustments to the NPV Estimations Due to Project Timing

Adjustments to the NPV calculated by the shortcut method are required if the actual project timing is different than that assumed for that project type. By way of example, assume that project startup occurs one year later than the assumptions shown in Table 4-6 for that project type. In this case, the calculated NPV must be discounted by multiplying by 1/(1 + discount rate). If the discount rate is 25%, then the calculated NPV must be multiplied by 1/(1 + 0.25) or 0.80. If project startup occurs two years later, then the calculated NPV must be multiplied by $(1/(1 + 0.25))^2$ or 0.64. In this way, differences in project timing are made by either discounting or compounding the calculated NPV.

4.12.5 Comments on NPV Method

The sensitivity of the NPV to each of the assumptions listed in Table 4-7 has been tested. With only a few exceptions, changing the assumption within reasonable limits does not have a significant impact on the calculations. One exception is the total of 4% to 6% of investment for annual maintenance, property taxes, and overheads. If this differs by more than two percentage points (for example, due to unusually high or low maintenance), there will be about a 10% error in the investment term. One way around this is to include an additional operating cost to account for maintenance above the default of 2% to 4%.

Another important assumption is the 4% inflation rate. If inflation differs by one percentage point from this assumption, the cost and working capital change terms will have errors of around 8%.

The other assumption that is critical is about the working capital change term. The working capital is assumed to be liquidated at the end of the project, which has a large impact on the NPV of the working capital change. Note that this is not an important assumption for the cost term, which automatically includes a small amount of working capital.

It is important to note that the term for investment includes everything that is factored off of investment, including maintenance and other operating costs, as well as working capital change associated with the change in cost. The contribution of the pure investment to the NPV is only around 67% of the investment term, with around 6% attributable to the startup and project liaison costs, 9% due to creep investment, and 18% due to maintenance, property taxes, and other costs factored from investment. On the other hand, although an automatic working capital change of two months of cash costs is included in the cost term, it only accounts for about 3% of the NPV of the cost term.

The accuracy of this shortcut method must always be considered. The calculations were done very accurately, but there is always uncertainty in the input data and the validity of the assumptions. Therefore, no more than two significant digits were used in the assumptions and the tabular output so as not to overstate the accuracy. Under the set of assumptions shown in Table 4-7, the equation is exact. Under reasonably close assumptions, the error is likely to be less than 5–10%.

Coefficient "a" (NPV of $1 of Investment)

Lives	10%	12%	15%	20%	25%	30%	35%	40%
	Discount Rates							
5	-0.85	-0.82	-0.77	-0.69	-0.63	-0.57	-0.52	-0.48
10	-1.0	-0.94	-0.86	-0.75	-0.67	-0.60	-0.54	-0.49
15	-1.1	-1.0	-0.92	-0.79	-0.69	-0.61	-0.55	-0.49
20	-1.2	-1.1	-0.96	-0.80	-0.69	-0.61	-0.55	-0.49
25	-1.3	-1.1	-0.98	-0.81	-0.70	-0.61	-0.55	-0.49
30	-1.3	-1.2	-1.0	-0.81	-0.70	-0.61	-0.55	-0.49

Coefficient "b" (NPV of $1 of Cost Minus Revenue)

Lives	10%	12%	15%	20%	25%	30%	35%	40%
	Discount Rates							
5	-2.0	-1.8	-1.6	-1.2	-0.97	-0.78	-0.63	-0.51
10	-3.8	-3.3	-2.7	-1.9	-1.4	-1.1	-0.84	-0.66
15	-5.1	-4.3	-3.3	-2.3	-1.6	-1.2	-0.90	-0.69
20	-6.1	-5.0	-3.8	-2.5	-1.7	-1.2	-0.92	-0.70
25	-6.9	-5.5	-4.0	-2.5	-1.7	-1.2	-0.92	-0.70
30	-7.5	-5.8	-4.2	-2.6	-1.7	-1.2	-0.92	-0.70

Coefficient "c" (NPV of $1 of Working Capital Change)

Lives	10%	12%	15%	20%	25%	30%	35%	40%
	Discount Rates							
5	-0.23	-0.25	-0.26	-0.26	-0.25	-0.23	-0.21	-0.19
10	-0.49	-0.50	-0.50	-0.46	-0.41	-0.36	-0.31	-0.27
15	-0.69	-0.68	-0.65	-0.56	-0.48	-0.40	-0.34	-0.29
20	-0.84	-0.80	-0.73	-0.61	-0.51	-0.42	-0.35	-0.29
25	-0.95	-0.89	-0.79	-0.64	-0.52	-0.42	-0.35	-0.29
30	-1.0	-0.95	-0.82	-0.65	-0.52	-0.42	-0.35	-0.29

Coefficient "d" (NPV of $1 of One-time Cost)

Lives	10%	12%	15%	20%	25%	30%	35%	40%
	Discount Rates							
For Any Life	-0.56	-0.54	-0.52	-0.47	-0.44	-0.40	-0.37	-0.35

Table 4-4. Shortcut NPV Equation Coefficients—Large Projects

Coefficient "a" (NPV of $1 of Investment)

Lives	Discount Rates 10%	12%	15%	20%	25%	30%	35%	40%
5	-0.85	-0.82	-0.78	-0.73	-0.67	-0.63	-0.58	-0.55
10	-0.99	-0.94	-0.88	-0.79	-0.72	-0.66	-0.61	-0.56
15	-1.1	-1.0	-0.93	-0.82	-0.73	-0.67	-0.61	-0.57
20	-1.2	-1.1	-0.97	-0.84	-0.74	-0.67	-0.61	-0.57
25	-1.2	-1.1	-0.99	-0.84	-0.75	-0.67	-0.61	-0.57
30	-1.3	-1.2	-1.0	-0.85	-0.75	-0.67	-0.61	-0.57

Coefficient "b" (NPV of $1 of Cost Minus Revenue)

Lives	Discount Rates 10%	12%	15%	20%	25%	30%	35%	40%
5	-2.4	-2.2	-2.0	-1.6	-1.4	-1.1	-0.97	-0.83
10	-4.3	-3.8	-3.2	-2.5	-1.9	-1.5	-1.2	-1.00
15	-5.7	-4.9	-3.9	-2.9	-2.1	-1.7	-1.3	-1.10
20	-6.8	-5.6	-4.4	-3.1	-2.2	-1.7	-1.3	-1.10
25	-7.6	-6.2	-4.7	-3.1	-2.3	-1.7	-1.3	-1.10
30	-8.2	-6.5	-4.8	-3.2	-2.3	-1.7	-1.3	-1.10

Coefficient "c" (NPV of $1 of Working Capital Change)

Lives	Discount Rates 10%	12%	15%	20%	25%	30%	35%	40%
5	-0.28	-0.31	-0.33	-0.35	-0.36	-0.35	-0.33	-0.32
10	-0.56	-0.58	-0.60	-0.59	-0.55	-0.51	-0.47	-0.42
15	-0.77	-0.78	-0.76	-0.70	-0.63	-0.56	-0.50	-0.45
20	-0.92	-0.91	-0.86	-0.76	-0.66	-0.58	-0.51	-0.45
25	-1.0	-1.0	-0.92	-0.79	-0.68	-0.59	-0.52	-0.46
30	-1.1	-1.1	-0.95	-0.80	-0.68	-0.59	-0.52	-0.46

Coefficient "d" (NPV of $1 of One-time Cost)

Lives	Discount Rates 10%	12%	15%	20%	25%	30%	35%	40%
For Any Life	-0.60	-0.59	-0.57	-0.55	-0.52	-0.50	-0.49	-0.47

Table 4-5. Shortcut NPV Equation Coefficients — Midsize Projects

Coefficient "a" (NPV of $1 of Investment)

	Discount Rates							
Lives	10%	12%	15%	20%	25%	30%	35%	40%
5	-0.83	-0.82	-0.81	-0.78	-0.76	-0.74	-0.72	-0.70
10	-0.93	-0.91	-0.88	-0.83	-0.80	-0.77	-0.74	-0.71
15	-1.0	-0.97	-0.92	-0.86	-0.81	-0.78	-0.75	-0.72
20	-1.1	-1.0	-0.95	-0.87	-0.82	-0.78	-0.75	-0.72
25	-1.1	-1.0	-0.96	-0.88	-0.82	-0.78	-0.75	-0.72
30	-1.1	-1.1	-0.97	-0.88	-0.82	-0.78	-0.75	-0.72

Coefficient "b" (NPV of $1 of Cost Minus Revenue)

	Discount Rates							
Lives	10%	12%	15%	20%	25%	30%	35%	40%
5	-2.6	-2.4	-2.2	-1.9	-1.6	-1.4	-1.3	-1.1
10	-4.5	-4.1	-3.5	-2.8	-2.3	-1.9	-1.6	-1.4
15	-6.0	-5.3	-4.4	-3.3	-2.6	-2.1	-1.7	-1.4
20	-7.2	-6.1	-4.9	-3.5	-2.7	-2.1	-1.7	-1.5
25	-8.0	-6.6	-5.2	-3.6	-2.7	-2.1	-1.8	-1.5
30	-8.7	-7.0	-5.4	-3.7	-2.7	-2.2	-1.8	-1.5

Coefficient "c" (NPV of $1 of Working Capital Change)

	Discount Rates							
Lives	10%	12%	15%	20%	25%	30%	35%	40%
5	-0.30	-0.33	-0.37	-0.41	-0.43	-0.44	-0.43	-0.43
10	-0.59	-0.63	-0.66	-0.68	-0.67	-0.64	-0.61	-0.57
15	-0.81	-0.84	-0.84	-0.81	-0.76	-0.71	-0.65	-0.60
20	-0.98	-0.98	-0.95	-0.88	-0.80	-0.73	-0.67	-0.61
25	-1.1	-1.1	-1.0	-0.91	-0.81	-0.73	-0.67	-0.61
30	-1.2	-1.1	-1.1	-0.92	-0.82	-0.74	-0.67	-0.61

Coefficient "d" (NPV of $1 of One-time Cost)

	Discount Rates							
Lives	10%	12%	15%	20%	25%	30%	35%	40%
For Any Life	-0.60	-0.59	-0.57	-0.55	-0.52	-0.50	-0.49	-0.47

Table 4-6. Shortcut NPV Equation Coefficients — Small Improvement Projects

Project Type →	Large Project	Midsize Project	Small Improvement Project
Project Characteristics	New Process Area Entire System > ± $20 Million		Existing Process Area Improvements Pieces of Equipment ≤ ± $5 Million
Key Assumptions			
1. Years until startup	3	2	1
2. Project spendout (% years 1, 2, &3)	0, 40, 60	0, 100, 0	100, 0, 0
3. Startup costs (as % of proj. investment)	10	6	4
4. Annual creep investment (as % of proj. investment)	2	2	1
5. Annual maintenance (as % of proj. investment)	4	3	2
6. Yr. 1 operating cost/savings (as % of ongoing annual)	50	100	100
7. One-time cash cost (% years 1 & 2)	0, 100	100, 0	0,100
8. One-time working capital adjustment (% years 2, 3, 4, & 5)	0,0,50,50	0,100,0,0	100,0,0,0

Other assumptions common to all project types:
1. Annual costs and revenues are for a typical year.
2. All dollars are expressed in today's dollars.
3. U.S. tax and depreciation rates.
4. 4% escalation rate for all cash flows.
5. Automatic working capital change of 2 months of cash costs.
6. Only terminal value is working capital liquidation.
7. End of year cash flows.
8. 2% of investment for annual property taxes and overheads.

Table 4-7. Economic Assumptions by Project Type

4.12.6 An Example

A pollution-prevention alternative requires $5 million in new investment (small improvement project), expressed in today's dollars. Compared to the base case, it results in annual pretax savings of $2 million in disposal fees, expressed in today's dollars. What is the NPV for this alternative, compared to the base case?

First, the two primary cash flows are calculated: investment (I) is $5 million; cost (C) is a negative $2 million. Because cost will decrease relative to the base case, C is negative.

Assume for the moment that the appropriate life is 10 years and that the appropriate discount rate is 25%. This might be the case for a product that is nearing the end of its life cycle and has capital constraints. The NPV is then

$$
\begin{aligned}
NPV_{25} &= -\ 0.80 * I + (-2.3) * (C - R) \\
&= -\ 0.80 * 5M + (-2.3) * (-2M) \\
&= +\ \$0.6M \text{ or } \$600{,}000
\end{aligned}
$$

4.13 Waste-Management Economics: A Balanced Approach

In the past, process R&D, process design, and waste treatment were dealt with independently. Because treatment costs and concerns were relatively low, this approach yielded systems close to optimum. Today, however, waste treatment costs and concerns have become much more important. If we continue as in the past to develop the chemical process and then to separately address waste streams, we will no longer arrive at the best overall solutions. The new paradigm, then, is to integrate process R&D, process design, and waste treatment design to find the best overall solutions to a waste problem.[3]

In DuPont, we are finding that although selection of optimum end-of-pipe treatment can reduce investment and costs by 30%, *integrated design* can achieve a 50% to 100% reduction in the cost of this treatment. This section presents an approach to move closer to true integrated design.

4.13.1 The Path to Better Solutions

The best solution to any waste problem is invariably a blend of approaches. These may include (1) improved/new processes to reduce waste generation/introduction; (2) segregation plus direct reuse, or treatment followed by recycle of the waste media (water and air); (3) end-of-pipe treatment; and (4) safe disposal. The blend may also involve other plants on a site, other sites, suppliers, and even customers. The payoff will come from *quickly* finding the *optimum* blend.

Items 1 and 2 given in the preceding paragraph are generally referred to as *process renewal.* Process renewal has many hidden benefits, such as improved public acceptance, less potential for future liabilities (fines, penalties, cleanup costs, obsolescence due to tightening regulations, etc.), and lower permitting, recordkeeping, and report writing costs, and is, therefore, at the top of the pollution prevention hierarchy. With traditional independent design approaches, however, waste solutions tend to have too little process renewal, and instead rely heavily on treatment and disposal—yielding costly solutions.

Figure 4-33 highlights a path to more economic integrated solutions. It involves an intermediate step of using the incremental costs of waste treatment to drive a business toward process renewal. The remainder of this section will review incremental treatment costs for new treatment facilities and how they can be used as an intermedi-

Emission Type	1997 Investment ($ per scfm)	1999 Cash Operating Cost ($/yr per scfm)
Hydrocarbon VOCs	40	9.5
Chlorocarbon VOCs	65	10
Chlorocarbon VOCs+ Particulates	75	13

Table 4-8. Minimum Incremental Costs for Air Emissions Abatement.

	Without Water Treatment Cost	With Water Treatment Cost
1997 Investment ($)	1,300,000	2,400,000
1999 Cash Operating Cost ($/yr)	300,000	550,000
1997 Net Present Cost ($, 10 years, at 25%)	1,500,000	2,800,000

Table 4-9. Comparison of Investment, Cost, and Economics for Water Scrubbing With and Without Incremental Costs of Wastewater Treatment Included.

ate step to reach integrated or holistic designs. The danger in remaining in the current state is suboptimization of process and waste treatment designs, and hence a nonoptimum overall solution.

4.13.2 Incremental Costs Of Waste Treatment

The incremental costs of waste treatment define the incentive for process renewal. The incremental costs presented below are *minimum* values, apply only to *new* (yet to be built) treatment facilities, and represent the incremental investment and cash operating cost (COC) to treat an additional unit (i.e., 1 gpm, 1 scfm, 1 lb) of waste at the end of the pipe. For example, if investment is $1000 for a proposed 10-gpm treatment plant and $1050 for an 11-gpm treatment plant, the *incremental* investment for 1 gpm capacity is defined as $50, or $50/gpm.

Tables 4-1 and 4-8 present the minimum incremental costs for new aqueous waste treatment (above-ground activated sludge biological treatment) and air-emissions abatement facilities, respectively. In some cases, actual incremental costs will be as much as five times or more of the minimums presented in Tables 4-1 and 4-8.

4.13.3 How Should You Use These Incremental Costs?

Example 1: Pump Seal Flush. A project engineer wishes to know the impact of a pump seal water flush on the planned new wastewater treatment facility. For this analysis, we will consider the impact of the incremental water flow only (no organics) on the treatment facility. Using the incremental costs for aqueous waste treatment from Table 4-1, the minimum added investment and cash operating cost of treatment for the additional 1/2 gpm is:

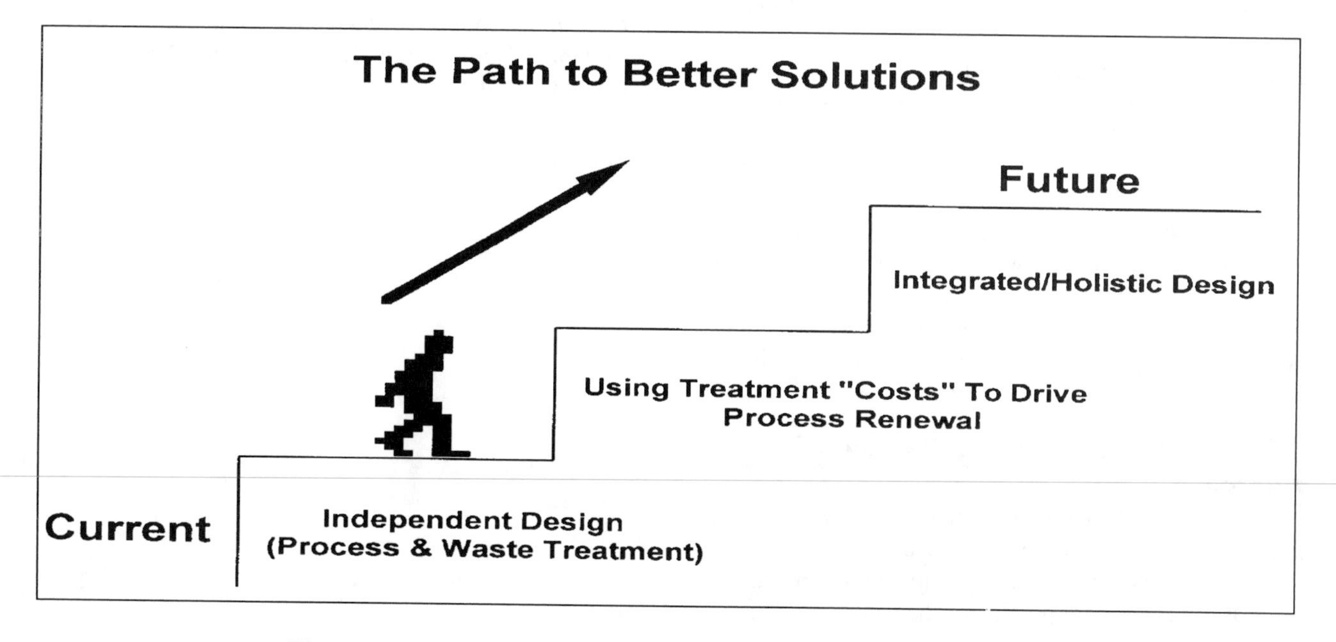

Figure 4-33. The Path to More Economic Integrated Solutions

Minimum Added Investment

$3000/gpm * 1/2 gpm = $1500 per pump

Minimum Added Operating Cost

$300/yr/gpm * 1/2 gpm = $150/yr per pump

In comparison, investment for a pump seal ranges between $700 and $4500 versus the additional $1500 for the new wastewater treatment plant. The minimum additional NPC of treating the pump seal flush is $1300 (10 years at 25% discount rate, 1997 dollars). Should you select a different pump seal?

Example 2 — Water Scrubber

A water scrubber is proposed to remove organics from a process stream. The scrubber will produce an aqueous waste containing 100 lb/h organics and 170 gpm water. The investment, operating costs, and economics of the water scrubber alone are:

1997 Investment ($):	1,300,000
1999 cash operating cost ($/yr):	300,000
NPC (1997, 10 years, at 25%):	1,500,000

What is the real cost of water scrubbing, however? To answer this question, we must consider the incremental costs for wastewater treatment:

Minimum Added Investment

Due to flow 170 gpm * $3000/gpm	= $500,000
Due to organic 100 lb/h * $6000/(lb/hr)	= $600,000
Total	$1,100,000

Minimum Added Operating Cost

Due to Flow 170 gpm * $300/yr/gpm =	$ 50,000/yr
Due to Organic 100 lb/h * $2000/yr/(lb/h) =	$200,000/yr
Total	$250,000/yr

If we include these incremental treatment costs in the economic analysis of the water scrubber, we obtain the results shown in Table 4-9.

The real cost of water scrubbing is therefore almost twice as much when wastewater treatment is included. Should another option be considered?

4.13.4 Characteristics of a Balanced Approach

In addition to using the incremental costs just presented, a balanced approach will have other characteristics that must be incorporated into our thinking when addressing waste-management problems. They are:

- Understand the business drivers for change, for example, cost reduction, corporate goal, regulation
- Involve diverse viewpoints and skills, that is, R&D, technologists, site process, evaluators, regulatory, business
- Take a holistic process/waste management perspective
- Identify the economic drivers early, for example, factors critically affecting the economics of treatment
- Brainstorm to identify options
- Know the cost of conventional treatment early
- Screen the options early
- Further define, demonstrate, and evaluate only the promising options
- Quickly identify the optimum blend

Literature Cited

1. Dyer, J. A., and K. L. Mulholland. February 1994. "Toxic Air Emissions: What is the Full Cost to Your Business?" *Environmental Engineering: A Special Supplement to February 1994 Chemical Engineering.* 101 (Suppl.): 4–8.
2. Keller, R. A., and J. A. Dyer. January 1998. "Abating Halogenated VOCs." *Chemical Engineering.* 105 (1): 100–105.
3. Dyer, J. A., and W. C. Taylor. June 1994. "Waste Management: A Balanced Approach." *Proceedings of the Air & Waste Management Association's 87th Annual Meeting and Exhibition*, 94-RP122B.05, Cincinnati, OH.

CHAPTER 5

Case Study

5.1 Introduction

The traditional approach to process design has been to first engineer the process and then engineer the treatment and disposal of waste streams. With increased regulatory and societal pressures to eliminate waste emissions to the environment, the total system must now be analyzed—process plus treatment—to find the most economic option. In general, processes that minimize the amount of waste generated at the source are the most economical. Even in existing facilities, waste generation can be reduced by more than 30% on average, while at the same time lowering operating costs and new capital investment.

This chapter presents a case study of a chemical manufacturing facility that was faced with the task of reducing air and wastewater emissions when the plant was relocated to a new location. The study takes the information from the previous chapters and illustrates both the "how to" of a pollution-prevention program as well as the tools and techniques that help engineers and scientists identify process improvements that reduce waste generation inside the pipes and vessels. This case is indicative of what can be done to reduce waste generation and emissions from existing manufacturing facilities. Given the right situation—in this case, relocation of the existing facility—significant strides can be made in protecting the environment, while making more money for the business.

5.2 Background of Case Study

A chemical intermediate is manufactured by a 30-year-old process that generates an unacceptable level of waste compared to current standards of environmental performance. The facility is located within city limits, and new local environmental regulations require that an expensive end-of-pipe treatment device be installed on a gaseous emission source unless more economical source reduction measures can be identified and implemented. The local community is unaware that the process emits noxious compounds—so, faced with increased public scrutiny due to new regulations, the business has elected to relocate the manufacturing process to a new site.

The separation steps involved in the existing process include: (1) aqueous scrubbing of a reactor off-gas to condense and capture the reactant and product; (2) extraction of the reactant and product into a benzene solvent; and (3) high-purity distillation of the reactant and product. Benzene is recovered for recycle, and the aqueous raffinate stream is stripped of benzene before discharge to a biological wastewater treatment plant.

The business decision to relocate the process, coupled with the investment associated with end-of-pipe treatment of gaseous and wastewater emissions, gives process and design engineers the chance to critically review the process for optimal design. The business team has elected to use the pollution prevention program methodology outlined in the center column of Figure 5.1 to analyze the process before investing in a new facility.[1] The review will target major process changes that can achieve significant environmental and financial gains over the existing manufacturing process. Potential benefits include reduced capital investment, lower operating costs, waste reduction, process simplification, and elimination of benzene handling.

The methodology shown in Figure 5.1 is applicable at all phases of a project. In the case study, the program will be implemented at the reengineering phase of an existing process. The same methodology should be used, however, when the process is first conceived in the chemist's laboratory and at periodic intervals through startup and normal plant operation.

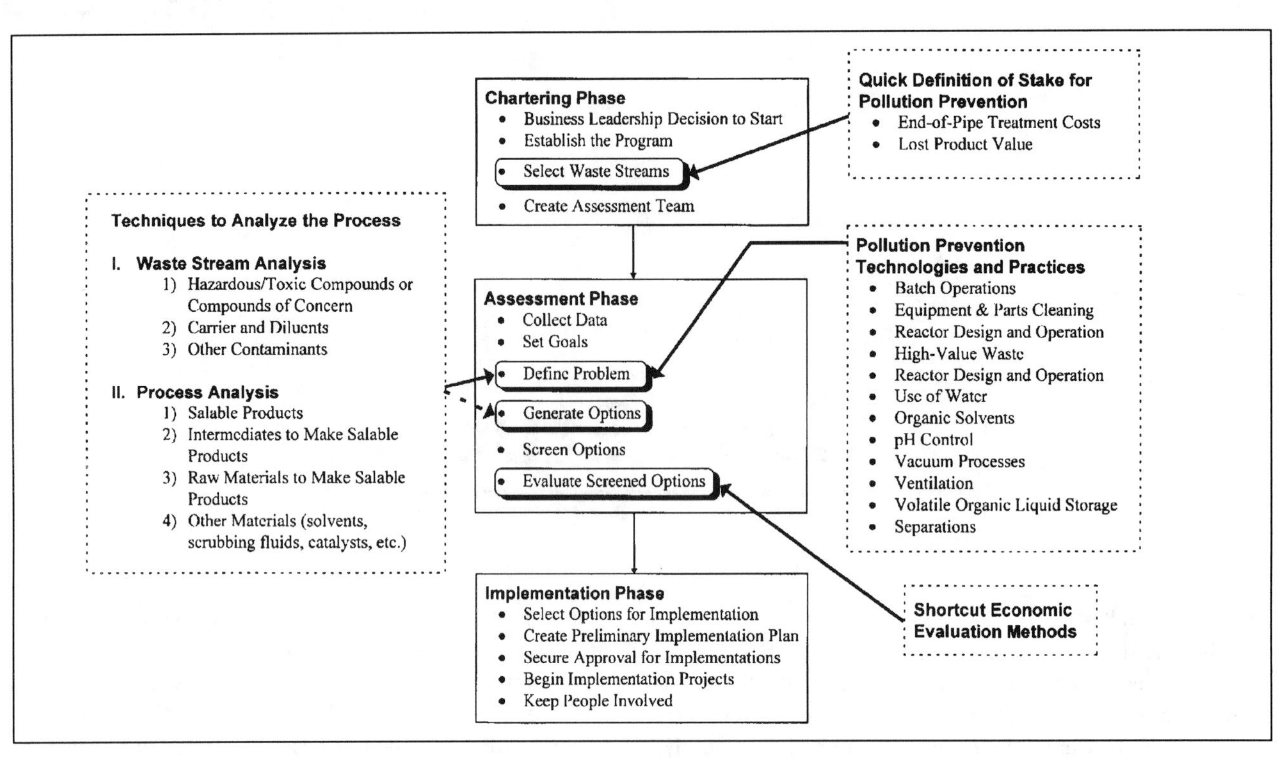

Figure 5-1. The Path to Pollution Prevention

5.3 Chartering Phase

The chartering phase discussed in Chapter 3 consists of the business leadership decision to start; establishing the program; selecting the waste streams; and creating a core assessment team.

The business decision to begin a pollution-prevention program for the case study process is triggered by excessive air toxic emissions as well as a nitrogen content (total nitrogen, which includes organic nitrogen, ammonia, nitrite, and nitrate) in the wastewater treatment outfall that is higher than allowed at the new location. The business leadership has chosen a project engineer to lead the core assessment team and to establish a program to explore and implement ways to reduce air toxic emissions and the nitrogen content in the wastewater outfall. The core assessment team will consist of the lead project engineer, a process engineer and chemist from the existing manufacturing plant, and a process engineer and chemist from the new site for a total of five members.

An examination of the process (Figure 5-2) reveals two major waste streams, the exit gas stream from the acid scrubber (stream A10) and the exit liquid stream from the benzene extractor (stream A22). Two other waste-stream emissions are also present: the stream stripper vent (stream A40) and fugitive emissions from the benzene storage and handling facilities. However, the team has decided to concentrate on the two major waste streams (A10 and A22).

As illustrated in Chapter 4, there are several ways to look at the incentive for pollution prevention. The choice will depend on the particular circumstances, that is, is an existing waste treatment or abatement system available, or is a new treatment system required? Three approaches are available to the core assessment team to determine the incentive:

- New end-of-pipe treatment
- Raw material costs
- Cost of manufacture

To further support the business leadership's decision, the core assessment team will identify the investment for end-of-pipe treatment to control emissions from the two major waste streams. In addition, to fully define the incentive for pollution prevention, the team will also determine the lost product value from the existing process.

5.3.1 Incentive Based on End-of-Pipe Treatment

Stream A10 is a 10,000-standard-cubic-feet-per-minute (scfm) gas stream containing hydrocarbons and sulfur- and nitrogen-containing compounds. The preferred abatement technology in this case would likely be a thermal oxidizer followed by a caustic scrubber to remove the oxides of sulfur and nitrogen. (In this case, no particulates are present in the gas stream.) Figure 5-3 shows the investment required for thermal oxidizers and acid gas scrubbers.[1-2] Here, the base investment cost is \$1.6 million, calculated with the values for halogenated volatile organic compounds (HVOC) gas abatement from Figure 5-3:

$$10{,}000 \text{ scfm} \times 160 \text{ \$/scfm} = \$1{,}600{,}000$$

Because stream A10 contains sulfur-containing compounds, the sulfur dioxides generated from oxidation of the VOCs must be scrubbed. In this case, the same acid gas scrubbing technology as for HVOCs is applicable.

The aqueous stream from the benzene extractor (stream A22) must be steam stripped to remove benzene before this 35 gallon per minute (gpm) stream can be sent to wastewater treatment. Figure 5-4 shows the range of capital investment that would be required to steam strip an aqueous stream followed by incineration of the condensed and decanted benzene/steam vapor stream. For a high-volatility VOC, the investment would be \$1.1 million, while for a low-volatility VOC, it would be \$1.9 million.

Thus, the minimum total capital investment for end-of-pipe treatment for the two major waste streams will range from \$2,700,000 to \$3,500,000 (calculations later will use an average of \$3 million). These numbers do not include any additional investment for pretreatment to lower the high nitrogen content of the wastewater prior to biological treatment or for the addition of a denitrification step in the existing wastewater treatment facility.

5.3.2 Lost-Product Value Incentive for Pollution Prevention

The cost of manufacture accounts for all of the relevant operating costs, including raw-material costs, and it should be cast in the form of dollars per pound of a key raw material. Another key measure that is known is the product selling price in dollars per pound of product. Depending on the current state of the business, either the cost of manufacture or product selling price is used to determine the incentive for pollution prevention in terms of lost-product value.

For a business with excess production capacity, the incentive for pollution prevention is the product of the cost of manufacture (\$/lb raw material) and the amount (in lb) of raw material that goes to waste, either directly or as a byproduct of reaction. For a sold-out business, every additional pound produced can be sold; therefore, the incentive for pollution prevention is determined by multiplying the product selling price (\$/lb product) by the additional

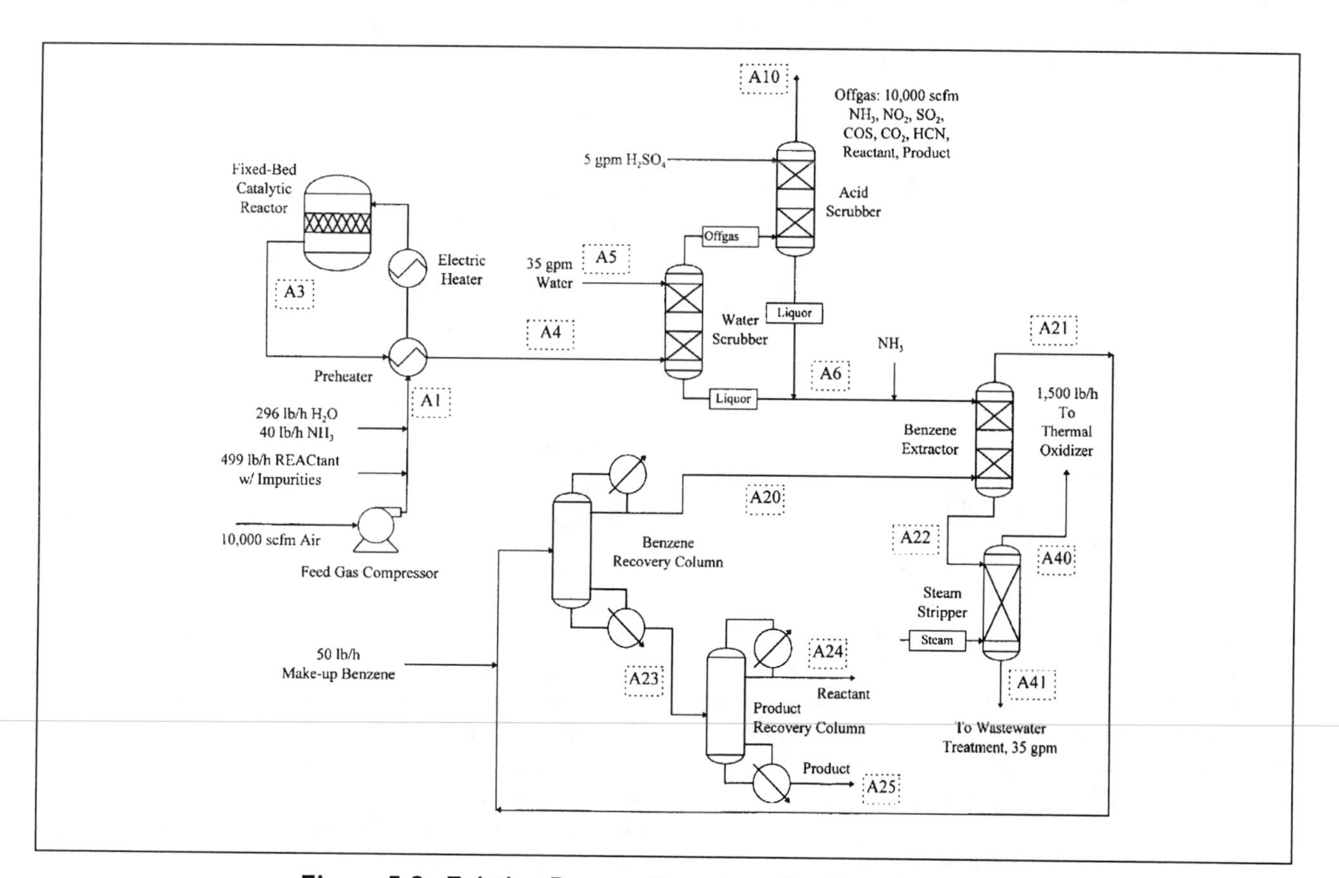

Figure 5-2. Existing Process Flowsheet for Case Study Plant

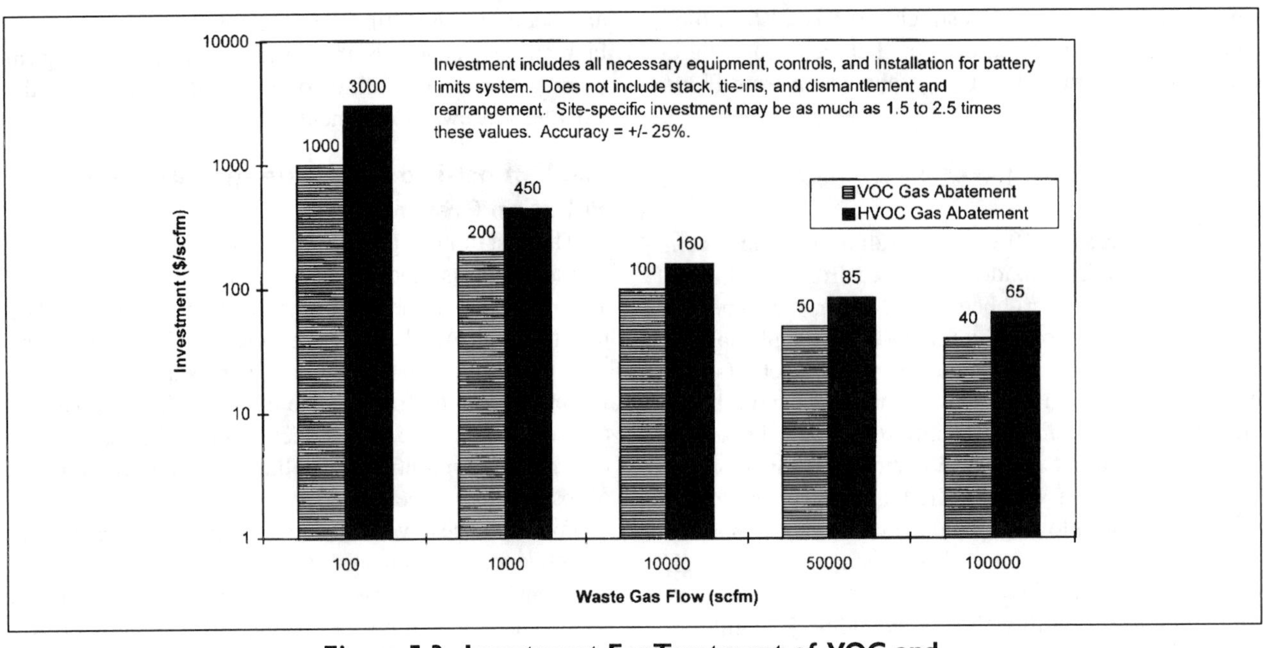

Figure 5-3. Investment For Treatment of VOC and HVOC Waste Gas Streams (U.S. dollars, ENR=5700)

amount (in lb) of product that can be sold.

For the case study process, the cost of manufacture is \$2/lb reactant (REAC), and 98 lb/h REAC is either decomposed or lost as product (PROD) in the waste streams. Therefore, for a business with excess capacity, the incentive for pollution prevention is:

\$2/lb x 98 lb/h = \$196/h or \$1.6 million/yr
(for 8000 h/yr of operation)

The product selling price is \$5/lb PROD. Therefore, for a sold-out business, the incentive is:

\$5/lb x 98 lb/h x 1.1 lb PROD/lb REAC =
\$544/h = \$4.4 million/yr

The total incentive for pollution prevention is then defined as the sum of the end-of-pipe treatment incentive and the lost-product-value incentive. For the case study, in the case of excess capacity, this is \$4.6 million (\$3 million average investment for end-of-pipe treatment plus \$1.6 million for lost-product value). For a sold-out business, it is \$7.4 million (\$3 million plus \$4.4 million).

These incentives for pollution prevention are very rough order-of-magnitude calculations. However, they do reflect the significant savings and additional revenue that are possible through waste reduction.

5.4 Assessment Phase

The assessment phase is the heart of the pollution-prevention program. It consists of tasks that will help the team to understand how the target waste streams are generated and how they can be reduced or eliminated, preferably at the source.

5.4.1 Collect Data

The amount of information that must be collected will depend on the complexity of the waste stream and the process that generates it. Mass and energy balances and process flow diagrams are a minimum requirement for most assessments. For the case study, the data transmitted to the brainstorming participants will include:

- A process diagram (Figure 5-2)
- A mass and energy balance (Table 5-1)
- Process chemistry
- Description of the various process steps and streams (Figure 5-5)
- Selected component properties (Table 5-2)
- Major process streams (Table 5-3)

5.4.2 Set Goals

Setting goals helps the team to analyze the drivers for pollution prevention and to develop the appropriate criteria to screen the options that will be generated during the upcoming brainstorming session. The goals for the case study are to reduce the nitrogen content in the plant outfall; minimize air toxic emissions; and lower the capital investment and cost of manufacture for the new facility.

5.4.3 Define the Problem

Defining the problem helps the team to understand the targeted waste streams and the manufacturing process steps that generate them.

Two techniques for parsing the waste-generation problem are the waste-stream analysis and the process analysis, as discussed in Chapter 3. Whichever technique is used, the goal is to frame the problem such that pertinent questions arise. When the right questions are asked, the more feasible and practical solutions for pollution prevention become obvious. Figures 3-4 and 3-5 summarize these two analysis techniques, which are discussed again below as they relate to the case study.

Waste-Stream Analysis. The best pollution-prevention options cannot be implemented unless they are identified. To uncover the best options, one approach is to focus on the waste streams themselves, using the following four steps:

1. List all components in the waste stream, along with any key parameters. For instance, for a wastewater stream these could be water, organic compounds, inorganic compounds (both dissolved and suspended), pH, and so on. A waste-stream selection form such as the one shown in Figures 5-6 and 5-7 (completed for waste streams A10 and A22, respectively) can be used for this purpose.

2. Identify the compounds triggering the concern, for example, compounds regulated under the Resource Conservation and Recovery Act (RCRA), hazardous air pollutants (HAPs), or carcinogenic compounds. Determine the sources of these compounds within the process, then develop pollution-prevention options to minimize or eliminate the generation of these compounds.

The compounds of concern in the case study process are HCN and the nitrogen- and sulfur-containing compounds in stream A10, and benzene and nitrogen-bearing compounds in the aqueous waste (stream A22).

3. Identify the highest volume materials (often these are diluents such as water, air, a carrier gas, or a solvent), because these materials or diluents often drive the investment and operating costs for end-of-pipe treatment of the waste streams. Determine the sources of these diluents within the process, then develop pollution-prevention options to reduce their volume or eliminate them.

The volumetric flow of the acid gas scrubber vent (stream A10) is controlled by the nitrogen and unreacted oxygen flows that originate in the reactor. The volumetric flow of the benzene extractor bottoms (stream A22) is set by the water flow from the off-gas scrubbers.

4. If the compounds identified in Step 2 are successfully minimized or eliminated, identify the next set of compounds that has a large impact on investment and

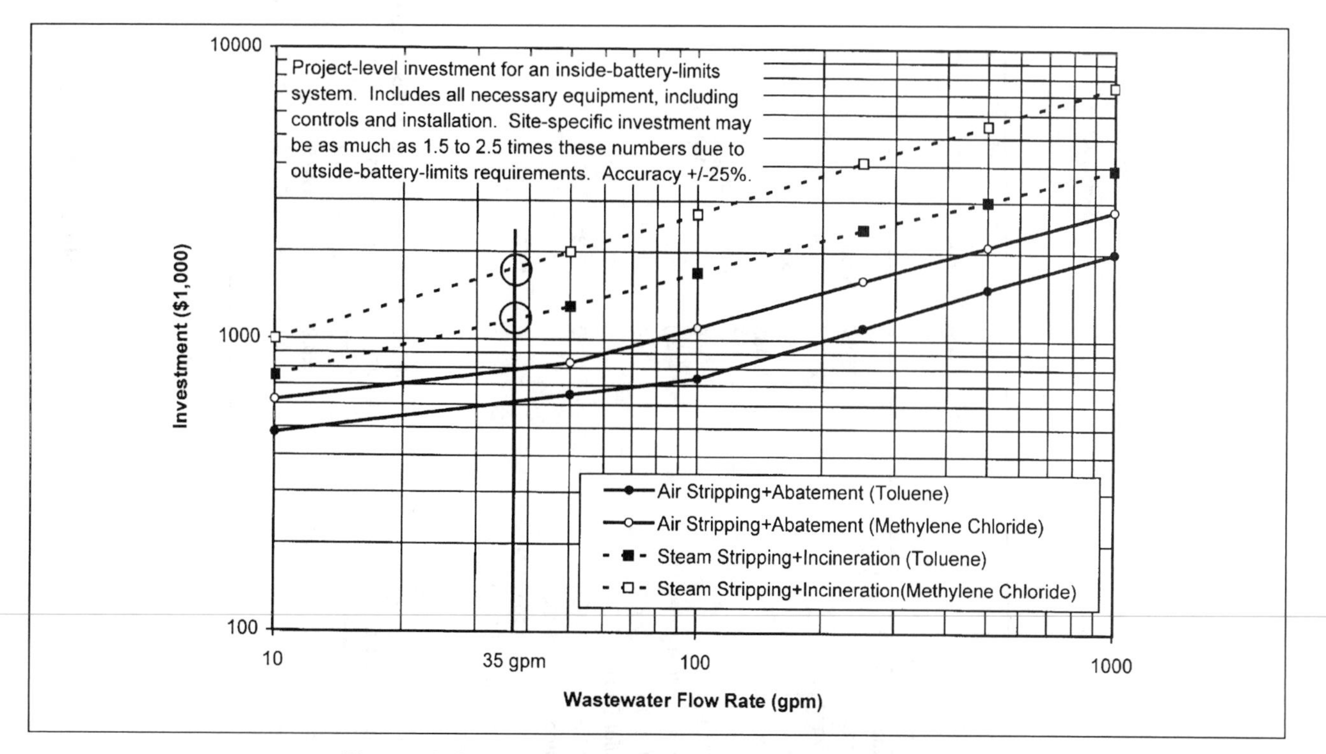

Figure 5-4. Investment For Air and Steam Stripping (Concentration Implicit in Cost Curves, ENR=5700)

Stream Number	A1	A3	A4	A5	A6	A10	A20	A21	A22	A23	A24	A25	A40	A41
Phase	Vapor	Vapor	Vapor	Liquid	Liquid	Vapor	Liquid	Liquid	Liquid	Liquid	Liquid	Liquid	Vapor	Liquid
Component							**Mass Flow (lb/hr)**							
Benzene (C_6H_6)	0	0	0	0	0	0	4949.6	4949.6	33.4	0.0089	0.0089	0	33.4	0
Nitrogen (N_2)	36952	36952	36952	0	0.22	36952	0.22	0.22	0.0024	0	0	0	0.0024	0
Oxygen (O_2)	11298	11064	11064	0	0.053	11064	0.053	0.053	0.00018	0	0	0	0.00018	0
Water (H_2O)	296	433.3	433.3	17500	14821	3112.2	15.9	15.9	14820	0	0	0	1500	16331
REACtant	483	205.8	205.8	0	202.8	3	2.7	203.1	1.7	200.4	199.3	1	1.7	0.0009
PRODuct	1	207.7	207.7	0	207.7	0	1.4E-05	206.6	1.1	206.6	1	205.6	0.054	1
IMPURity	15	15	15	0	2.4	12.6	0.52	2.4	0.013	1.9	1.9	0.0016	0.013	0
Ammonia (NH_3)	40	8	8	0	0	0	0.13	0.13	7.9	0	0	0	7.8	0.03
Nitric Oxide (NO_2)	0	14	14	0	2.3	11.7	0.21	0.21	2.1	0	0	0	2.1	0.004
Sulfur Dioxide (SO_2)	0	39.1	39.1	0	2.9	36.2	0.27	0.27	2.6	0	0	0	2.6	0.002
Carbonyl Sulfide (COS)	0	18.3	18.3	0	0.0018	18.3	1.6E-04	1.6E-04	0.0016	0	0	0	0.0016	0
Carbon Dioxide (CO_2)	0	120.9	120.9	0	0.025	120.9	0.0023	0.0023	0.022	0	0	0	0.022	0
Hydrogen Cyanide (HCN)	0	16.5	16.5	0	4.3	12.2	0.4	0.4	3.9	0	0	0	3.5	0.34
Sulfuric Acid (H_2SO_4)	0	0	0	0	26	0	0	0	0	0	0	0	0	0
Ammonium Sulfate ($(NH_4)_2SO_4$)	0	0	0	0	30	0	0	0	65	0	0	0	0	65
Total	49085	49094.6	49094.6	17500	15299.7	51343.1	4970.00	5378.89	14937.7	408.909	202.209	206.602	1551.19	16397.4
Temperature (°C)	266.5	400	352	30	54.2	48.6	-65	53.9	49.7	139.6	31.7	136.3	105.4	108.9
Pressure (lb/in^2 absolute)	75	31	29	25	18	18	9.5	14.7	14.7	10	0.5	0.65	18	20
Molecular Weight (g/g-mole)	28.9	28.9	28.9	18	18.4	27.9	77.3	78.8	18.1	104.3	99	109.9	18.4	18
Density (lb/ft^3).	0.21	0.069	0.069	62.9	61.8	0.081	60.2	53.7	61.9	64.9	66.1	69.7	0.046	58.9

Table 5-1. Mass and Energy Balances for Case Study

Compound	Molecular Weight (g/mole)	Normal Boiling Point (°C)	Critical Temperature (°C)	Critical Pressure (atm)	Critical Volume (cm^3/mole)	Critical Compress-ibility	Melting Point (°C)
Benzene	78	80	289	48	260	0.2730	5.5
Nitrogen	28	-196	-147	33.5	90	0.2910	-210
Oxygen	32	-183	-119	49.8	76.4	0.3014	-218
Water	18	100	374	218	55.4	0.2277	0
REACtant	99	132	355	60	250	0.2933	-30
PRODuct	110	235	482	58	276	0.2589	60
IMPURity	85	116	343	76	186	0.2796	-20
Ammonia	17	-33.4	132	111	72		-77.7
Nitric Oxide (NO_2)	46	21.3					-9.3
Sulfur Dioxide	64	-10	157	78	123		-76
Carbonyl Sulfide (COS)	60	50	105	61			-138
Carbon Dioxide	44	-78	31.1	73	96		-57
Hydrogen Cyanide (HCN)	27	26	183	55	135		-14

Table 5-2. Component Properties for Case Study

operating cost (or both) in end-of-pipe treatment. For example, if the aqueous waste stream was originally a hazardous waste and was incinerated, eliminating the hazardous compound(s) may permit the stream to be sent to the wastewater treatment facility. However, if doing so overloads the biochemical oxygen demand (BOD) capacity of the existing wastewater treatment facility, it may be necessary to identify options to reduce organic load in the aqueous waste stream.

Process Analysis. For either a new or existing process, the following steps are taken:

1. List all raw materials reacting to salable products, any intermediates, and all salable products. This is "List 1."

For the case study process, the salable compound is PROD, and the raw materials are REAC, oxygen, and ammonia (Figure 5-8).

2. List all other materials in the process, such as nonsalable byproducts, solvents, water, air, nitrogen, acids and bases. This is "List 2."

For the case study, these other components (water, nitrogen, excess oxygen, reactor byproducts, benzene, etc.) are not salable or do not make a salable product (Figure 5-8). They are only necessary because of the chemistry and engineering technologies that were chosen in the development of the original product and process.

3. For each compound in List 2, ask "How can I use a material from List 1 to do the same function of the compound in List 2?" or "How can I modify the process to eliminate the need for the material in List 2?"

4. For those materials in List 2 that are the result of producing nonsalable products (i.e., waste byproducts), ask "How can the chemistry or process be modified to minimize or eliminate the wastes (for example, 100% reaction selectivity to a desired product)?"

5.4.4 Generate Ideas

When the core assessment team has a good understanding of the process and the source and cause of each waste stream, it should convene to brainstorm for ideas. To generate *all* of the possible ideas for process improvement and waste reduction, the core team needs to involve additional talents and diverse points of view.

For the case study, the extended brainstorming team will include, in addition to the core team, a separations specialist, a pollution prevention and environmental specialist, a reaction engineering specialist, a chemist, and a business expert.

To educate the engineering community within the corporation and to facilitate the identification of source reduction options during brainstorming, a DuPont team has developed a set of pollution-prevention engineering technologies and practices covering the areas in the middle box on the right side of Figure 5-1. The team transformed a great deal of process- and industry-specific information on pollution-prevention successes (which was primarily in the form of case histories) into generalized knowledge. The information is organized in a "unit operations" format that can be more easily implemented by project teams and existing manufacturing facilities. The use of these technologies and practices results in faster identification and implementation of preferred pollution-prevention solutions.[1] Chapters 6 through 18 present these generalized technologies and practices.

For this case study, the brainstorming session produced

Process Chemistry

The raw material, REAC, reacts with ammonia and oxygen to substitute a triple-bonded nitrogen for the hydrogen atoms on the methyl group. The oxygen reacts with the displaced hydrogen atoms to form water. Unfortunately, at high temperatures (greater than 360°C), the compounds REAC and PROD are thermally decomposed (i.e., oxidized) to CO_2, NO_x, SO_2, COS, HCN, etc.

$$\underset{\text{(REAC)}}{\overset{\text{CH3}}{\overset{|}{\text{R-N-C-S}}}} + 3/2\ O_2 + NH_3 \xrightarrow[\text{STEAM}]{\text{CATALYST}} \underset{\text{(PROD)}}{\overset{\text{CN}}{\overset{|}{\text{R-N-C-S}}}} + 3H_2O$$

Feed to Reactor

- Air is compressed to 75 psia (60 psig) by a gas compressor.
- Liquid reactant (REAC) is then added to the hot compressed air and vaporized.
- Steam and ammonia (NH_3) are added next and the entire stream is heated to approximately 300°C by process-to-process heat exchange with the hot reactor exit gas.

Reactor

- A fixed-bed catalytic reactor is used. The catalyst is an alumina honeycomb substrate coated with a precious metal catalyst.
- Inlet to the catalyst bed is about 340°C and 75 psia.
- The reacted gases exit at greater than 400°C and 51 psia.
- REAC reacts to PROD and water.
- Some REAC and PROD are converted to undesirable byproduct gases such as SO_2, COS, CO_2, NO_x and HCN.

Process-to-Process Heat Exchange (Preheater)

- Reactor exit gas enters the heat exchanger at 440°C and 51 psia and is cooled to 400°C at 49 psia.
- Reactor feed gas enters at about 265°C and is heated to about 300°C.

Electric Heater

Reactor feed gas is heated from 300°C to 340°C, the temperature required to initiate the reaction.

Figure 5-5. Process Description for Case Study

Water Scrubber

- 30 gpm fresh water is fed to the top of the scrubber.
- The exiting gas contains some REAC, PROD, and IMPUR, together with byproduct gases made in the reactor.
- The exiting gas also contains nitrogen and unreacted oxygen.
- The exiting water stream contains the majority of the REAC, PROD and IMPUR and some dissolved gases.

Acid Scrubber

- 5 gpm dilute sulfuric acid is added to the scrubber to remove the final traces of PROD and REAC from the gas stream. This will also remove any residual ammonia.
- The exiting gas stream contains primarily nitrogen, oxygen, and reactor byproduct gases.

Feed to Benzene Extractor

The water and acid scrubber aqueous streams are mixed together and neutralized with ammonia to a pH greater than 7.0. This is to prevent corrosion in downstream pieces of equipment, which are fabricated from carbon steel.

Benzene Extractor

Benzene is used to extract PROD and REAC from the water. The exiting water stream contains ammonium salts, sodium salts from the soft water fed to the scrubber, some REAC and PROD, benzene, and dissolved gases.

Steam Stripper

- Benzene is stripped from the wastewater by steam, and the overhead benzene/steam stream is sent to a thermal oxidizer to destroy the benzene.
- The stripped water is sent to wastewater treatment.

Benzene Recovery Column

- Distills the benzene from the PROD and REAC.
- Benzene is recycled back to the extractor.

Product Recovery Column

- Separates PROD and REAC by distillation.
- REAC is recycled back to the front-end of the process.

Figure 5-5. Process Description for Case Study

over 100 ideas for process improvement. Some of the ideas are listed in Figure 5-9.

Three key questions were asked during the brainstorming session that proved to be the "catalyst" to improve the process.

Why Use Air as the Source of Oxygen? The nitrogen and excess oxygen act as heat sinks to prevent hot spots in the fixed catalyst bed. Steam was suggested as an alternative heat sink; however, subsequent testing showed that the activity of the catalyst would have been negatively affected by the excess steam. Another suggestion was to use a fluidized-bed technology to improve the reactor temperature profile and minimize hot spots.

Why Use Water to Remove the Product and Reactant from the Gas Stream? Water is a cheap, universal solvent and heat-transfer fluid. The water serves two functions in the existing process: to cool the hot reactor gases and to scrub the condensed reactant and product from the gas stream. When the original process was developed, the impact on wastewater treatment was not taken into consideration.

What Are the Physical Properties of the Product and Reactant? The product solidifies at 50–60°C. The reactant has a normal boiling point at above 120°C and a freezing point below -20°C. These properties, together with increased concentrations at reduced air flow, will allow for direct condensation of the bulk of the product.

5.4.5 Screen Ideas

A first-cut screening of the ideas is performed during the brainstorming session to take advantage of the experience and "gut feel" of the participants in the room. No idea is ever truly eliminated completely, but it is important to make an initial ranking to help the core assessment team focus on those ideas with the highest likelihood of success. The core team, however, is responsible for reconsidering all ideas in a separate session. Because the brainstorming session participants are normally weary by the time the first-cut screening is done, simple methods are used to rank the ideas. Some methods that have been used successfully include:

- judge yes or no;
- rank high, medium, or low against specific criteria
- divide the number of ideas by ten and allow everyone to vote using self-adhesive colored dots

Criteria mentioned in Section 5.4.2 can be used for ranking options for the case study. Other criteria can be found in Section 3.6.2.

During the first-cut screening, the number of ideas is normally cut in half; then in a subsequent meeting, the core assessment team revisits the options generated during the brainstorming process. Typically, a second-cut screening is made to further narrow the number of "feasible"

Feed Streams	Flow/Constituents
Water	30 gpm
Dilute Acid	5 gpm
Reactor Feeds	10,000 scfm air
	40 lb/hr NH_3
	296 lb/hr steam
	483 lb/hr REACant
Make-up Benzene	50 pph
Benzene Recirculation Rate	11 gpm
REAC Recirculation Rate	199 lb/hr
Waste Streams	**Flow**
Scrubber Off-gases (N_2, O_2, NO_x, SO_x, COS, toxics)	10,000 acfm
Wastewater to Treatment (NH_4 salts, Na salts, dilute toxics, high nitrogen load)	35 gpm
Steam Stripper Overheads to Thermal Oxidizer (H_2O, NH_3, benzene, HCN, others)	1550 lb/hr

Table 5-3. Major Process Streams for Case Study

FORM: Waste Stream Selection

Date: January 15, 1998

Process: Intermediates Manufacture

Waste Stream ID: Stream A10 — Acid Scrubber Vent

Waste Description: Gas Stream Containing Air Toxics

DISPOSITION
Is this stream a "quick hit"?
___Yes _X_ No
If not check one:
___ Do assessment now
___ Do assessment later
___ Do nothing at this time
___ Info required

WASTE TREATMENT
Off-site Disposal? ___ Yes _X_ No
Disposal Medium _X_ Air Emission ___ Incineration
___ Landfill ___ Wastewater
___ Other________________
Regulatory Exposure? _X_ Yes ___ No

WASTE GENERATION
High Toxicity? _X_ Yes ___ No
Special Safety Hazard? ___ Yes _X_ No

Stream

____
(Annual Rate)

____
(Rate per Unit of Production

Waste Compounds and Composition in Order of Importance

Compound	Wt., **Vol.**, or Mol%	Waste Origin (Present Knowledge)
Nitrogen	Carrier (> 80%)	Air to process
Oxygen	19%	Air to process
Water	9%	From scrubber
Impurity	80 ppmv	Feed to process
NO2, SO2, COS	50 ppmv	Reactor byproducts
CO2, HCN	86 ppmv	Reactor byproducts

WASTE COST AND OTHER INFORMATION
Yield loss: ______33%______
Treatment: ____________
Packaging: ____________
Transportation: ____________
Additional Materials: ____________
Taxes: ____________
Cost of Manufacture: ______$2/lb REAC______

List nonquantifiable "soft-costs":
____
____
____
____
____
____

Total Estimated Cost: ____________

NOTICE: This form is intended for use as a preassessment worksheet. It is not a permanent record of waste generation or treatment. The information it contains may be speculative.

Figure 5-6. Completed Waste Stream Selection Form for Vent Scrubber Off-Gas (stream A10)

FORM: Waste Stream Selection

Date: January 15, 1998

Process: Intermediates Manufacture

Waste Stream ID: Stream A22 — Benzene Extractor Bottoms

Waste Description: Wastewater Stream Containing Benzene and Nitrogen-Bearing Compounds

DISPOSITION
Is this stream a "quick hit"?
___ Yes _X_ No
If not check one:
___ Do assessment now
___ Do assessment later
___ Do nothing at this time
___ Info required

WASTE TREATMENT
Off-site Disposal? ___ Yes _X_ No
Disposal Medium ___ Air Emission ___ Incineration
___ Landfill _X_ Wastewater
___ Other ______________
Regulatory Exposure? _X_ Yes ___ No

WASTE GENERATION
High Toxicity? _X_ Yes ___ No
Special Safety Hazard? ___ Yes _X_ No

Waste Compounds and Composition in Order of Importance

Stream	Compound	Wt., Vol., or Mol%	Waste Origin (Present Knowledge)
16 Mgal/yr	Water	Carrier (> 80%)	Scrubber water feed
(Annual Rate)	REAC & PROD	190 ppmw	Feed and reactor product
______	Benzene	0.22%	Benzene extractor
(Rate per Unit	Ammonia	530 ppmw	pH control
of Production	NO2, SO2	320 ppmw	Reactor byproducts
	HCN	210 ppmw	Reactor byproducts

WASTE COST AND OTHER INFORMATION
Yield loss: 33%
Treatment: ______
Packaging: ______
Transportation: ______
Additional Materials: ______
Taxes: ______
Cost of Manufacture: $2/lb REAC

Total Estimated Cost: ______

List nonquantifiable "soft-costs":
______

NOTICE: This form is intended for use as a preassessment worksheet. It is not a permanent record of waste generation or treatment. The information it contains may be speculative.

Figure 5-7. Completed Waste Stream Selection Form for Benzene Extractor Bottoms (stream A22)

FORM: Process Constituents and Sources

Process: Intermediates Manufacture   Date: January 15, 1998

List 1

Constituent:	Source (Feed, Reactor, Unit Operation):
Salable Products	
PROD	Reactor
______	______
______	______
______	______
Intermediates (result in salable products)	
______	______
______	______
______	______
______	______
______	______
Essential Raw Materials(only those constituents used to produce the intermediates and salable products)	
REAC	Feed
Ammonia	Feed
Oxygen	Air Feed to Reactor
______	______
______	______
______	______

List 2

Constituent:	Source (Feed, Reactor, Unit Operation):
Other Materials (Nonsalable byproducts, solvents, water, air, nitrogen, acids, etc.)	
Nitrogen	Air Feed to Reactor
IMPURity	Feed
Water	Feed to Scrubbers
Benzene	Extractor Solvent
NO_2, SO_2, COS, CO_2, HCN	Reactor Byproducts
______	______
______	______
______	______
______	______
______	______

Figure 5-8. Process Constituents and Sources Form for Case Study Process

options that will be carried forward and evaluated in further detail. Experience shows that about 10–15% of the original ideas make it through the second-cut screening.

For the case study, once the idea-generation phase was complete, the ideas were then reexamined and ranked using a yes/no rating method and the following criteria:

- Technical feasibility
- Economic viability
- Waste-reduction potential

Figures 5-10 through 5-12 briefly summarize the process improvement ideas that survived the first-cut screening. The ideas are categorized as those requiring low capital investment, high capital investment, or additional research and development. More details on each of these ideas can be found in Appendix B. For each screened idea, the Appendix includes a brief description of the process, chemistry, or technological changes, concerns, estimated capital investment and/or operating cost, benefits if implemented, resources required, timing, and probability of success.

5.4.6 Evaluate the Screened Ideas

This stage of the pollution-prevention program entails a more detailed and thorough technical and economic evaluation of the screened ideas. The first two levels of screening just described minimize the number of feasible alternatives that need to be evaluated in detail.

In general, revised mass and energy balances, process flow diagrams, and operating requirements are generated for each screened idea. Based on these inputs, the investment, operating cost, and net present value for each option can then be determined and compared against the existing process.

5.5 Implementation Phase

The tasks listed in Figure 5-1 will turn the options identified by the team into the projects that will actually accomplish the waste reductions. Although not discussed here, this is a very important phase of the pollution-prevention program. This was one good reason to have a project engineer lead the core assessment team for this case study.

Equipment and Streams Eliminated
Reactor preheater
Greatly reduced reactor feed gas compressor size
Benzene extraction column
Two absorption columns (scrubbers)
Benzene recovery column
Benzene handling facilities (tanks, pumps, etc.)
Much smaller thermal oxidizer
No water feed
No dilute acid feed
Equipment Added
Product cooler/condenser downstream of reactor
Spray contactor and chiller using cold REAC as absorbent
REAC/water azeotropic recovery column to separate water of reaction.
Increased size of product recovery column

Table 5-4. Equipment and Streams Eliminated or Added to the Redesigned Process

	Cost or (savings), $ Million
Compressor rating reduced from 900 hp to 50 hp	(0.9)
Extraction equipment eliminated	(2.1)
Scrubber size reduced	(0.2)
Benzene handling facilities eliminated	(3.1)
Stream stripper eliminated	(0.4)
Thermal oxidizer size reduced	(1.6)
Product condenser required	0.5
Investment savings	7.8
NPV of savings (12%, 10 yr)	12.9

Table 5-5. Investment Savings and Net Present Value Savings Achieved by the New Design

Acid Scrubber Off-Gas Stream (A10)

Use pure oxygen in the reactor instead of air to eliminate N_2.

Recycle the acid scrubber off-gas (stream A10) to the reactor and add pure O_2 as make-up.

Use a new or improved catalyst in the reactor.

Replace the existing fixed bed reactor with a fluidized-bed reactor to eliminate hot spots that lead to unwanted byproduct formation.

Add a second heat exchanger in series with the preheater to improve energy recovery.

Use a different heat sink instead of nitrogen, such as CO_2 or steam.

Use an indirect-contact heat exchanger to cool the reactor off-gas and condense the product and reactant.

Wastewater Stream from Benzene Extractor (A22)

Combine the acid and water scrubbers to reduce water consumption.

Recycle the water stream from the benzene extractor (A22) to the water scrubber.

Recycle the water stream from the steam stripper (A41) to the water scrubber.

Use caustic soda instead of ammonia for pH control of the water stream to the benzene extractor (stream A6) to reduce total nitrogen load to wastewater treatment.

Replace benzene with a different extraction solvent, such as toluene or xylene.

Freeze-crystallize REAC and PROD from the water to improve product recovery and reduce total nitrogen load to wastewater treatment.

Use a multieffect evaporator to concentrate stream A41.

Use the chilled reactant (REAC) to scrub itself from the reactor off-gas.

Use a condenser and decanter to separate and recycle the benzene from the steam stripper overheads instead of a thermal oxidizer.

Figure 5-9. Process Improvement Ideas Generated During the Brainstorming Session

5.6 Results

Figure 5-13 shows the flow diagram for one possible redesigned process.

Pilot testing of a fluidized-bed reactor revealed that a 70% reduction in the total gas flow around the reactor was possible. The lower gas flow, coupled with the ability to maintain a high-pressure recirculating gas loop around the reactor, reduced compressor energy requirements by more than 85%. The fluidized bed and the use of make-up oxygen obviated the need for large quantities of air, which resulted in significant savings in the gas compressor.

More importantly, reactant and product concentrations increased with the decreased air flow, allowing for direct condensation of the bulk of the product. Due to the boiling-point differences between the reactant and product, a product-rich condensate is removed in this step (80%), bypassing the absorption step and part of the distillation train. The remainder of the reactant and product are scrubbed using chilled reactant (i.e., reactant in place of water) in a considerably (60%) smaller-diameter unit.

The change in scrubbing medium has profound implications in further separations processing, in that the extraction step and raffinate stripping operations are completely eliminated. This substitution was arrived at in a effort to eliminate the large aqueous waste load imposed upon the process by aqueous scrubbing of moderately soluble organics. The elimination of the extraction step, benzene handling, and raffinate stripping operations provided significant project investment savings and lower operating costs. Table 5-4 summarizes the equipment pieces and process/waste streams that were eliminated from or added to the redesigned process.

Emission reductions include: vent gas flow was reduced from 10,000 scfm to less than 50 scfm; waste-

Recycle Vent Stream and Add Additional O_2 or Enriched Air as Needed

Cost:	$200,000 investment, $500,000/yr operating cost
Benefit:	Reduced end-of-pipe investment of $700,000 Reduced operating cost of $70,000/yr
Waste Minimization:	Reduced gas flow to be treated
Energy Conservation:	Reduced electricity requirements
Probability of Success:	90%

Install Condenser to Remove Product

Cost:	$500,000 investment
Benefit:	Unknown
Waste Minimization:	No change
Energy Conservation:	Reduced separation requirements
Probability of Success:	90%

Recycle Benzene From the Steam Stripper Overhead Stream

Cost:	$100,000 investment, $10,000/yr operating cost
Benefit:	Reduced operating cost of $90,000/yr
Waste Minimization:	Reduced gas rate to thermal oxidizer
Energy Conservation:	No change
Probability of Success:	90%

Use Steam Stripper Bottoms as the Source of Water for the Water Scrubber

Cost:	$300,000 investment, $10,000/yr operating cost
Benefit:	Water conservation and reduced treatment cost
Waste Minimization:	Reduced wastewater treatment load (35 gpm to 1 gpm)
Energy Conservation:	No change
Probability of Success:	90%

Figure 5-10. Low-Capital-Investment Process Improvement Ideas from First-Cut Screening

Use Superheated Steam (as the inert gas) and Pure O_2 Instead of Air

Cost:	\$2,000,000-\$6,700,000 investment, \$800,000-\$3,500,000/yr operating cost
Benefit:	Reduced end-of-pipe investment of \$800,000 Reduced operating cost of \$300,000/yr
Waste Minimization:	Reduced gas flow to be treated
Energy Conservation:	Reduced electricity and steam requirements
Probability of Success:	30%

Replace the Water Scrubber With a Solvent Scrubber

Cost:	\$1,300,000 investment, \$100,000/yr operating cost
Benefit:	Reduced operating cost of \$700,000/yr
Waste Minimization:	Reduced wastewater treatment load (35 gpm to 1 gpm)
Energy Conservation:	Reduced steam requirements
Probability of Success:	90%

Use Freeze Crystallization to Separate REACtant, PRODuct, and Water

Cost:	\$3,500,000 investment, \$1,500,000/yr operating cost
Benefit:	Reduced operating cost of \$700,000/yr
Waste Minimization:	No change
Energy Conservation:	No change
Probability of Success:	40%

Purify REACtant and PRODuct Directly From the Water Phase, Rather Than Using Extraction

Cost:	\$10,500,000 investment, \$4,500,000/yr operating cost
Benefit:	Reduced operating cost of \$700,000/yr
Waste Minimization:	Reduced wastewater treatment load (35 gpm to 1 gpm)
Energy Conservation:	No change
Probability of Success:	70%

Figure 5-11. High-Capital-Investment Process Improvement Ideas from First-Cut Screening

water flow was reduced from 35 gal/min to 1 gal/min; and all benzene fugitive emissions were eliminated. In addition, thermal oxidizer emissions were lowered by more than 90% due to the smaller vent gas flow, and energy consumption was significantly reduced. Personnel exposure problems and administrative controls required with benzene are also no longer an issue with the revised process.

Changes in the distillation train are required; however, no measurable increase in capital investment results. In addition, the overall distillation train has a lower operating cost now that the benzene solvent recovery step is eliminated. The thermal oxidizer required to treat off-gases is also greatly reduced in size and cost. Table 5-5 details the financial benefits of the revised manufacturing process.

Use Fluidized-Bed Reactor (reduces air volume)

Cost:	$500,000 investment, $500,000/yr operating cost
Benefit:	Reduced end-of-pipe investment of $800,000 Reduced operating cost of $40,000/yr
Waste Minimization:	Reduced gas flow to be treated
Energy Conservation:	Reduced electricity requirements
Probability of Success:	80%

Use New Catalyst With Better Selectivity/Conversion

Cost:	Unknown
Benefit:	Unknown
Waste Minimization:	Reduces waste generation by 50% for a 5% increase in yield
Energy Conservation:	None
Probability of Success:	50%

Change Air-to-Feed Ratio to Reactor to Reduce COS Generation

Cost:	Unknown
Benefit:	Unknown
Waste Minimization:	None
Energy Conservation:	None
Probability of Success:	50%

Recover and Recycle Tars From REACtant/PRODuct Splitter

Cost:	Unknown
Benefit:	Unknown
Waste Minimization:	Eliminate hazardous waste incineration
Energy Conservation:	None
Probability of Success:	70-80%

Figure 5-12. Process Improvement Ideas from First-Cut Screening Requiring Research and Development

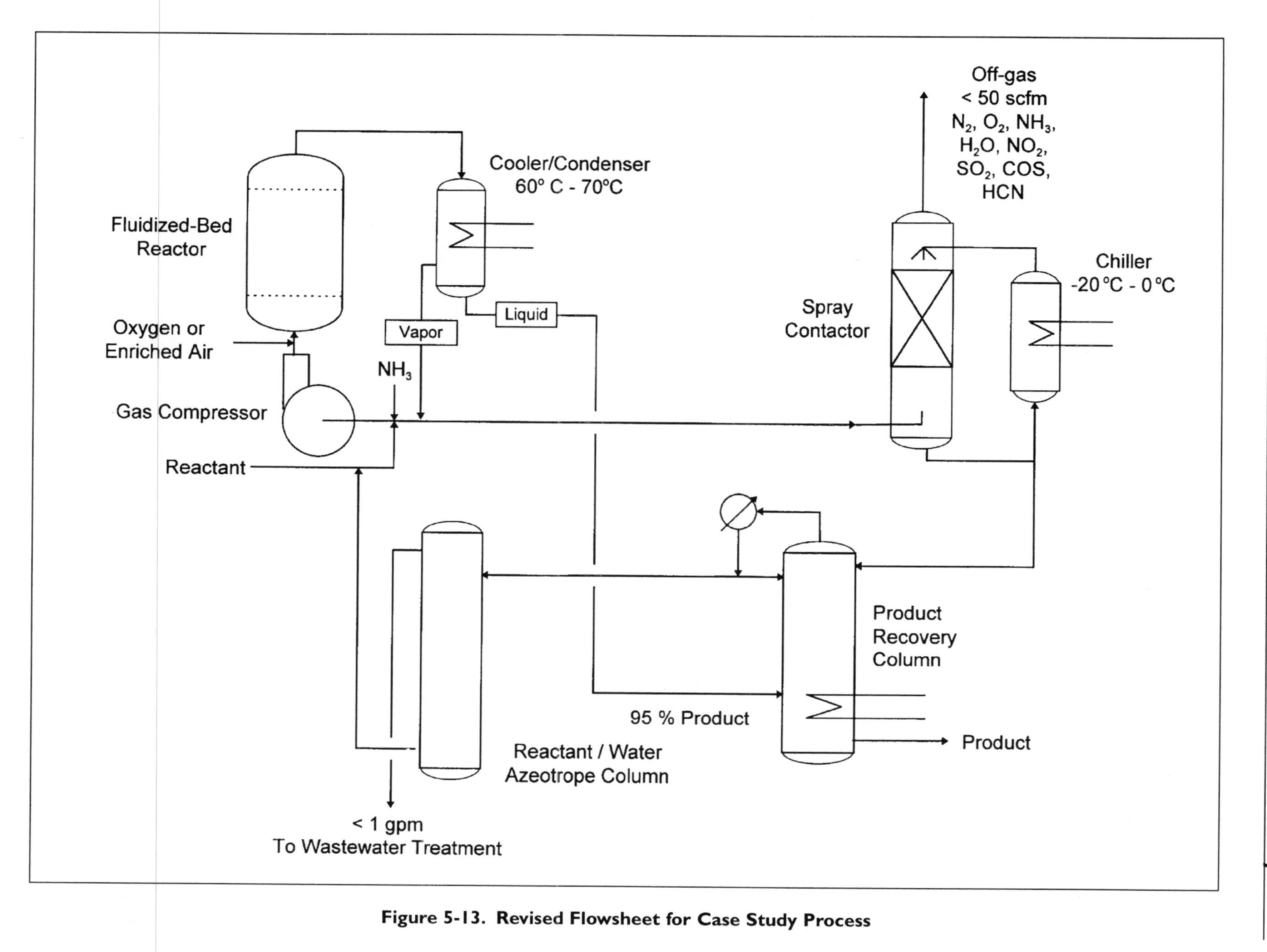

Figure 5-13. Revised Flowsheet for Case Study Process

Literature Cited

1. Dyer, J. A., and K. L. Mulholland, January 1998, "Follow This Path to Pollution Prevention," *Chemical Engineering Progress*, 94: 34–42.

2. Dyer, J. A., and K. L. Mulholland. February 1994, "Toxic Air Emissions: What is the Full Cost to Your Business?" *Environmental Engineering: A Special Supplement to February 1994 Chemical Engineering,* 101 (suppl.): 4–8.

CHAPTER 6

Pollution-Prevention Technologies and Practices

6.1 Introduction

Some of the arguments used by engineers and scientists unfamiliar with the methodology, strategies, and true wisdom of pollution prevention or waste minimization are that it is too process-specific; that we cannot leverage the learnings from one plant or industry to another; or, to use a cliché, that the view is not worth the climb. In the first five chapters we have attempted to dispel these myths. In fact, the wisdom of pollution prevention is quite the opposite—by dissecting the process into its component parts, asking the right questions, and applying good engineering and generalized pollution-prevention technologies and practices, improved process yield and uptime become synonymous with improved environmental performance.

The remaining chapters of this book present the "meat" of pollution prevention—the engineering technologies and practices. To facilitate the rapid generation of process improvement ideas during brainstorming (whether in a large or small group, or even on an individual basis), the extensive databases of pollution-prevention case histories available within the DuPont Company and the public domain were creatively transformed into fundamental pollution-prevention knowledge that can be applied across all businesses and industries. In most of these chapters, a pollution-prevention continuum is presented that captures (in a hierarchical format) the fundamental pollution-prevention strategies that apply for that particular engineering technology, operating practice, or unit operation. The continuums are a snapshot of the breadth of options that are available to engineers and scientists for minimizing emissions, beginning at the process source and working down through treatment, in some cases. The pollution-prevention scale shown in each continuum is relative, and should not be taken too literally. In some cases, a continuum is not presented, because that technology or practice did not lend itself to this presentation format.

This chapter introduces the engineering technologies and operating practices discussed in detail in Chapters 7 through 18. In addition, a case study is presented that exemplifies the need for and applicability of generalized pollution-prevention technologies and practices.

6.2 Engineering Technologies, Operating Practices, and Unit Operations

The following is a list of categories—some of which are unit operations, some of which are engineering technologies, some of which are operating practices, and some of which do not fit nicely into any one classification—that define the collection of pollution-prevention technologies and practices presented in Chapters 7 through 18. Included with each technology or practice is a brief description of the key environmental emission sources and their respective causes.

Batch Operations. Emissions from batch operations result during charging of raw materials (e.g., vapor displacement, dust generation), reactor operation (e.g., formation of unwanted byproducts, nitrogen padding of vessels, sampling), reactor discharge (e.g., solvent flashing, product accumulation in lines, and pressure transfer), and vessel cleaning (e.g., during product changeovers).

Equipment and Parts Cleaning. Cleaning of equipment and parts is an important preparatory step in many industrial processes, particularly in batch processes. With the use of multipurpose batch equipment growing in high-value, specialty chemicals processes, a chief concern becomes cross-contamination of products in addition to environmental concerns, such as air emissions, wastewater treatment, and hazardous waste disposal. In many processes, the waste associated with cleaning represents the majority of the waste generated and/or the peak load on a treatment or emissions control operation. The business incentive to reduce or eliminate this peak load becomes substantial.

High-Value Waste. Waste generation sometimes results by not recognizing the opportunity for certain waste streams to be reclassified and sold as a product or converted into a usable form. In one DuPont process, the waste esters from an intermediates process have become more valuable than the commodity product; therefore, there is now an incentive to increase the yield of the "waste esters."

Reactor Design and Operation. Waste generation in any chemical manufacturing process can almost always be traced back to the heart of the process, the reaction system. Emissions result from poor solvent selection, impurities in raw materials, improper equipment design, not understanding the reaction chemistry, product accumulation that promotes side reactions, inadequate mixing, poorly designed heat-transfer equipment, improper addition of raw materials, inadequate process control and instrumentation, and so on. The reaction system represents the key to pollution prevention in many processes, and yet it is also the process area that encounters the most resistance in brainstorming sessions.

Use of Water as a Solvent and Heat-Transfer Fluid. Water is used as the universal solvent and heat-transfer-fluid-of-choice in the manufacturing industries. Even though water is not a contaminant by itself, it is the major component of many waste streams (often greater than 98% by weight). For this reason, water use often drives the cost of end-of-pipe treatment for many manufacturing processes. Emissions result from the simple act of contacting water with raw materials, solvents, and products. The challenge is how to minimize the degree of contact in the first place and, when contact is essential, to maximize the number of times the same water stream is reused, so that freshwater use is minimized.

Organic Solvents. Organic solvents are used to dissolve materials for easier processing, to extract materials from a mixture, to create or coat films, and to provide a medium for liquid-phase reactions. Emissions result from high temperatures, vessel leaks, indiscriminate use of inert gases, selection of solvents of high volatility, hot-air dryers, and ventilation of manufacturing areas. The problem is magnified when highly regulated, highly toxic solvents are chosen early in the design of the reaction chemistry. More than any other pollution-prevention technology or practice, solvent selection must be addressed very early in the product life cycle.

pH Control. The pH of an aqueous system can affect byproduct formation in chemical reactions, excess alkali consumption and inorganic salt formation in acid gas (e.g., HCl, HBr) scrubbers, the solubility of organic acids and bases in wastewater streams, and air emissions of volatile organics from wastewater treatment systems. Neutralization with mineral acids and bases produces inorganic salts that are often removed by water washing. During washing, the water almost invariably becomes contaminated with other organic compounds, resulting in the generation of a "salty" aqueous waste.

Vacuum Generation. Vapor emissions result from the evaporation of volatile organic compounds in vessels operating under vacuum, the magnitude of which can be drastically increased by high levels of inerts (such as nitrogen) dissolved in the liquid, air inleakage through flanges, valve packings, and so on. In vacuum systems using steam as the motive fluid, liquid entrainment due to high vent-line gas velocities can lead to increased organic load on wastewater treatment systems.

Ventilation of Manufacturing Areas. Air emissions result from contamination of room dilution ventilation air with process gases, fugitive solvent emissions, and dusts and fumes. For most processes built prior to the 1970s, the atmosphere was viewed as an infinite sink; therefore, air was freely used to ventilate manufacturing areas to minimize operator exposure to harmful vapors and fumes. Nowadays, these large-volume air streams (tens and hundreds of thousands of standard cubic feet per minute) represent a significant business liability if they must be treated.

Volatile Organic Liquid Storage. Vapor emissions are caused by vapor displacement during tank filling (working losses) and daily temperature changes (breathing losses). The problem is magnified when the tank is purged with an inert gas. High-turnover tanks storing volatile solvents are the most likely candidates for pollution prevention.

Separation Technologies. Most manufacturing processes, especially in the chemical industry, comprise two major sections: a reaction section that transforms a feed(s) into a product(s), and a separations section that separates a mixed stream(s) into its individual constituents. The selection of the right separation technology to use in the first place, along with the ongoing operation of that technology, dictates the amount of waste that is generated. Excessive environmental emissions result from improper technology selection, poor/inadequate design (e.g., insufficient theoretical stages in a packed-bed scrubber, wrong feed location in a distillation column, an operating temperature in a decanter that is too high), lack of good vapor–liquid equilibrium data, fouled components, and so on.

Equipment Leaks. Fugitive emissions are defined as unintentional air emissions resulting from the effects of malfunctions, age, lack of proper maintenance, operator errors, improper equipment specifications, use of inferior technology, and externally caused damage. Equipment leaks represent a specific subcategory of fugitive emissions, and can account for a substantial amount of the reportable air emissions at a plant site. Equipment leaks can occur at valve stems, flanges, pumps, pipe fittings, manhole seals, transfer lines, and so on.

6.3 Dyelate Process Case Study

The Mulholland Company manufactures a herbicide named Dyelate using a batch process. The herbicide is formulated as a dry flowable granule and a wettable powder. It is used on vegetable crops to control weed growth. The product has an attractive cash flow, but is a large waste generator for the company. Waste from this herbicide process accounts for 65% of the waste load on a multiproduct site. In fact, the wastewater treatment facility is operating at maximum capacity, and cannot accept any additional organic load. This is a negative for the Mulholland company, because it precludes new products from being manufactured at the site.

The owner of the company has asked the Dyelate business to investigate ways to reduce waste generation from their process. A new product line is being developed in the R&D laboratory, and the owner would like to avoid an estimated $10,000,000 to expand and upgrade the existing wastewater treatment facility on the site. The lead process engineer for the Dyelate process has been assigned the task of designing a pollution-prevention program to reduce waste generation by at least 50%. The process engineer's current mindset is that the Dyelate process is unique, and there is no other operation just like it in the world. However, a pollution-prevention consultant hired by the Mulholland company seeks to show the process engineer how the company's seemingly unique waste-generation problems can be transformed into problems that are common across many types of processes.

Figure 6-1 presents a block flow diagram of the Dyelate process. The process consists of seven key steps: condensation, ring closure, extraction, acidification, halogenation, filtration and drying, and storage and packaging. Figure 6-2 is a more detailed flow diagram showing the key process and waste streams.

The subsections below briefly describe each process step, outline the key waste-generation issues, pose questions that can be framed in terms of one or more of the generalized pollution-prevention technologies and practices, and, finally, offer potential solutions to minimize waste generation at the source.

6.3.1 Condensation

Condensation is the first in a series of reaction steps to produce the Dyelate compound. In this first step, Reactant A (a six-carbon, nitrogen-bearing organic compound) and Reactant B (a seven-carbon, oxygenated hydrocarbon containing hydroxyl and carboxylic acid functional groups) combine in Pot I to form an intermediate, labeled "Int1." The reaction is performed in a moderate-volatility, boiling hydrocarbon solvent (Solvent) using a mineral acid (Acid) as a catalyst. In the condensation reaction, water is produced as a byproduct, and must be vacuum distilled from the reactor to drive the reaction to completion (approximately 100 mm Hg and 90°C). Vacuum is used to reduce the boilup temperature to minimize formation of undesirable side products. The vacuum is provided by two stages of steam jets. Water and solvent vapors from the reactor are condensed in a series of vent condensers, and then decanted to produce a recyclable organic-solvent phase and a water phase containing small amounts of residual solvent. The reaction mass is then cooled under vacuum and batch-transferred to Pot II for the next step, Ring Closure.

The key waste-generation issues in this step of the process are the formation of side products due to the 90°C reactor temperature, and the production of a significant wastewater stream containing small amounts of organic solvent. The wastewater stream is the steam condensate from the vacuum jet system. Questions that could be asked from a pollution-prevention perspective include:

- Can the reactor be operated at higher vacuum (i.e., lower absolute pressure) to reduce the reactor operating temperature and, hence, side-product formation?
- Can the solvent-contaminated wastewater stream be steam or air stripped to remove residual solvent? Can colder decanter temperatures be used to improve the solvent/water separation?
- Is there an alternative solvent that boils at a lower temperature than the existing solvent, while also possessing lower solubility in water?

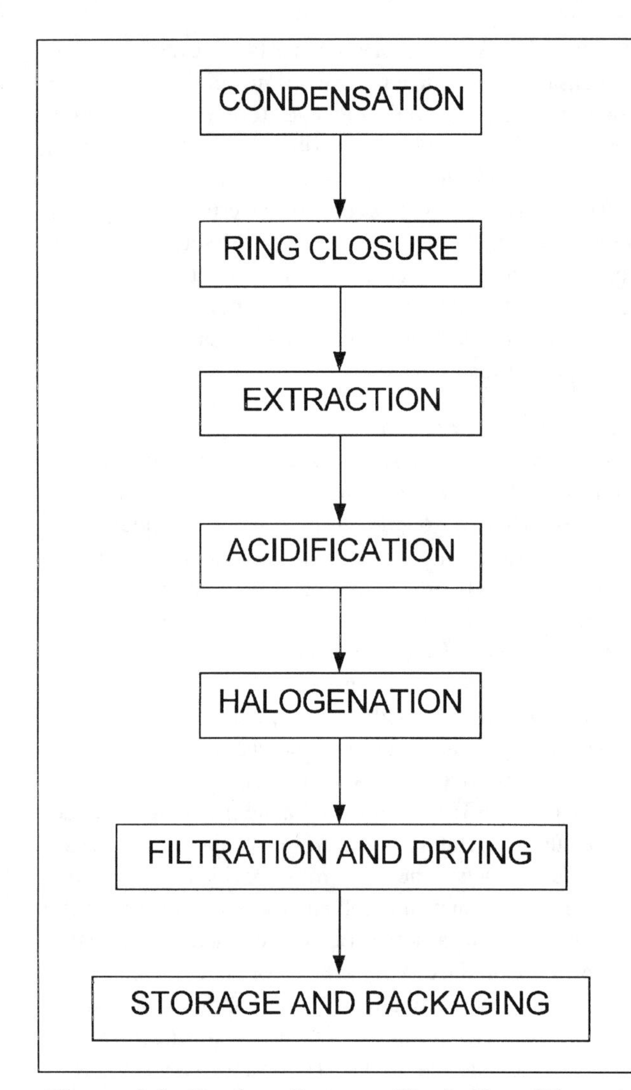

Figure 6-1. Dyelate Process Block-Flow Diagram

- Can the vacuum system be improved to minimize wastewater generation (e.g., vacuum pumps instead of steam jets)? Are the steam jets oversized?
- Can a heterogeneous catalyst be used instead of a homogeneous one?

Now let us look at this process step from the standpoint of generalized pollution-prevention engineering technologies and practices. Looking beyond the "unique" aspects of this process, it is apparent that a number of the unit operations, engineering technologies, and operating practices described in Section 6.2 will be appropriate for this case study: batch operations, reactor design and operation, use of water as a solvent and heat transfer fluid, organic solvent selection, and vacuum generation. A generalized set of pollution-prevention engineering technologies and practices, then, will prove useful in improving the performance of this process.

6.3.2 Ring Closure

In this step, the intermediate (Int1) rapidly reacts with the sodium salt of an alcohol (Na-OR) in Pot II to form the sodium salt of the intermediate (Int1-Na). A byproduct of this reaction is a high-volatility alcohol (Volatile Alcohol). The Na-salt of the alcohol (Na-OR) is added to the process as an aqueous solution. The volatile alcohol is removed from the reactor by heating the contents of Pot II using the reactor steam jacket.Volatile Alcohol and vaporized Solvent are then separated in a packed column. The volatile alcohol is condensed and stored for reuse. Following distillation of the volatile alcohol, water is added to Pot II to lower the viscosity of the reaction mass before batch-transfer to the extractor.

Two sources of waste generation in this step are the introduction of water with Na-OR and the use of water for viscosity control at the end of the batch cycle. Can the volatile alcohol byproduct be used as the solvent for Na-OR instead? In other words, can a constituent already in the process be used in place of a new solvent (i.e., fresh-water)? Would semibatch addition of the Na-OR reactant make larger salt particles and, thereby, reduce the viscosity of the reaction mass, eliminating the need for water addition?

A third source of waste generation in Ring Closure is the formation of side products in the steam-jacketed batch reactor. The steam jacket causes hot spots to develop along the reactor wall (i.e., a nonuniform temperature profile), enhancing thermal degradation of the desired intermediate product. Would a pumparound loop with an external heat exchanger improve mixing and provide a more uniform temperature profile throughout the batch reactor? Alternatively, because this is a very fast reaction, a pipeline or tubular reactor might provide the best combination of yield, conversion, and heat-transfer efficiency.

In this step, therefore, the generalized pollution-prevention technologies and practices that apply are reactor design and operation and use of water as a solvent.

6.3.3 Extraction

The purpose of the extraction step is to use water to separate the intermediate sodium salt (Int1-Na) from Solvent. Caustic soda (50% NaOH in water) is added to the extractor to improve the separation (pH > 11). The Int1-Na and water settle to the bottom of the extractor, while the solvent rises to the top. The Int1-Na/water phase is pumped from the bottom to an intermediate hold tank. Additional water and caustic soda are then fed to the extractor, mixed, and allowed to phase-separate again. After this settling period, the Solvent layer is pumped to a solvent-recovery system. There is no pH control on the system.

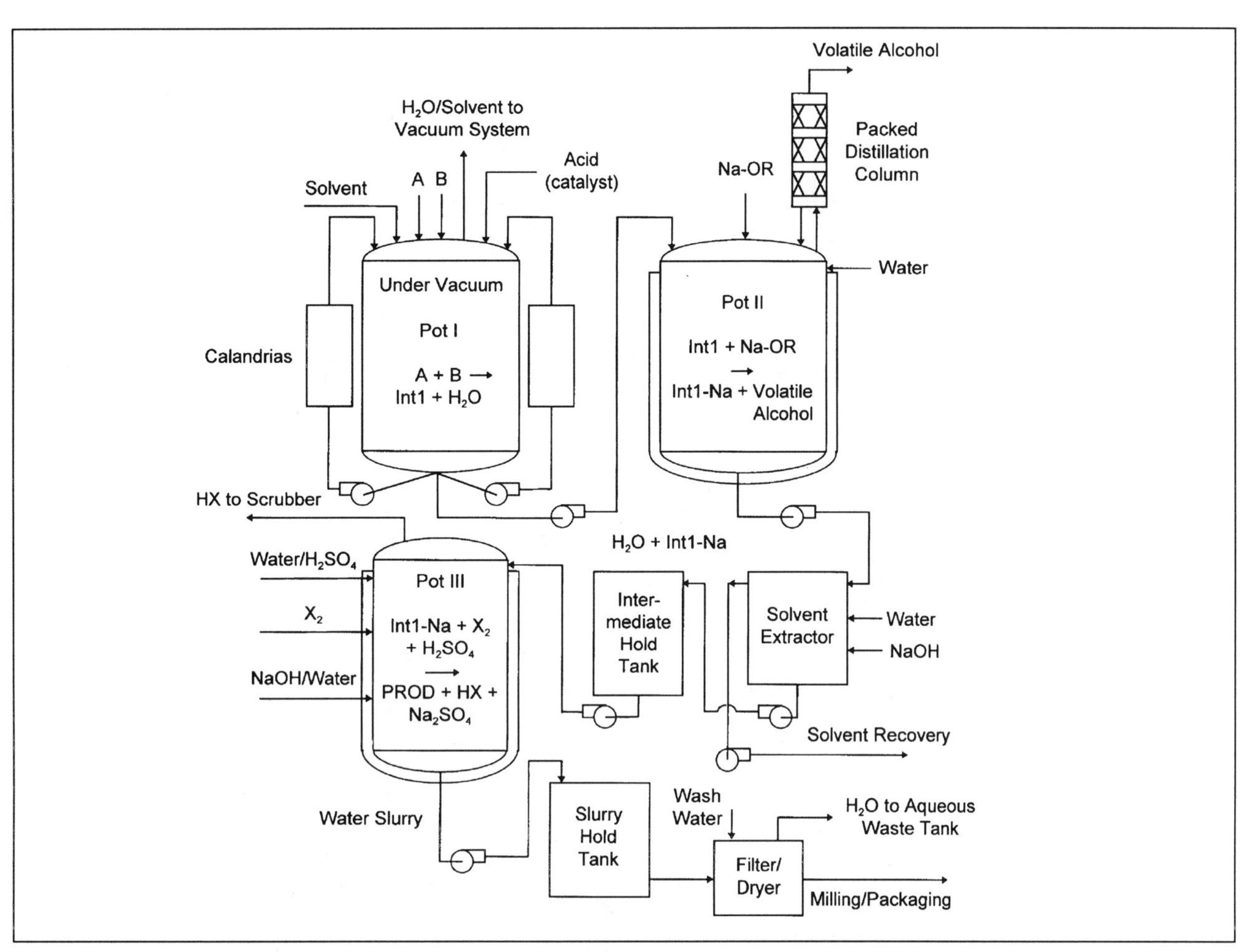

Figure 6-2. Detailed Flow Diagram for Existing Dyelate Process

The key waste-generation issue in this step is the use of water for extraction. Once again, freshwater is added to the process. The use of NaOH also introduces more Na^+ to the system that will ultimately leave the process as a dissolved sodium salt.

Questions that come to mind are:

- How does extraction efficiency vary with pH?
- Would a pH controller improve extraction efficiency, while also reducing NaOH consumption, and hence inorganic salt generation downstream?
- Can the extraction step be eliminated entirely by reconsidering when to add water to separate the product for recovery from the solvent?
- Is there an alternative separation technology to extraction?

The desired generalized pollution prevention technologies and practices for this step are thus pH control and separation technology.

6.3.4 Acidification and Halogenation

In Pot III, three distinct reaction steps occur: acidification, halogenation, and neutralization. In the acidification step, water, sulfuric acid (H_2SO_4), and the Int1-Na/water charge are fed to the reactor. The sulfuric acid reacts with Int1-Na at pH > 10 to form Int1 and Na_2SO_4.

In the next step, the pH is lowered to 2.0 and the reactor temperature is raised before the addition of X_2 for halogenation of Int1. The halogenation reaction produces the desired product (PROD) and byproduct acid (HX). The HX is vented to a caustic scrubber where it is neutralized and sent to wastewater treatment. The product precipitates from solution as it is formed, producing a water slurry.

In the neutralization step, the pH of the reaction mass is raised to 7–8 using 50% NaOH. The product slurry is then pumped through a shear mixer (breaks large agglomerates into smaller particles) and cooler before entering the Slurry Hold Tank. The pK_a of the product is approximately 8.0; therefore, above pH 8, the solubility of PROD increases dramatically (i.e., the deprotonated form of the

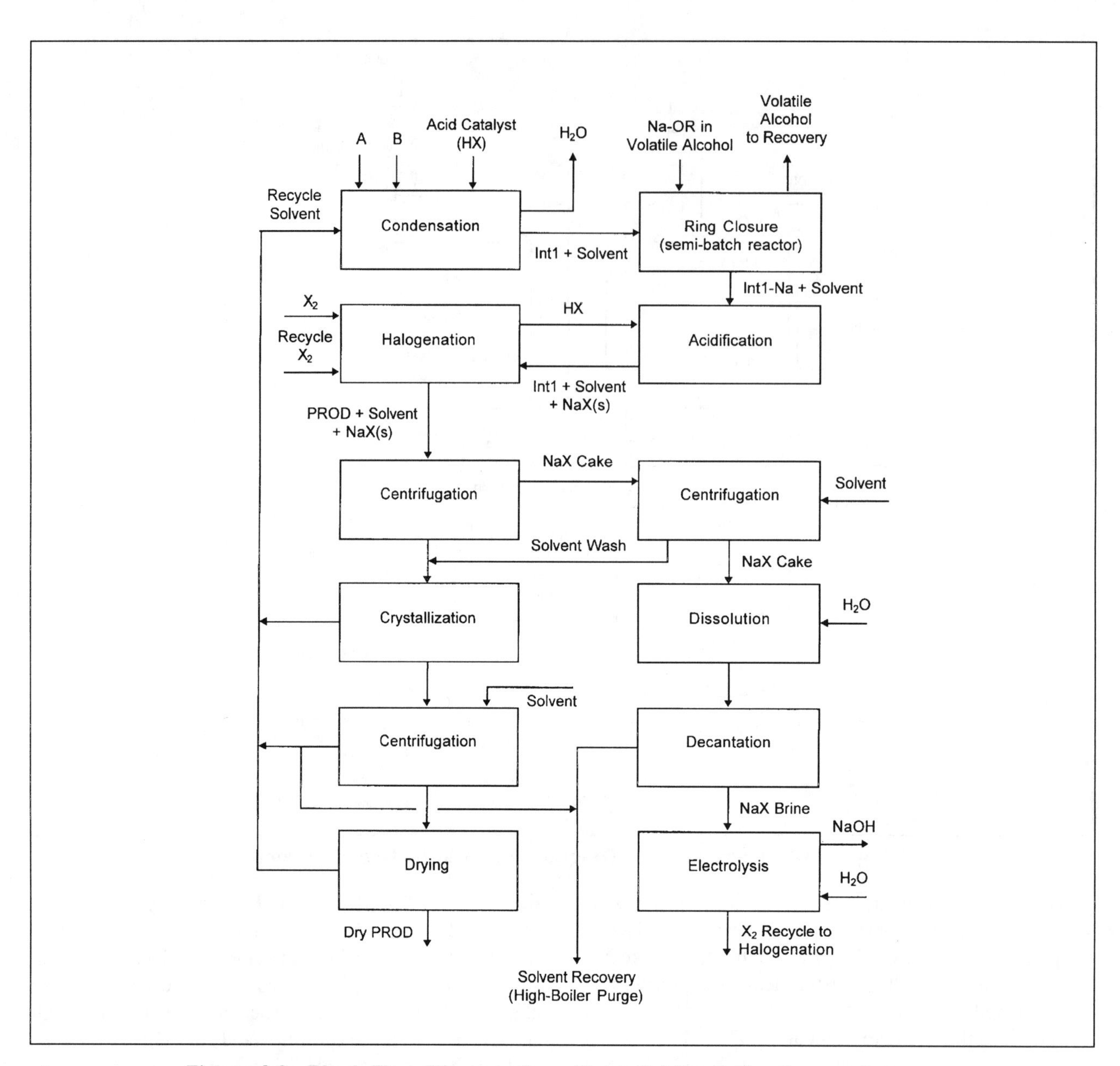

Figure 6-3. Block Flow Diagram for a Potential Semi-Continuous Process Alternative to the Existing Dyelate Process

product is water-soluble, while the protonated form is not). At times, the reactor pH rises above 8.0 because of inadequate operating procedures.

Waste generation in these three steps results primarily from the introduction of a mineral acid (H_2SO_4), which leaves the process as Na_2SO_4, the generation of byproduct HX, which is vented, scrubbed, and neutralized, and poor pH control, which can dramatically increase the dissolved PROD concentration in the product/water slurry.

An engineer or chemist participating in a brainstorming of this process might ask:

- Can the byproduct HX be recycled to the reactor and used in place of the H_2SO_4? This would eliminate the introduction of a new constituent to the process that ultimately results in the generation of a wastewater stream from the caustic scrubber.
- Can the pH control system and operating procedures be improved to avoid high-pH excursions that lead to high PROD solubilities in the reactor water? This water is ultimately purged from the process in the filtration/drying system and sent to wastewater treatment for disposal. A high PROD concentration means lower yield,

- In the ring closure step, replace the Na-OR base (electron pair donor) with ammonia or a solid base.
- Upgrade the extraction system to better control the organic/aqueous phase split. Consider changes to the piping and interface detection system to improve the separation.
- Complete a water balance and an inorganic ion balance (Na^+, $SO4^{2-}$, etc.) for the process.
- Evaluate an ion exchange process for removing Na^+ after the extraction step, but before the halogenation step, to reduce Na^+ contamination of the final product.
- Look at the reactant mole ratios in Pot III to avoid overcharging sulfuric acid, which in turn will decrease caustic consumption for neutralization.
- Use hydrogen peroxide to regenerate X_2 from HX in the halogenation step. Alternatively, consider an electrolytic route.
- Complete a crystallization study of how to make larger PROD crystals in Pot III, so as to reduce the volume of aqueous waste generated in the filtration and water washing steps.
- Modify the acidification/halogenation batch recipe to minimize the amount of Na^+ moving downstream with the product slurry. For example, after completing the acidification step, drain the bulk of the water/acid solution from Pot III, add fresh, sodium-free water, then agitate and halogenate.
- Use aqueous waste from the filtration step to flush piping between vessels instead of recycle water of higher quality.
- Redesign the equipment layout in drying/size reduction/packaging to more fully utilize gravity, thereby reducing the number of conveying operations and their associated dry product losses.
- Switch from a batch process to a semi-continuous process.

Table 6-1. Additional Process Improvement Ideas for the Dyelate Process

lost revenue, and higher organic load on the biological treatment unit.

Generalized pollution-prevention concepts on pH control and reactor design and operation would yet again prove valuable.

6.3.5 Filtration and Drying

The PROD is separated from water using a continuous vacuum belt filter. The product slurry enters the filter at 15% solids and leaves at 85–93% solids.Vacuum is applied with a water-sealed vacuum pump. After bulk water removal on the first-half of the filter belt, the wet cake is then washed with freshwater to remove salt impurities. The water that is removed from the slurry, including any wash water, is held in an aqueous-waste tank before being sent to wastewater treatment. The PROD is at its solubility limit at the filter operating temperature and slurry pH. After filtration, the wet cake is belt-conveyed to a rotary dryer where a closed-loop, hot, air–nitrogen gas stream removes more water from the product cake. The dry product is then screw-conveyed to a hold bin for temporary storage, size reduction, and packaging. Any product entrained in the air–nitrogen conveying gas is captured in a bag filter in the conveying loop.

The vacuum filter is old and is responsible for a higher-than-desired yield loss. In addition, the water removed in this step of the process represents a large aqueous load on wastewater treatment. Possible process improvement questions include:

- Can the wash water be collected and reused in the process?
- Would countercurrent, multistage water washing, including reuse of the water from the latter wash stages, substantially reduce aqueous-waste generation and improve product quality?

- Can an alternative solid–liquid separation technology, such as centrifugation, improve cake dewatering and reduce product yield losses?
- Would lowering the slurry feed temperature improve product recovery by reducing its solubility in water?

Once again, these questions point to the benefit that would be gained from the following generalized pollution-prevention technologies and practices: separation technology and use of water as a solvent.

6.3.6 Observations and Potential Process Improvement Options

A key observation from this case study is the benefit that can be gained by framing the waste-generation problem in the context of generalized pollution-prevention engineering technologies and operating practices. In this way, successful approaches (based on science and experience) from other chemical processes, plants, and even industries can be brought to bear on the problem at hand. This, too, is the benefit that unit operations have brought to the chemical engineering profession. "The unit-operation concept is this: by studying systematically these operations themselves—operations which clearly cross industry and process lines—the treatment of all processes is unified and simplified."[1]

Another observation is the advantage that a detailed mass balance—in particular, a water balance—would bring in brainstorming improvements to this process. Water is a major player throughout the process—entering with mineral acids, reactants, caustic soda, and being fed as freshwater for dilution, extraction, washing, and heat transfer. Substantial waste reductions will be achieved only by understanding the true water balance for this process.

A pollution-prevention team that reviewed this manufacturing process generated many of the same questions and possible alternatives discussed earlier. Some additional process improvement ideas are listed in Table 6-1. One of the more far-reaching ideas was to convert from a batch to a semicontinuous process. A possible semicontinuous process is shown in Figure 6-3. The overall goal of this alternative concept was to devise a process that prevented contact between PROD and water. Some of the highlights of this alternative process concept are:

- Use a new halogenation-resistant solvent in place of the existing hydrocarbon solvent.
- Replace the mineral-acid catalyst in the condensation step with HX to avoid introducing a new species to the process.
- Replace H_2SO_4 in the acidification step with recycle HX from the halogenation reactor.
- Change the ring-closure step to a semibatch process to hopefully produce larger salt particles, reduce the viscosity of the reaction mass, and eliminate water addition.
- Use the volatile alcohol as the solvent for the Na-OR reactant.
- Instead of recovering PROD solids from an aqueous stream, recover NaX solids from a solvent stream, thereby eliminating the extraction step.
- The NaX solids are centrifuged and washed with solvent to recover any PROD, dissolved in water, and decanted to remove residual solvent. NaX could then be converted by electrolysis to X_2 and NaOH to recover the halogen value.
- PROD is crystallized from the new solvent, centrifuged, solvent washed, and then dried.
- High boilers would be purged in the solvent recovery-recycle system. The goal is to supply 100% of the solvent added in the Condensation Step with recycle solvent.

A substantial amount of research-and-development effort and new capital investment would be required to implement this concept; therefore, a more detailed engineering evaluation is needed to assess the economic viability of this process concept, assuming it is feasible.

The authors hope that this case study has shown the value that generalized technologies and practices bring to the practice of pollution prevention. In addition, this case study should give the reader a better feel for how to ask the right questions when looking at a seemingly unique chemical process.

Literature Cited

1. McCabe, W. L., and J. C. Smith. 1976. *Unit Operations of Chemical Engineering*, p. 4. New York: McGraw-Hill.

CHAPTER 7

Pollution Prevention in Batch Operations

7.1 Introduction

Batch processes are a preferred operating method for manufacturing small volumes of high-value products. Examples include many pharmaceutical, agricultural, and specialty chemical products. Unfortunately, batch operations are also associated with the production of unacceptably high amounts of waste per unit of product. In the past, many pharmaceutical and agricultural products businesses could "afford" this wastefulness, because the high value of the final product often overshadowed the cost of treating the waste. In addition, waste streams from these high-value-product businesses often contain toxic compounds which are toxic, biorefractory, and previously unregulated.

However, in recent years, two things have changed. First, businesses are becoming aware of the *true cost* of waste production, which is always greater than just the cost for disposal or treatment. The true cost of waste production also includes the monetary value of lost product, the cost of solvent purchases, the fees for permitting and monitoring emissions, and the increased exposure to safety and environmental risks.

Second, with increased regulation of emissions to air, water, and land, a good number of plant sites can no longer handle the often concentrated, toxic-waste streams (especially wastewater streams) emanating from batch processes in existing control equipment. As a result, businesses relying on batch processes are now being faced with significant new capital investment to pretreat concentrated wastewater before existing biological treatment plants, to expand biological wastewater treatment capacity, to install state-of-the-art hazardous-waste incinerators for liquid wastes, and to provide point-source treatment of air emissions.

The *true cost* of waste generation provides an incentive for batch operations to engage in pollution-prevention solutions. In this chapter we discuss techniques and technologies that engineers can use to reduce waste generation at the source in batch operations.

7.2 Nature and Sources of Emissions

A batch operation is by definition a discontinuous process. Process variables such as temperature, pressure, and concentration vary throughout the batch cycle. Likewise, waste emissions associated with these cycles are seldom considered constant.[1] For example, a batch reaction step may generate a nitrogen purge stream whose volatile organic compound (VOC) concentration varies with time, as shown in Figure 7-1. The emission rate peaks in the first hour, and then decreases with time as the reaction nears completion.

Such time-dependent behavior makes it difficult to effectively specify a continuous gas-abatement system. To meet permit limits, the abatement equipment must be sized to treat the maximum achievable waste flow and concentration. For some emissions, this maximum may last only a few minutes. It is easy to see how uneconomical the "end-of-pipe" approach can be to a business. For this reason, source reduction of these emissions by pollution-prevention strategies can reduce the dependence on such costly control technologies.

Every stage of a batch cycle has some amount of waste emissions associated with it. Noncondensibles (such as nitrogen and air) escape the process during reactor charging and discharging, carrying with them high concentrations of VOCs. Reactor operation may generate unwanted byproducts that contribute to product losses. Most noticeable are the large amounts of cleaning wastes generated during vessel cleaning between batch cycles and product

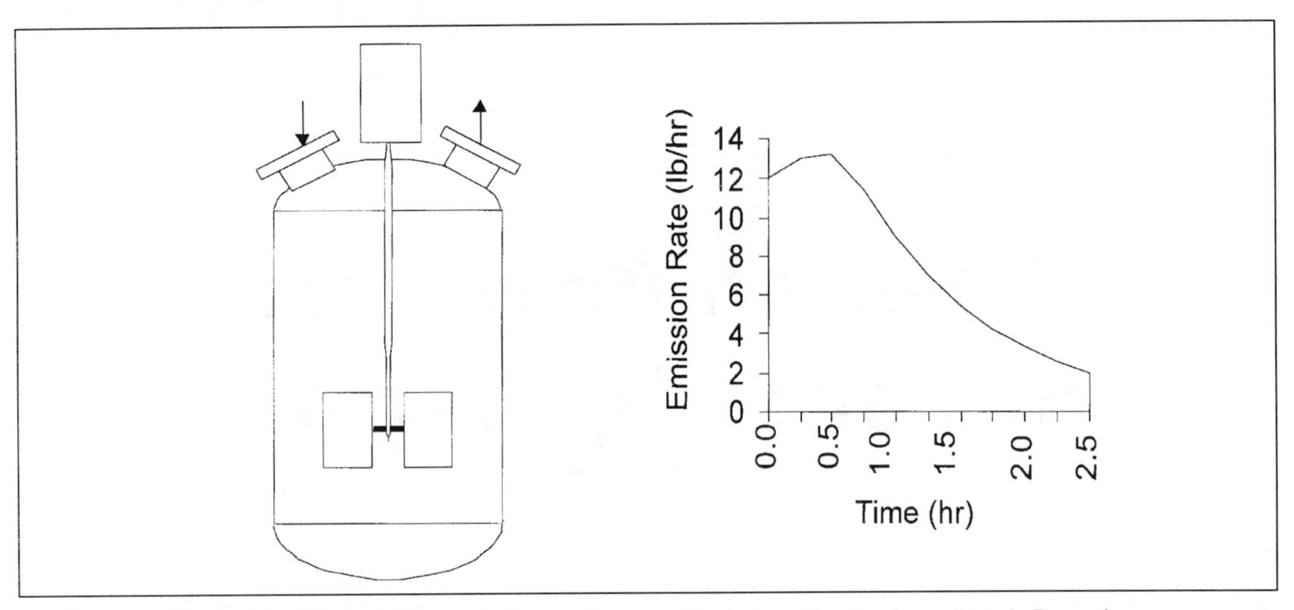

Figure 7-1. Typical Nitrogen Purge Stream Emission Profile for a Batch Reaction

campaigns. The pollution-prevention strategies discussed in the next section offer the potential for making substantial reductions in emissions for all steps of batch operations.

7.3 Pollution-Prevention Strategies

Most waste generation in batch operations occurs in the reactors. For this reason, pollution-prevention technologies and practices described in this chapter focus on the batch reactor. Four key reactor process steps are discussed: reactor charging, reactor operation, reactor discharging, and reactor cleaning.

7.3.1 Reactor Charging

Solvent vapor losses during raw-material addition occur for two primary reasons: leaks from the process equipment, and vapor displacement as the reactor is filled with liquids or solids. To circumvent these two problems, consider the following pollution-prevention strategies:

- *Gravity-introduce solvents* into the reactor vessel instead of using a centrifugal, positive-displacement, or diaphragm pump to reduce the pressure buildup associated with pumping.[2] This can often be achieved in new designs by changing the location and elevation of raw material feed tanks to be above the reactor to take advantage of gravity instead of beside it.
- *Install closed-loop vapor recycling systems for pumping operations.*[3] As shown in Figure 7-2, this would involve piping the displaced vapor from the reactor (as it fills) back to the vessel containing the feed material that is being added to the reactor, for example, a solvent storage tank. Be careful of materials compatibility, however.
- *Minimize solvent displacement from reactors by charging solids before liquid solvents.*[4] A more general strategy is to reconsider the sequence of raw-materials addition to minimize unwanted off-gassing due to vapor displacement and/or chemical reactions.
- *Charge solids using lock hoppers* instead of rotary valves or manual dumping through open lids. Lock hoppers isolate the reactor from the open atmosphere, so that organic vapor emissions are minimized. In contrast, rotary valves often allow vapors to escape as they rotate from the reactor vapor space back to the solids feed bin. Lock hoppers are typically used when solids must be added to a reactor containing a volatile solvent that is at an elevated temperature and/or pressure. In addition, they can be designed to eliminate air infiltration into a reactor for cases where oxygen and moisture would create an unsafe or undesirable condition. Figure 7-3 shows a typical lock-hopper arrangement where two sliding-gate valves isolate the solids feed bin (open to the atmosphere) from an inerted reactor operating at elevated temperature and pressure.
- *Consider cut-in hoppers* when bags of solids must be manually charged to a reactor.[5] Rather than dumping bags of solid raw material through an open lid (which can lead to significant dusting and solvent vapor losses), cut-in hoppers "cut open" the bags inside a closed feed hopper to contain any dust that is generated.
- *Eliminate manual charging of dry solids* (i.e., manual addition) by introducing solids in slurry form[3] or using dense-phase conveying. Consider using a raw material or intermediate already in the process to serve as the

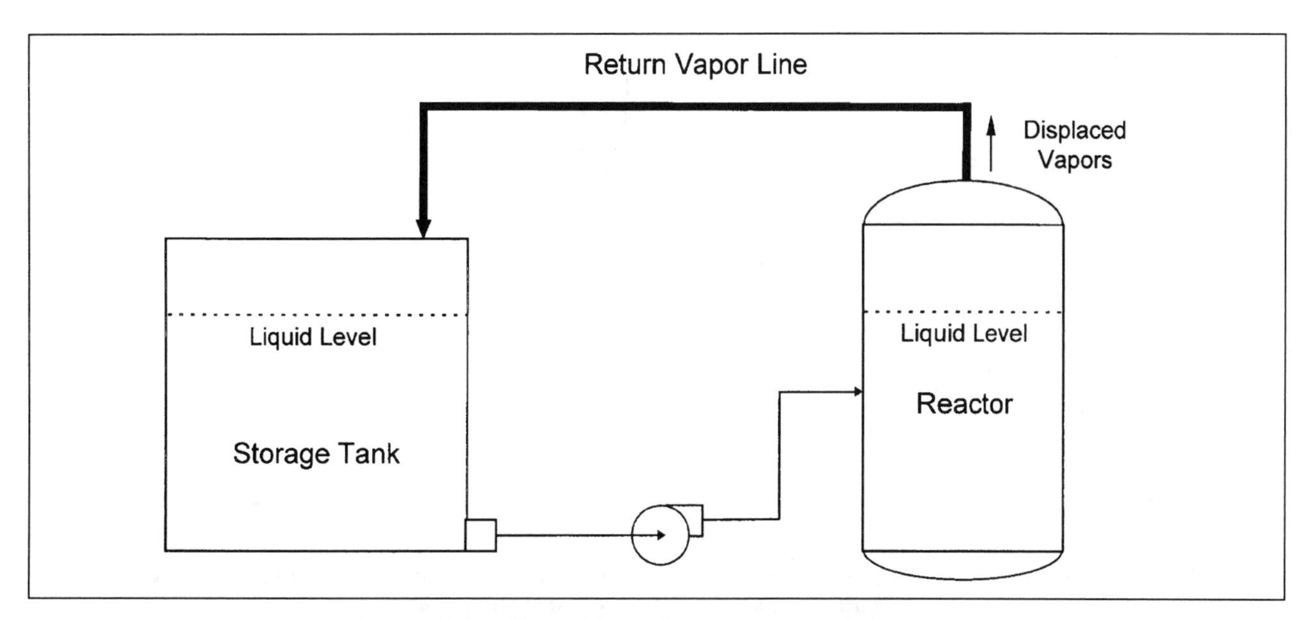

Figure 7-2. Closed-Loop Vapor Recycling System

carrier fluid for the slurry, rather than introducing a new chemical to the process. Dense-phase conveying uses less conveying gas to move the solids, thereby reducing the amount of solvent evaporated from the reactor as the conveying gas is vented.

- *Vent displaced vapors through a refrigerated vent condenser* to recover and recycle condensable solvents.[6]
- *Consider solvents with a lower vapor pressure* to minimize evaporation and facilitate recovery by condensation. Be careful with liquid-solvent mixtures where a small weight percentage of a volatile solvent can account for the majority of the vapor emissions. For example, a flammable solvent mixture used for cleaning parts is shown in Table 7-1. Even though the liquid mixture contains only 10% acetone by weight, the saturated vapor phase is more than 40% acetone at 26°C and 1 atm total pressure. Removing or reducing the weight fraction of acetone in the solvent mixture would cut vapor emissions significantly.

7.3.2 Reactor Operation

The pollution-prevention ideas that follow focus on the inert blanketing, sampling, sealing, control, sequencing, and heating of batch reactors. Chapter 10 contains a more exhaustive discussion of reactor design and operation for both continuous and batch chemical reactors.

- *Reduce vapor losses by enclosing open-air tanks.*[3] Use lids with gaskets that seal tightly if manual access is a necessity. Otherwise, consider a closed reactor with a permanent top and closed raw-material addition systems as described earlier.
- *Reduce vapor losses by improving seals on agitators and lids.*[1] In some cases, the agitator can be replaced with a jet mixing system, which consists of a jet nozzle submerged in the reactor fluid, a pump, and a liquid recirculation loop (see Chapter 10).
- *Sequence the addition of reactants and reagents* to optimize yields and lower emissions. For example, a base, such as sodium hydroxide, is added at the end of a batch reaction cycle to neutralize the reaction mass before feeding the next process vessel; the pH in the

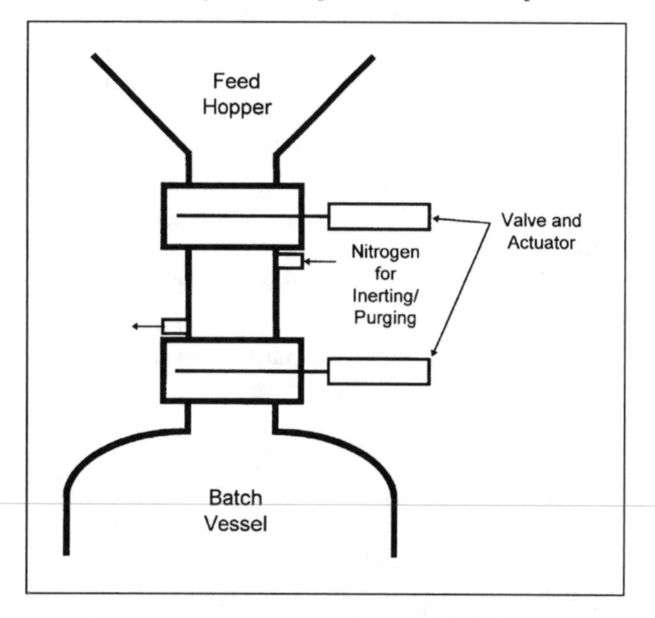

Figure 7-3. Charging Solids to an Inerted, Batch Reactor Using a Lock Hopper

Compound	Liquid Composition (Weight %)	Vapor Composition (Volume or Mole %)
Acetone	10	44
Methyl Ethyl Ketone	30	43
Toluene	30	10
Xylene	30	3

Table 7-1. Composition of Saturated Vapor Phase in Equilibrium With a Liquid Cleaning Solvent Mixture at 26°C and 1 atm Total Pressure

reactor is allowed to vary with reaction progress. Should the sodium hydroxide be added throughout the reaction sequence to maintain a constant pH to improve yield and minimize byproduct formation?

- *Control nitrogen purge rates* with automatic flow control devices instead of manual throttling valves.[1] Too often, the manual valves are found wide open, wasting nitrogen and dramatically increasing VOC losses.
- *Use a nitrogen-pad system* (i.e., pressure control of nitrogen in the reactor vessel headspace) instead of a purge based on flow control.[4] Figure 7-4 shows a typical piping diagram for a nitrogen-pad system where the nitrogen-supply pressure is greater than 60 psig. Dual pressure regulators are employed to improve low-pressure control in the reactor headspace.
- *Optimize existing reactor design* based on reaction kinetics, mixing characteristics, and other parameters to reduce byproduct formation.[3] Refer to Chapter 10 for more details on reactor design and operation.
- *Collect and recycle excess reactants and solvents* used in the reaction.[7] Can solvent purification equipment be added or upgraded to allow recycle?
- *Use a minimum number of intermediate-stage process samples* to determine the reaction endpoint.[4] Also, strongly consider automatic, in-line sampling,[6] rather than manual sampling from vessel openings, to minimize vapor emissions from openings and the amount of sampling waste. Evaluate the impact of the time lag associated with manual samples analyzed in the site laboratory on reactor yield and waste generation. In many instances, the additional turnaround time required for laboratory analyses leads to excess waste production. For example, in a process for making an agricultural intermediate, manual sampling of a batch, caustic-scrubbing system for "% NaOH" led to unnecessary safety margins in the operating procedures for purging the spent caustic-scrubbing solution. Because of the long turnaround time from the laboratory, the spent scrubbing solution contained 3–5% unused NaOH. A viable alternative to this approach would be to use an in-line pH meter to regulate caustic addition to minimize the amount of NaOH being wasted.
- *Use statistical process control techniques* to regulate the reaction, instead of relying on intermediate sampling.[6]
- *Eliminate excessive reactor boilup* by replacing direct steam-jacket heating with an external, recirculating, heating loop, consisting of a pump and noncontact heat exchanger.[4] This may also prevent unwanted tar formation in reactions where a hot vessel wall leads to product degradation.
- *Eliminate fugitive emissions* from pressure-relief valves by using an upstream rupture disk in series with the relief valve.[4]

7.3.3 Reactor Discharging

- *Replace nitrogen blowcasing with a pump.* The pressurized nitrogen used to push the reaction mass from the reactor must be vented downstream. This nitrogen will leave saturated with solvent and, possibly, valuable product and reactants.

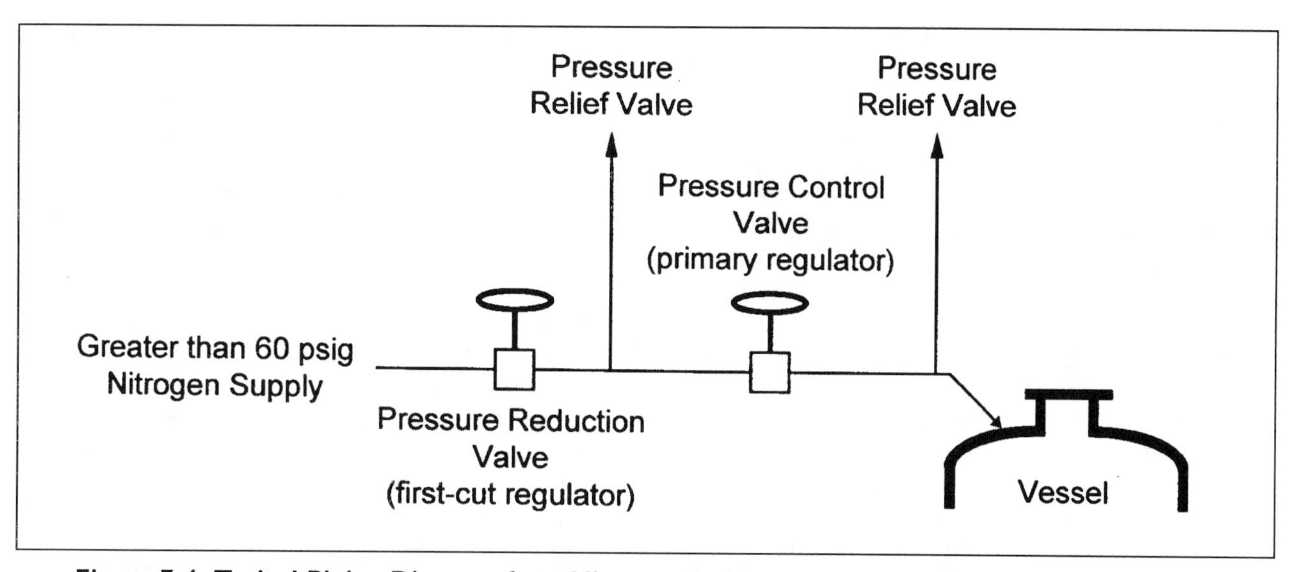

Figure 7-4. Typical Piping Diagram for a Nitrogen-Padded Reactor with a Gas Supply Pressure Greater Than 60 psig (this figure illustrates the concept behind a nitrogen pad system; it should not be used for construction details in individual applications)

- *Design and install discharge lines on an incline* to take advantage of gravity flow to the downstream equipment. Avoiding a pump lowers investment and operating costs as well as fugitive emissions from leaks.
- *Reduce the reactor batch temperature before discharge* to lower the vapor pressure of volatile organic compounds, which can include solvents and, in some cases, reactants and products.[1]

7.3.4 Reactor Cleaning

- *Optimize the product manufacturing sequence and product campaigns* to minimize washing operations. This may mean structuring product campaigns to manufacture clear or light-colored products before heavily pigmented products or products with loose quality specifications before products with tight quality specifications. Within a single product campaign, it may mean understanding the chemistry that leads to equipment fouling better, and then changing the reaction sequence to minimize tar formation, equipment scaling, polymer buildup, and so on.
- *Maximize production runs* to decrease the frequency of washes.[6] Check with customers to see if product specifications can be loosened. If some customers can accept higher levels of impurities, fill their orders later in the product campaign, if possible.
- *Minimize the number of solvent rinses* used to clean reactors by understanding the level of acceptable contamination.[4] Can off-spec product be used to flush the reactors instead of clean solvent?
- *Increase the smoothness of vessel internals* to reduce the quantity of cleaning solvent used.[3] Consider non-stick, nonporous linings, eliminating baffles, and minimizing other difficult-to-access points where material can accumulate and harden.
- *Use high-pressure rotary nozzles* to reduce the quantity of cleaning solvent used.[2]
- *Position drain valves at the lowest point* on the reactor vessel to improve removal of residual solvent and product before drying or the next product campaign.
- *Use plastic or foam "pipe pigs"* to clean the inside of pipes.[2]
- *Replace volatile chlorinated solvents with lower-volatility, nonchlorinated solvents.*
- *Replace solvent-based cleaning with aqueous-based cleaning* using detergents or other surfactants, if necessary, to increase the cleaning effectiveness. How does pH affect cleaning?
- *Collect the final wash rinse for reuse as first-pass rinse* during the next cleaning cycle.[6] Can lower quality solvent or water from elsewhere in the plant be used as the first rinse?

7.4 Modeling Batch Processes

Changes in standard operating conditions can affect the amount of waste generated by a batch process. However, few plants can accept the risks associated with making experimental batch runs. Developments in computer modeling software now allow changes in batch operations to be analyzed to a greater extent than ever before.

Most of the major chemical process flowsheet simulation companies, such as Aspen Technology, Inc. (Cambridge, Massachusetts), Simulation Sciences, Inc. (Brea, California), Hyprotech, Ltd. (Calgary, Alberta, Canada), and OLI Systems, Inc. (Morris Plains, New Jersey), have already developed and/or are greatly expanding their capability to simulate batch unit operations and overall batch processes. Some examples discussed below include Aspen Technology Inc.'s ASPEN PLUS, BATCH PLUS, and Aspen Custom Modeler (formerly SPEEDUP simulation packages, OLI System, Inc.'s DynaChem software, and Hyprotech Ltd.'s BatchDesign-Kit (BDK). This is not an all-inclusive list, but is based on the authors' familiarity with these simulation packages.

ASPEN PLUS, which is a well-known software package for continuous process flowsheet modeling, can be used to simulate some batch unit operations. These include batch distillation and chemical reaction kinetics. In addition, nonideal vapor/liquid/liquid systems are easily studied using ASPEN PLUS's physical property system.

BATCH PLUS is a new batch simulation package offered by Aspen Technology, Inc., that contains more than 40 shortcut batch and semicontinuous unit operations. The software program also has access to the standard ASPEN PLUS physical property and thermodynamics package. BATCH PLUS evaluates process alternatives, generates material balances, performs cycle time calculations, and develops cost estimates.

Aspen Custom Modeler is an equation-based dynamic simulator that includes an extensive library of unit operations, but also has the flexibility to allow users to build custom dynamic models. Another useful feature is that any physical properties taken from ASPEN PLUS can be used directly in Aspen Custom Modeler without any changes.

DynaChem is the dynamic simulation module of the environmental simulation program (ESP) developed and marketed by OLI Systems, Inc. ESP and DynaChem are specifically designed to model the thermodynamics and kinetics of complex aqueous systems, that is, systems containing greater than 65 mol % water. This includes the ability to model interphase equilibria between aqueous, organic liquid, vapor, and multiple solid phases; intraphase equilibria, including redox and speciation reactions, biochemical reactions, reaction kinetics, and other phenomena, such as ion exchange, coprecipitation, and sorption. ESP and DynaChem access a database covering the aqueous inorganic chemistry of 78 elements and the aqueous organic chemistry of more than 2500 organic components.

Developed at the Massachusetts Institute of Technology's Department of Chemical Engineering, the BatchDesign-Kit is a computer-aided system designed to support process development and design activities in the batch manufacturing industries. Using reaction network analysis, graphical process flowsheets, batch recipe sheets, and economic assessments, the software can interactively define a complete, optimized batch process flowsheet, including complete material balances and estimates of waste generation. The pollution-prevention aspect of BDK comes from its unique ability to define the zero avoidable pollution (ZAP) level for a batch reaction, and then to optimize the trade-offs between cost and environmental impacts to achieve the minimum avoidable pollution (MAP) batch process. BDK is also capable of helping in the selection of the most appropriate reaction pathway and process solvent. The software includes an expert system capable of characterizing waste streams and recommending the lowest-cost treatment options.[8]

7.5 Batch vs. Continuous Operation

For some processes, the conversion from batch to continuous operation is an effective way of reducing waste generation as well as increasing total throughput.[3] Cleaning waste is most significantly reduced by switching to a continuous process. Continuous process operations are rarely interrupted for cleaning, because the equipment is dedicated to the production of one or only a few products. Therefore, switching to a continuous process may mean practically eliminating cleaning waste. In addition, solvent recovery in a continuous process is feasible, because it becomes more economical to install dedicated distillation equipment to handle the necessary chemical separations. Recovered solvent can then be reused in the process, instead of being incinerated or disposed of as a hazardous waste.

7.6 Case Studies

The case studies that follow illustrate the benefits of successful pollution-prevention practices in batch operations.

7.6.1 Replace an Organic Solvent with an Aqueous Solvent for Cleaning

The small lots manufacturing area (SLM) at a site undertook a conversion to aqueous cleaning in order to reduce solvent waste generation. SLM is an agricultural product manufacturing area using batch organic synthesis processing involving reaction vessels in series.

Past operations involved several rinses with various organic solvents as a way to clean the vessels between product campaigns. Once spent, these organic wastes were discarded off-site as hazardous wastes, not recycled. In order to reduce organic-solvent waste production, SLM attempted aqueous cleaning by replacing a number of organic-solvent flushes with detergent flushes and water rinses. The reduction in solvent usage per cleaning cycle totaled 6000 gallons, a 60% reduction over organic-solvent cleaning. Moreover, the resulting aqueous waste can be processed in the site's wastewater treatment facility (it was important to select a biodegradable detergent and to ensure that the additional organic and hydraulic load on the biotreatment basin could be handled).

In addition to the environmental benefits, other significant benefits were realized in converting to aqueous cleaning. The business realized a $168,000 per year savings in solvent waste disposal and a $70,000 per year savings in solvent raw-material costs. Cleaning effectiveness, measured by the amount of material carrying over into the next product, was increased tenfold. The increase in production capability caused by reducing the cleaning time is estimated to save as much as $750,000 per year.

7.6.2 Methylene Chloride Emissions Reduction

Methylene chloride is used as a coating and cleaning solvent in the manufacture of several graphic arts and electronic photopolymer films. In the late 1980s, a site released more than 3 million pounds of chlorinated solvents into the air.

The coating-solution preparation areas accounted for the majority of the air emissions. Coating solutions are prepared in batches in agitated vessels using a blend of polymers, monomers, photoinitiators, pigments, and solvents. These atmospheric mix tanks were not well sealed, resulting in large fugitive and point-source emissions of methylene chloride.

In an effort to enclose the batch vessels as much as possible, the mix tanks were fitted with bolted, gasketed lids. The vessels were also designed with pressure/vacuum conservation vents to allow the vapor pressure to rise to 3 psig before the tanks breathed. By sealing the process, the site was able to reduce air emissions by 40% and save $426,000 per year in methylene chloride costs.

7.6.3 Conversion from Batch to Continuous Operation

The manufacturers of a herbicide intermediate were unable to meet demand when operating their batch process at full capacity. To overcome the production shortfall, they purchased the intermediate from a competitor at a price higher than their cost of manufacturing.

Duplication of the existing batch process seemed the conservative approach to increasing production. However, the business team saw an opportunity to meet the expansion objectives with minimal investment by changing from batch to continuous reaction technology. The team persevered through many intensive technology and project reviews, and concluded that the continuous process appeared inherently safer and technically viable. Within eleven months from concept to startup, the business team achieved a 240% increase in production capacity when compared to the original batch process. In addition, the continuous process demonstrated a 29% reduction in methanol emissions per pound of product when compared to the batch technology.

Literature Cited

1. Chadha, N., and C. S. Parmele. January 1993. "Minimize Emissions of Air Toxics via Process Changes." *Chemical Engineering Progress.* 89(1): 37–42.

2. U.S. Environmental Protection Agency. June 1990. *Guides to Pollution Prevention: The Paint Manufacturing Industry.* EPA/625/7-90/005. Cincinnati, Ohio: U.S. EPA, Risk Reduction Engineering Laboratory.

3. Chadha, N. November 1994. "Develop Multimedia Pollution Prevention Strategies." *Chemical Engineering Progress.* 90(11): 32–39.

4. Jacobs Engineering Group, Inc. October 1994. *Compendium of Source Control Options for Reduction of Volatile Organic Compound Emissions from Bulk Pharmaceutical Production Facilities*: 11–13. Pasadena, CA: Jacobs Engineering Group, Inc.

5. U.S. Environmental Protection Agency. February 1990. *Guides to Pollution Prevention: The Pesticide Formulating Industry.* EPA/625/7-90/004. Washington, D.C.: U.S. EPA, Office of Research and Development.

6. U.S. Environmental Protection Agency. October 1991. *Guides to Pollution Prevention: The Pharmaceutical Industry.* EPA/625/7-91/017. Washington, D.C.: U.S. EPA, Office of Research and Development.

7. Benforado, D. M., G. Ridlehoover, and M. D. Gores. September 1991. "Pollution Prevention: One Firm's Experience." *Chemical Engineering.* 98(9): 130–133.

8. Linninger, A. A., A. A. Shahin, E. Stephanopoulos, C. Han, and G. Stephanopoulos. 1994. "Synthesis and Assessment of Batch Processes for Pollution Prevention." *AIChE Symposium Series,* 90(303): 46–57.

CHAPTER 8

Equipment and Parts Cleaning

8.1 Introduction

Cleaning equipment and parts is an important preparatory step in many industrial processes. Cleaning removes dirt, oil, wax, oxides, residual products, monomers, and other contaminants that may adversely affect the final product quality or the operating efficiency of the equipment or part. Two general types of materials are removed: "soft" films, such as dirt, oil, and grease, and "hard" films, such as hardened polymeric films, scale, and paints. Until recently, solvent strippers, such as methylene chloride, were frequently used to remove both types of films from the surfaces being cleaned.

With the use of multipurpose equipment growing in batch manufacturing operations, a chief concern is cross-product contamination (in addition to environmental concerns, such as air emissions and hazardous-waste disposal). In this chapter we discuss pollution-prevention engineering practices and technologies that help to avoid the need to clean equipment, and, where cleaning is necessary, the approaches that are available to replace or reduce the use of organic solvents.

8.2 Nature of Emission Sources

Cleaning, degreasing, and coating removal technologies generally involve the application of an organic solvent(s) to the material being removed. During the process, air emissions are often generated from the use of the solvent(s), and after the cleaning process, other waste streams—solvent- and metal-containing wastewater and sludges—remain that require disposal. In particular, the wastewater streams and sludges often contain materials that classify them as a Resource Conservation and Recovery Act (RCRA) hazardous waste.[1-2]

8.3 The Pollution-Prevention Continuum

The pollution-prevention continuum, shown in Figure 8-1, contains the relative merits of options that are available to process and project engineers to eliminate or reduce the frequency of equipment cleaning or to minimize the waste generated during the cleaning process. The decision of how far to move toward a zero-waste and -emissions design will depend on a number of factors, including corporate and business environmental goals, economics, technical feasibility, and applicable regulations.

8.3.1 More Than 95% Pollution Prevention

In the design of a new facility, equipment should be designed without low spots, ledges, or other crevices that will collect material that later becomes difficult to remove. Properly designed vessels will contain sloping interior bottoms and piping arrangements with valved low points or valves that drain back into the main vessels. For existing equipment, drain valves can be installed at strategic "low-point" locations on the process equipment. If the "low points" are too low for standard tanks and pumps, then design a special portable collection vessel to collect and transport the material to a tank (for more details, refer to the Case Study, "Minimize the Need for Cleaning" in this chapter.)[3]

In addition, the equipment can be dedicated to a single product line, thus eliminating the need to wash the equipment interior between production campaigns.

Alternatively, flush contaminated equipment with the salable product or a process intermediate, and then recycle the flush back to the process where it can be purified in existing separations equipment.

A continuous process inherently generates less waste, because there is no need to shut down, drain, and clean the equipment.

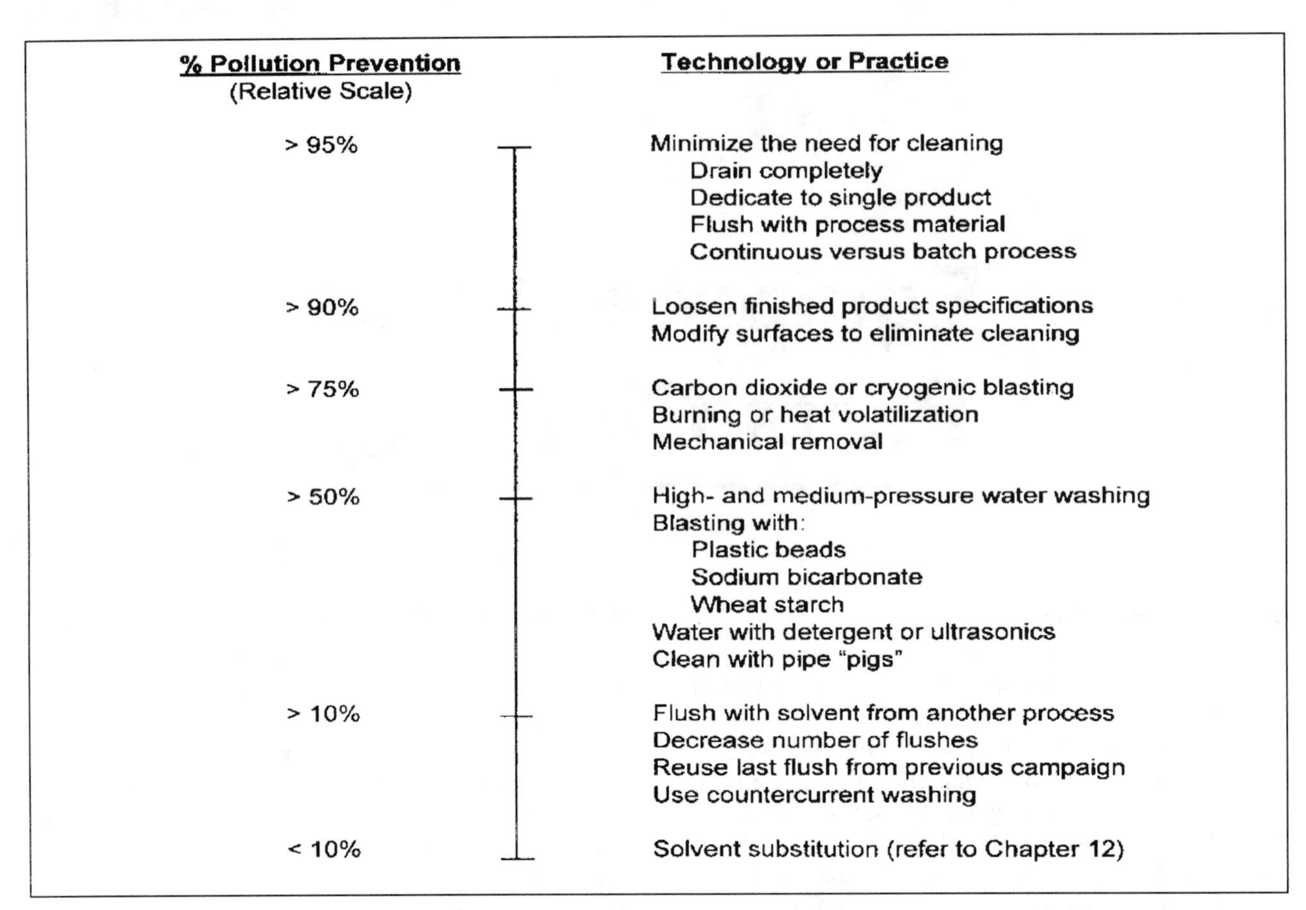

Figure 8-1. The Pollution-Prevention Continuum for Equipment Cleaning

8.3.2 More Than 90% Pollution Prevention

Loosening finished-product specifications allows for higher levels of impurities in the final product or cross-contamination of products that reduces or eliminates the need for solvent washes between product campaigns. Many times, specifications for products manufactured in the same equipment are different. One set of specifications may be more stringent than another. Through careful planning and inventory control, product changeovers can be made from products with tighter specifications to those with looser specifications.

Application of an antistick coating, such as Teflon or silicone, to equipment interior walls could enable easy drainage and removal of leftover residue. One example is coating the rollers in adhesive-film manufacturing facilities with a nonstick material. These coatings allow the adhesive-coated substrate to pass over the roller without sticking or leaving adhesive on the roller. In polymer drum coating/drying operations, nonstick coatings or the use of synthetic lubricants, such as Krytox, have been suggested as ways of preventing the buildup of polymer on the back of the doctor blades that apply and spread the polymer solution onto the heated drums.

8.3.3 More Than 75% Pollution Prevention

Carbon dioxide blasting systems consist of a refrigerated liquid CO_2 supply and equipment for converting the liquid CO_2 to solid pellets. The solid pellets remove coatings by a combination of impact, embrittlement, thermal contraction, and gas expansion. Because the CO_2 pellets sublime (go from solid to gas without liquefying), a wastewater or liquid waste is not produced, and a dry coating residue is collected.[1] A typical liquid CO_2 blast system is shown in Figure 8-2.

The liquid-nitrogen, cryogenic stripping process uses liquid nitrogen to cool the surface and to help propel the plastic-bead blasting media. The liquid nitrogen embrittles and shrinks the coating, and the high-velocity, nonabrasive pellets crack, debond, and break away the coating.[1]

Using heat generated by flames, lasers, or flashlamps to remove coatings results in the vaporization of the coated materials. The nonvolatile portions of the coating, such as metals, form particulate ash that is removed and collected by a vacuum air removal system.[1]

Where feasible, manual cleaning using scrapers or spatulas might eliminate the need for any subsequent solvent wash.

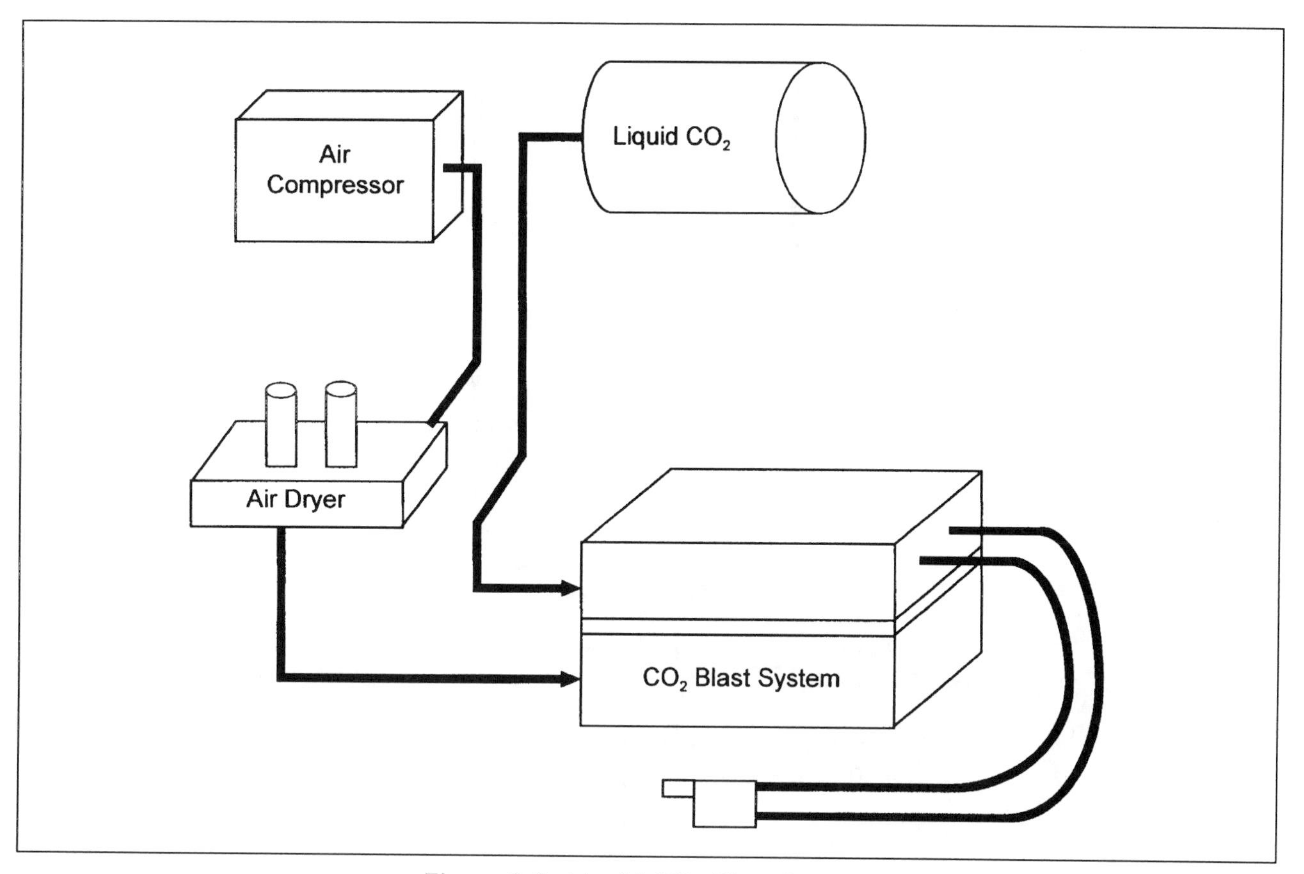

Figure 8-2. Liquid CO_2 Blast System

8.3.4 More Than 50% Pollution Prevention

High-Pressure Water Washing. High-pressure water washing (also called water-jet stripping) uses the impact energy of the water to remove coatings. The water is collected, filtered, and then recycled. The coating thickness and hardness dictate whether a high-pressure (15,000 to 30,000 psig) or medium-pressure (3,000 to 15,000 psig) waterjet is required. A typical high-pressure water-wash system for a process vessel is illustrated in Figure 8-3.

The high-pressure water lance is attached to a carriage, which is in turn affixed to the bottom of the vessel. A chain-drive moves the lance up and down the carriage as needed. A swivel joint at the base of the lance permits free rotation. The nozzle at the tip of the spinning lance has two apertures, which emit cone-shaped sprays of water at 10,000 psi with a combined flow rate of 16 gal/min. Operation of the lance is controlled from a panel well removed from the vessel. The process is designed such that no high-pressure spray leaves the interior of the vessel. These precautions assure operator safety during vessel washout.

Media Blasting. Several media can be used to strip the coatings away by a combination of impact and abrasion: plastic beads, baking soda, and wheat starch.[1] The plastic-bead process uses low-pressure air or centrifugal wheels to project nontoxic plastic media at the coated surface. The beads are collected, cleaned, and recycled. As the particles break down, they are discarded and replaced with new beads. The discarded material must be disposed of properly.

In the baking soda process, sodium bicarbonate is delivered by a wet blast system to remove the coating by impaction. As a result, the coating is shattered into a very fine particulate. The sodium bicarbonate medium is used only once, and is not recycled.

Wheat starch blasting uses low-pressure air to propel the particles at the coated surface. The coating is stripped away by a combination of impact and abrasion. The starch can be recycled and replaced as the particles breakdown. The fines plus coating material must be disposed of properly.

Aqueous Cleaning. Water with the appropriate pH or additives, such as detergents, can be used to remove "soft" coatings. The water solution is collected and recycled until it becomes saturated with the material being removed. However, consideration must be given

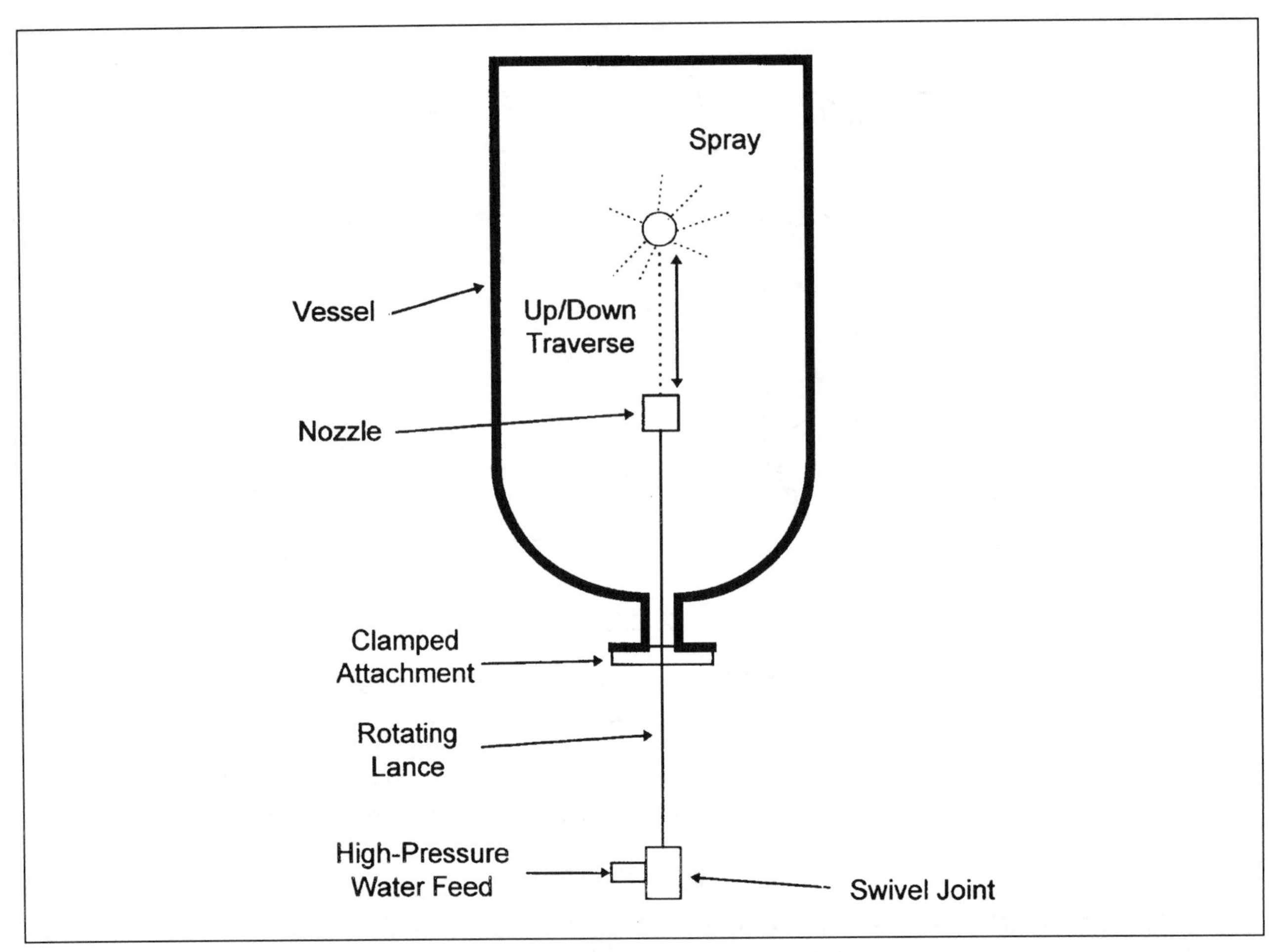

Figure 8-3. High-Pressure Water Wash System

to the impact of the detergents and other materials on the wastewater treatment facility.[4]

Ultrasonic cleaning makes use of cavitation in an aqueous solution for greater cleaning effectiveness. The high-frequency sound waves generate cavitation zones that exert enormous pressures (on the order of 10,000 pounds per square inch) and temperatures (approximately 20,000°F on a microscopic scale). These pressures and temperatures loosen contaminants and perform the actual scrubbing of the surface. Intricate surfaces can be cleaned effectively by the ultrasonic cleaning process.[5]

Pipe-Cleaning Pigs. "Pigs" are pipe-cleaning mechanisms made of any number of materials. They are actuated by high-pressure water, product, or air. Pigs remove residue buildup on pipe walls, thereby minimizing or eliminating subsequent washing.

8.3.5 More Than and Less Than 10% Pollution Prevention

In cases of more than 10% pollution prevention, instead of using a fresh solvent, a waste solvent from another process in the plant should be considered for use as the equipment flush. If feasible, this will reduce the plant's total waste load. Also, standard operating procedures that have been around since process startup may call for a certain volume of solvent to clean the equipment between batches or cycles. Challenging these old procedures will often reduce the amount of solvent used with no change in the resulting cleanliness of the equipment.

Normally, the final wash of a series of washes contains very few contaminants; therefore, save this final rinse and use it as the first wash for the next campaign. If spare tanks are available, then countercurrent washing should be evaluated. In a countercurrent wash system, the first wash is the second wash from the previous campaign, the second wash is the third wash from the previous campaign, and so on. The first wash of the present campaign is the only wash sent to recovery or waste treatment and disposal. As a result, solvent usage can be reduced to that needed for a single flush.

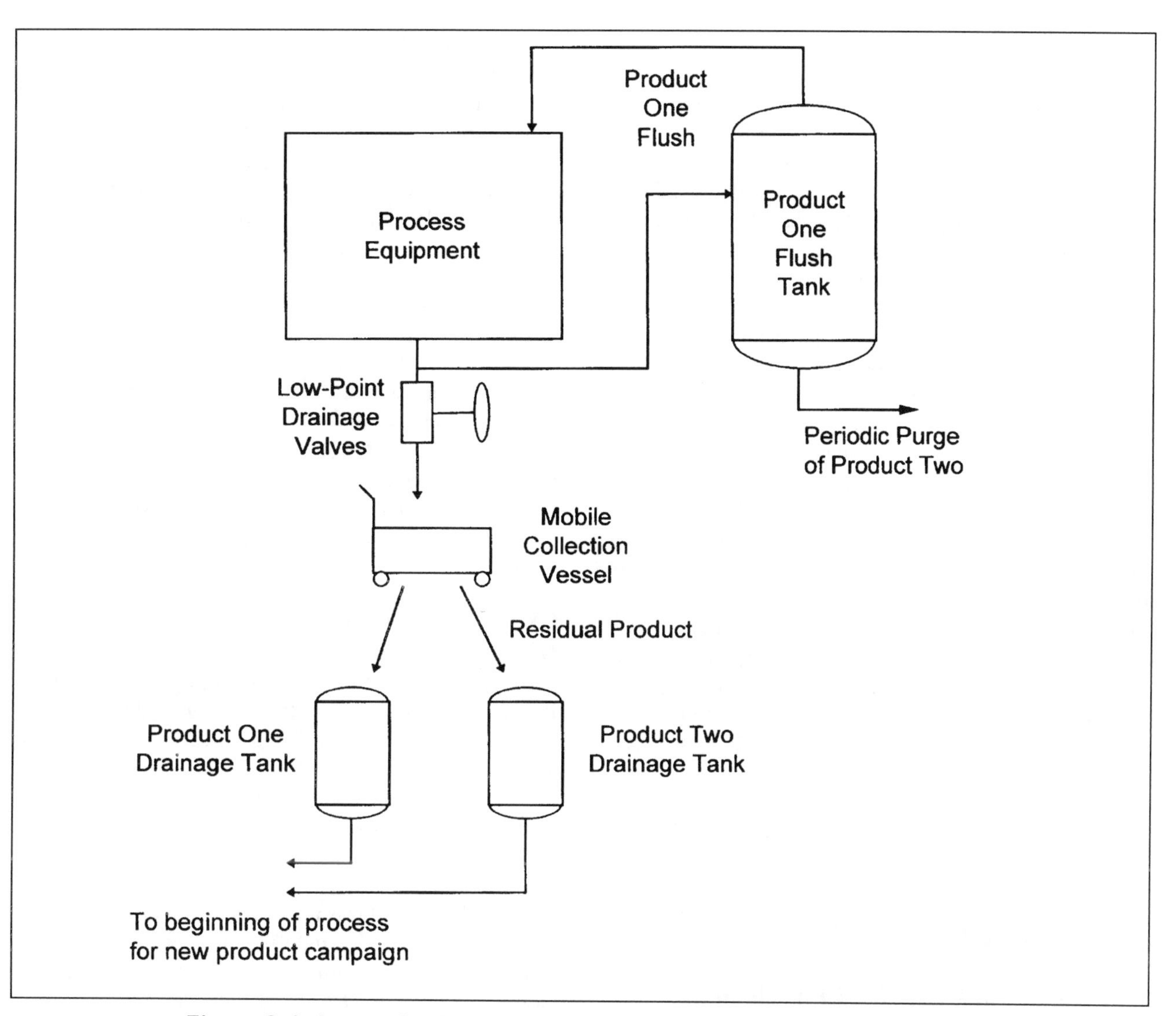

Figure 8-4. Low-point Drainage System with Portable Collection Buggy

For less than 10% pollution prevention, the cleaning solvent can be replaced with another solvent that has a lower negative impact on the environment or that is not regulated.

8.3.6 Employee Awareness

Another key aspect of equipment cleaning is employee awareness. Plant operators and maintenance personnel are in contact with the equipment every day. Training programs that help them to understand the sources and cause of waste generation, and its impact on the business and employee safety, will make a difference.

8.4 Cleaning Research

As an example, the Emission Reduction Research Center at the New Jersey Institute of Technology (NJIT) is supporting research programs at several universities whose goals are to better understand the adhesion of contaminants to vessel surfaces and to identify aqueous-based cleaning solvents that can replace traditional organic-based cleaning solvents. The research spans the range from fundamental surface studies to characterize surface/organic interactions that promote surface fouling, to the practical applications of designing effective aqueous cleaning solutions and developing a robotic water-jet system for batch vessel cleaning.

Professor M. Labib at NJIT has shown that the adhesion of organic materials to vessel surfaces is mainly caused by acid–base interactions between the organic material and the vessel wall.[6] In a number of cases, the residue material adhering to the vessel surface is different from the starting material and possesses different functional groups, such as aldehyde groups transformed into carboxylic acid groups. Professor Labib has found that organic acids and bases will interact with hydroxyl groups on the metal surface, that is, iron (Fe) and chromium (Cr) surface hydroxides for stainless steel. For this reason, the pH of the cleaning solution becomes critical, and must exceed the isoelectric point (point of zero charge) for the metal surface. Above the isoelectric point, the metal surface will be negatively charged. The goal, then, is to establish a net repulsion between the charged metal surface and the organic residue. For example, the isoelectric point for stainless steel is about pH 8.5; therefore, a pH greater than 8.5 will be needed to clean most organic compounds from stainless-steel surfaces. For organic bases, very high pH values are sometimes required.

Professor M. Leu at NJIT has led the development of an articulated robot arm to manipulate high-pressure water jets to achieve complete coverage of internal areas where fouling is known to occur. In addition, the NJIT research team has defined the jet properties (i.e., impingement conditions, composition, temperature, pressure) that ensure the desired cleaning efficiency.[7] For example, they have found that cleaning efficiency is a strong function of travel speed and the distance of the water-jet nozzle from the vessel surface. The goal is to maximize travel speed to minimize cleaning time and wastewater generation.

Professor Cannon at Pennsylvania State University has looked at the use of hydrogen peroxide (an oxidant) as a cleaning agent in aqueous cleaning solutions.[8] He has found that hydrogen peroxide, when safely handled and used, can aid in the dissolution of organic acid residues from vessel surfaces.

8.5 Case Histories

8.5.1 Minimize the Need for Cleaning

A site makes two types of chlorinated aromatic compounds that are sold as three different products: pure Product One, pure Product Two, and a mixture of the two products. Between product campaigns, large amounts of residual product remain that must be removed to prevent contamination of the product in the next campaign. In the past, this was done by flushing the equipment with an organic solvent. However, using an organic solvent had the following disadvantages: a large waste stream was produced, use of the solvent contributed to long turnaround times between campaigns, a large amount of off-spec product was made due to contamination of the initial product with the previous campaign's residue, and the solvent wash contained a significant amount of valuable product.

An innovative pollution-prevention team implemented an equipment drainage system to eliminate the need for the organic solvent. Valves were placed at the lowest elevation of the process, and the residual product was collected in a special buggy at the end of the campaign, so that it could be reused in the next campaign of that product (see Figure 8-4).

In deciding whether or not to implement the system, the pollution-prevention team had to consider the effect on product quality. After the equipment is drained with the new system, a small amount of the product residue still remains. For Product Two, the purity specifications are loose enough that equipment flushing before the next product campaign is not necessary. On the other hand, purity specifications for Product One are strict enough that flushing with a reserve amount of Product One is required. The flush tank shown in Figure 8-4 is designed for this purpose. The amount of Product Two within the flush tank is held constant by periodically drawing off small quantities of the flush for recycle back to the process or to make the mixed product.

Capital investment for the project was $10,000, but it resulted in a positive net present value of more than $2 million and a 100% reduction in waste generation.[3]

8.5.2 Procedural Changes and Relaxed Product Specifications

At a chemical manufacturing site, a series of distillation columns is used to purify different product crudes in separate campaigns. At the conclusion of each campaign, a portion of product crude was used to wash out the equipment. When the crude became too contaminated, it was sent for destruction in a hazardous-waste incinerator. First, an analysis of the washing procedure of a decant tank indicated that only one-tenth of the product-crude wash material was really needed to effect cleaning. Second, a dedicated pipeline for each crude was installed, thus eliminating the need to flush the line between campaigns. Third, an extended and improved drainage procedure was developed for a large packed-bed distillation column. Finally, the product specifications were relaxed, so that fewer washes were required to maintain product specifications. Capital investment for these process changes was $700,000; however, the project had a positive net present value of more than $3 million, and realized a 78% reduction in waste generation.[3]

8.5.3 High-Pressure Washing

At one chemical plant, cleaning kettles required 800 hours of labor and 110 tons of solvent annually. The cleaning process sometimes involved employees entering the kettles to scrape the walls, a process that took each employee 3 hours. Changing to a high-pressure rotating spray head reduced the amount of solvent needed, thereby reducing the time required for cleaning. In addition, employee safety was improved by eliminating the need for employees to enter the vessels. The change required a capital investment of $69,000, but it resulted in an annual savings of $61,500.

At another manufacturing site, several types of polymer are made in an agitated vessel that must be cleaned periodically to maintain product quality. The vessel was cleaned by washing with a flammable solvent. To eliminate the use of the solvent, a special high-pressure water jet was installed to clean the vessel. The water jet had a capital investment of $125,000; however, the project had a net present value of more than $2.5 million and a 98% reduction in waste generation.

In the production of polyvinyl chloride (PVC), polymer buildup on reactor surfaces, agitators, brackets, and other parts required cleaning after every batch. The manual cleaning operation was replaced by a high-pressure water-jet. To minimize the need for water jet cleaning, a proprietary additive was used to suppress the formation of polymer buildup on the walls. After each batch, a low-pressure water rinse removed sufficient material to prevent contamination of the next batch. Only after 500 batches was a high-pressure water-jet wash required.[9]

8.5.4 Waste Solvent Used for Flushing Equipment

In the manufacture of a cast-type product, methylene chloride was used as the process flush and cleaning solvent for 24 years. Two significant changes were made to the cleaning procedure: first, methylene chloride was replaced by a dibasic ester waste stream from another process; second, the waste load was further reduced by modifying the flushing procedure. Methylene chloride emissions were reduced by 97%, and a proposed $1.5 million project to control methylene chloride emissions to the air was avoided.

Literature Cited

1. U.S. Environmental Protection Agency. February 1994. *Guide to Cleaner Technologies: Organic Coating Removal.* EPA/625/R-93/015. Washington, D.C.: U.S. EPA, Office of Research and Development.
2. U.S. Environmental Protection Agency. February 1994. *Guide to Cleaner Technologies: Cleaning and Degreasing Process Changes.* EPA/625/R-93/017. Washington, D.C.: U.S. EPA, Office of Research and Development.
3. U.S. Environmental Protection Agency. November 1993. *DuPont Chambers Works Waste Minimization Project.* EPA/600/R-93/203. Washington, D.C.: U.S. EPA, Office of Research and Development.
4. U.S. Environmental Protection Agency. February 1994. *Guide to Cleaner Technologies: Alternatives to Chlorinated Solvents for Cleaning and Degreasing.* EPA/625/R-93/016. Washington, D.C.: U.S. EPA, Office of Research and Development.
5. Gavaskar, A. R. 1995. "Process Equipment for Cleaning and Degreasing." *Industrial Pollution Prevention Handbook.* H. M. Freeman, Ed. 467–481. New York: McGraw-Hill.
6. Deruijter, M. J., M. E. Labib, and P. J. Zanzucchi. 1994. "Adhesion of Organic Materials to Vessel Surfaces Used in Pharmaceutical and Chemical Industry." *Abstracts of Papers of the American Chemical Society.* 208(1): 165.
7. Meng, P., E. S. Geskin, M. C. Leu, and L. Tismenetskiy. 1996. "Waterjet In-Situ Reactor Cleaning." Proceedings at the *13th International Conference on Jetting Technology.* BHR Group Conference Series Publications, 21: 347-358.
8. Brant, F. R., and F. S. Cannon. 1996. "Aqueous-Based Cleaning with Hydrogen Peroxide." *Journal of Environmental Science and Health, Part A, Environmental Science and Engineering and Toxic and Hazardous Substance Control,* 31(9): 2409–2434.
9. Randall, P. M. 1994. "Pollution Prevention Strategies for the Minimizing of Industrial Wastes in the VCM-PVC Industry." *Environmental Progress.* 13(4): 269–277.

CHAPTER 9

High-Value Waste

9.1 Introduction

Wastes are unwanted materials that are regulated and, most often, must be destroyed or rendered harmless before being emitted to the environment. On the other hand, products are materials that someone will pay money to acquire, and are considered to have value to society. Many wastes often contain valuable materials that can be either recovered directly, or transformed into a product or recyclable raw material. In this chapter, we discuss the potential value of waste materials, and how several businesses successfully transformed waste into valuable products or raw materials.

9.2 Regulations

A potential barrier to waste transformation or reuse is the reclassification of a waste stream to a product. For example, before a Resource Conservation and Recovery Act (RCRA) waste can be classified as a product, it must be used directly without any reclamation. Experience has shown that to minimize any difficulties with the regulatory agencies, the following steps should be considered, as they relate to potential customers for the waste:

1. *A good safety, health, and environmental record.* Conduct an audit of each potential customer to ensure that there are minimal liability issues. The audit should consider past regulatory agency findings; community relations; past safety, health, and environmental incidents; groundwater protection and any needed corrective action; housekeeping; and in-place compliance systems.
2. *Easy tracking.* Find a customer that can take the waste material and transform it into a product at the same location. This way, the business can always show that the material is being used directly and any waste is properly treated and discarded. The business should stay away from companies that simply buy wastes, blend them, and then ship them throughout the country for different uses.

9.3 Conversion to a High-Value Waste

Turning a low-value waste stream into a high-value waste (or, ultimately, a product) requires an examination of how the waste stream can be modified to increase its value. Do trace carcinogens need to be removed? Are any radionuclides present? Do solids need to be removed? Does pH need to be adjusted?

9.3.1 Process and Waste-Stream Constituents

A waste stream has little value if it contains the wrong materials or if the concentration is too low for economical recovery. Initial efforts should focus on trying to change the waste stream into a higher-value waste using the following process analysis steps discussed in Chapter 3.

1. List the raw materials required for all chemical reactions in the process, all process intermediates produced, and all salable products. This is "List 1."
2. List all other materials, including solvents, water, air, nitrogen, and reaction byproducts that become waste. This becomes "List 2."
3. For any materials that are in List 2, but not in List 1, determine how to remove them from the manufacturing process. For example, can a constituent from List 1 replace a solvent in List 2?

9.3.2 Modify the Waste Stream

Subsequent efforts should focus on modifying the waste-stream size, composition, and physical or chemical properties. The waste-stream analysis techniques outlined in Chapter 3 can be used here also.

The best sale, recycle, or reuse option cannot be implemented unless it is identified. To uncover the best options, each waste-stream analysis should include:

1. A list of all materials in the waste stream; for example, for a wastewater stream, water, organics, inorganics, dissolved and suspended solids.
2. Identify the valuable compounds. Determine the source of these compounds and develop options to isolate or purify them.
3. Identify the highest-volume materials (e.g., water, air, carrier gas, solvent), because reducing the volume of these materials reduces the investment and operating cost for isolation and purification technologies.

By analyzing the process in this manner and applying good engineering and chemistry practices, a process that produces a reusable or salable high-value waste product can result.

9.4 The Waste-Value Continuum

The continuum depicted in Figure 9-1 shows the relative merits of actions to take to improve the value of a waste stream. The decision of how far to move up the continuum will depend on a number of factors, including corporate and business environmental goals, regulatory drivers, economics, the maturity of the process, and product life.

9.4.1 Eliminate from the Process

One valuable high-volume waste is solvent. The greatest value to the business would be to eliminate the use of the solvent altogether. Savings would result from avoided purchase costs, handling costs, and reprocessing costs for the solvent; moreover, the cost associated with the yield loss of valuable materials, reactants, intermediates, and products in any waste solvent would be eliminated.

Another valuable waste is the high-value material(s) in a high-volume waste stream. For example, if air is used as the source of oxygen for a process, nitrogen will leave the process with valuable materials that are often at too low a concentration to recover. Eliminating the nitrogen greatly reduces the loss of valuable materials.

9.4.2 Reuse in the Process

One common material reused in a process is the solvent. The opportunities to increase the value of waste solvent range from recovering solvents that were not previously recovered, to recovering more or higher-quality solvent than is currently the case.

Other reusable materials are chemical reactants and intermediates. Often, reactants and intermediates are present at low concentrations in a solvent, water, or gas stream. The first approach should be to increase their concentration in the waste stream; that is, to reduce the volume of the carrier, so that investment and operating costs for isolation and purification are minimized. A second approach is to modify the process, so that the valuable materials do not leave with the carrier fluid in the first place.

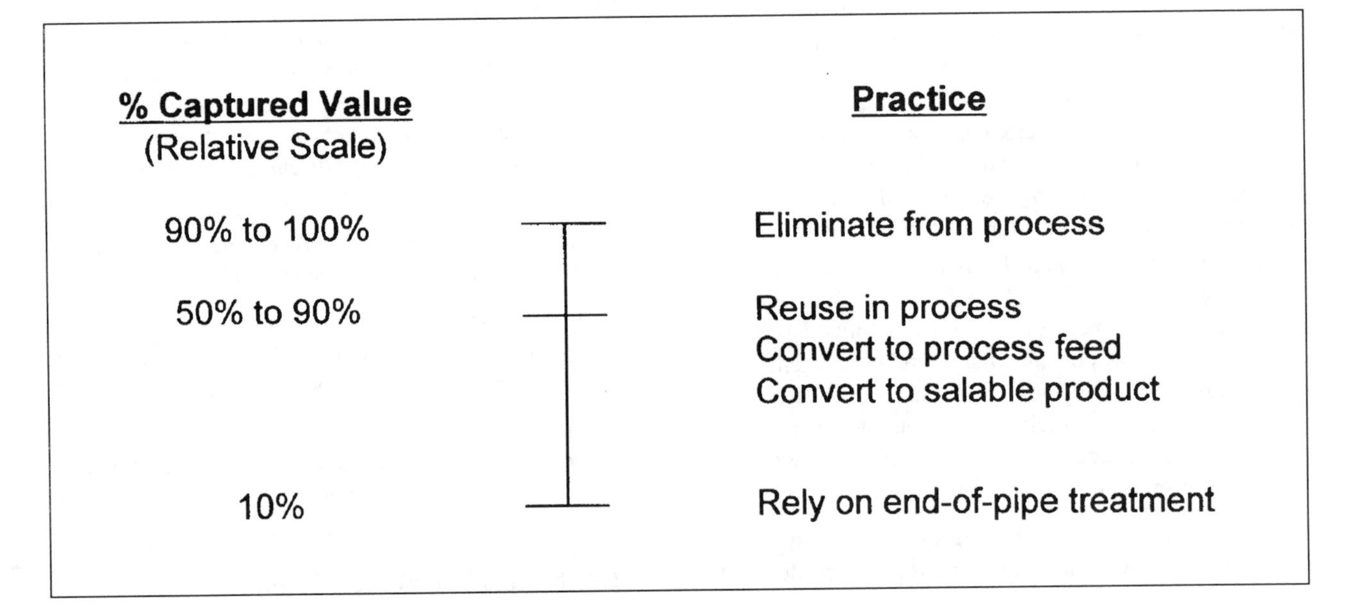

Figure 9-1. The Waste Value Continuum

9.4.3 Convert to Feed

In many chemical reactions, the byproducts are homologues of the products, intermediates, or original reactants. Often, the homologous byproducts can be converted back to the feed reactants in a separate reaction step. An example would be converting waste polyester film and fiber back into their monomer intermediates. If the byproducts cannot be converted back to the feed materials, then a process route to a salable byproduct should be pursued. An example would be purifying and selling the waste dibasic esters from a nylon intermediates process as a valuable raw material for cleaning solvents.

9.4.4 Convert to Product

Existing regulations, and often the existence of waste solvent mixtures, can preclude the economical reuse of the solvent directly within the process. However, do not rule out converting the waste solvents into a valuable product for external sale. Often, these solvents can be purified and sold in applications where higher levels of impurities can be tolerated. An example might be for use as a cleaning solvent.

When a waste solvent or byproduct is converted to a product, several items must be considered:

- If the new product results in a profitable business venture, then the manufacturing process could be intentionally operated to maximize production of byproduct rather than the original product.
- Even if a market is found, unless it generates equal or higher revenue than the original product, the search for higher-value-in-use markets should continue.
- If the new product is a former waste solvent, the buyer must be informed that the solvent supply could disappear in the future. This is because the supplier will often gain more business value by eliminating or reusing the solvent in their own process than by purifying and marketing it on the outside.

Another situation is when a market is found for a material that was once a waste. An example is selling waste ferric chloride solution from white pigments manufacture as a wastewater treatment flocculation and dewatering aid.

Finally, there are situations where valuable materials can be recovered from a waste stream, purified, and then sold as a product. An example would be filtering product solids from a liquid waste stream and selling them as a lower-grade product.

9.4.5 End-of-Pipe Treatment

A major step forward is to reduce the volume of the waste stream to minimize investment and operating costs for end-of-pipe treatment. In addition, attempts should be made to prevent materials that require a more costly treatment from entering the waste stream. One example is to remove a chlorinated organic from a waste-gas stream containing other nonhalogenated volatile organic compounds (VOCs). A second example is to segregate a low-flow, metals-containing wastewater stream from a larger-volume, metals-free wastewater stream.

9.5 Case Histories

9.5.1 Organic Solvents Eliminated from the Process

At DuPont, aromatic organic solvents used as catalyst carriers in the manufacture of intermediate monomers were incinerated as hazardous wastes. Alternative nonhazardous organic solvents were considered and rejected. However, the intermediate monomers were found to have the same dissolution capacity for the catalyst as did the original organic solvents, and could replace them. By utilizing existing catalyst feed and mixing equipment, realizing savings in ingredients recovery, and reducing operating and incineration costs for the waste solvents, the project realized a 33% internal rate of return (IRR) and a 100% reduction in the use of the original solvents.

9.5.2 Acid Scavengers Eliminated from the Process

A higher alcohol was washed with an aqueous solution containing inorganic chemical scavengers to remove residual acidity. Isopropyl alcohol was added to the aqueous solution to assist in separating the higher alcohol product from the wash water. The water-wash and isopropyl-alcohol addition steps were eliminated by directly neutralizing the residual acidity with a chemical agent instead. This process change reduced the aqueous waste load by 100% and realized a 93% IRR.[1]

9.5.3 Reusing Small Quantities of Solvents in a Process

Small quantities of solvents used for cleaning or processing tools and parts in the electronics, auto repair, and other smaller-scale industries, represent a large source of RCRA hazardous waste in aggregate, even though they are a small source per business or site. The U.S. EPA and Battelle Research Institute conducted a study on two technologies for on-site recovery of the solvents as well as a new process technology for reducing VOC emissions.[2] These technologies were:

- Atmospheric batch distillation
- Vacuum heat-pump distillation
- Low-emission vapor degreasing

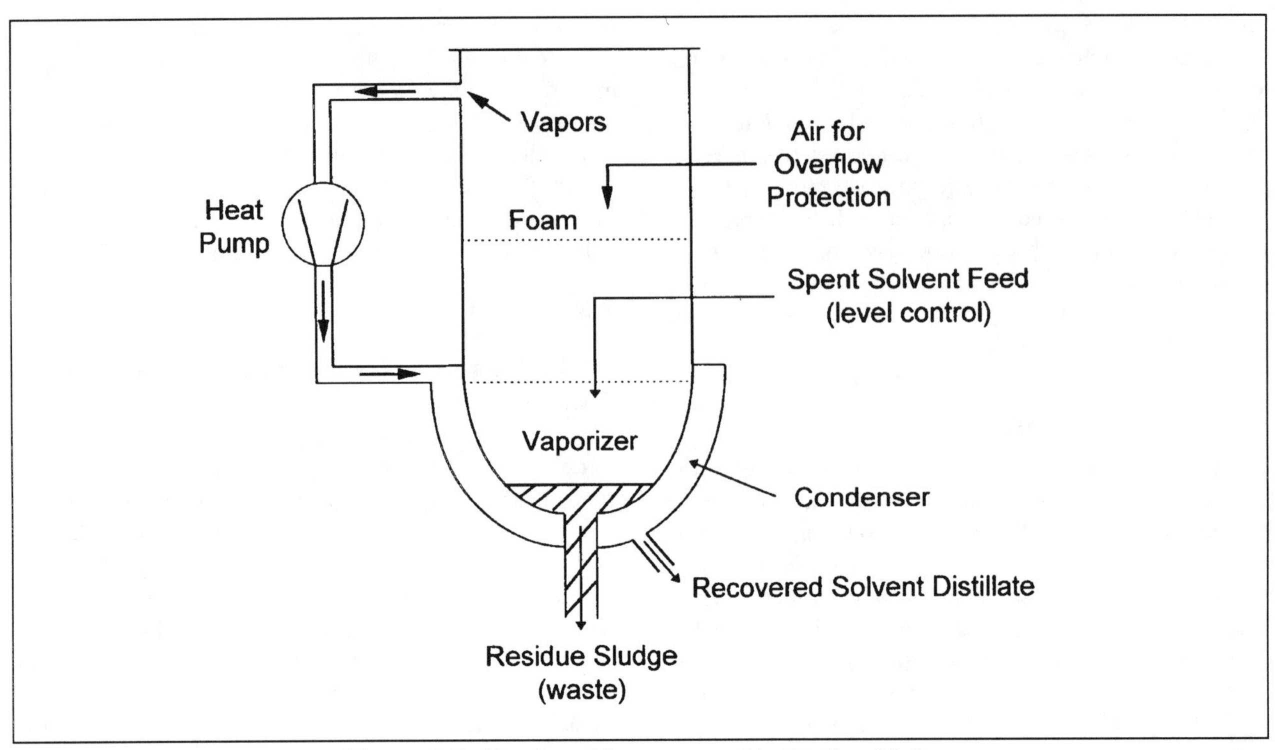

Figure 9-2. Vacuum Heat-pump Distillation Unit

Atmospheric batch distillation units can be purchased in 5 to 55 gallon-per-batch sizes. Some of the units can be modified for vacuum service when higher-boiling solvents (150°C or higher) are involved.

The vacuum heat-pump unit (Figure 9-2) is similar to the vacuum pump used with a conventional vacuum distillation system; however, in addition to drawing a vacuum, the pump functions as a heat pump. The single-stage rotary vacuum pump compresses the vapor and, in the process, heats the vapor. As the hot vapor condenses, it provides the energy to boil the unpurified liquid solvent under vacuum. Because of this heat exchange, this vacuum unit uses 50 to 75% less energy than a conventional vacuum distillation unit.

The low-emission vapor degreaser (LEVD) is a fully enclosed alternative to conventional open-top degreasing that reduces air emissions by more than 99%.

Compared to disposal, the two distillation units reduced operating costs significantly, resulting in a 2-year payback. The LEVD reduced solvent losses and labor costs with an estimated payback of 10 years.

9.5.4 Wash-Solvent Reuse in Paint Formulation

A large source of waste for the paint industry is the solvent used to clean tanks and equipment. The U.S. EPA funded Battelle Research Institute in a study of the feasibility of reusing wash solvents as a portion of the "fresh" solvent required to produce paint at Vanex Color, Inc.[3] The goal of the study was to evaluate product quality, the amount of wash solvent that could be reused, and cost.

The Vanex Color technical staff prepared a master list of paints, along with a list of which wash solvents could be used in the paint formulation. Product quality was tested by formulating one paint with 100% virgin solvent and another with 80% wash solvent. Test results showed that there was no difference in product quality between the two formulations. With improved cleaning procedures (such as mechanical scraping and better tracking of solvents used for washing), and the reuse of wash-solvent for paint formulation, the overall reduction in wash solvent waste generation was 80%. Overall economics showed a 12:1 payback; that is, for every $1000 per year of cost, $12,000 per year in earnings were realized.

9.5.5 Vinyl Acetate Recovery and Reuse

At the Union Carbide Seadrift Plant, the largest waste was a residue that contained high concentrations of vinyl acetate (VA) along with heavier components, such as poly-oils. The waste stream was characteristically ignitable, making it a hazardous waste under the RCRA. The plant averaged over 5 million pounds per year. The Seadrift site installed a VA recovery system, consisting of a feed tank and two distillation columns, to separate the heavy and light components from the VA, which was then recycled.[4] For a capital investment of $1.3 million, the Seadrift site realized annual savings of $570,000, and reduced the hazardous-waste load by 30%.

9.5.6 Reuse Sample Waste in the Process

At a DuPont site, a special adhesive is sampled prior to use. Previously, the spent samples were discarded as a hazardous waste; however, the site decided to consider the feasibility of recycling the spent samples to the process. Following testing and working with the manufacturing area, the change was made, resulting in a reduction in the waste load of 4000 pounds per year.

While the size of the sample waste stream was relatively small in the plant's overall waste picture, it represented significant savings on an annual basis—$17,600 in raw material and disposal costs. More importantly, it demonstrated the viability of reincorporating sampled materials into the production process. Following up on this success, the plant is now recycling many other samples back into their processes.

9.5.7 Reuse, Waste Elimination, and Conversion to a Feed Material for a Herbicide Process

As originally designed and installed, a DuPont herbicide process required two different mixed solvents to produce the product. The two spent solvent streams and a byproduct stream were incinerated as a hazardous waste. Over a 6-year period, 8,000,000 pounds of hazardous waste were reduced to zero by the following processes:

1. Recycling and reusing the first-step mixed-solvent stream for a capital investment of $2.5 million. This resulted in a net cost savings of $1.5 million/year because of reduced treatment costs and fresh-solvent purchases.
2. Finding a customer who would use the second solvent mixture directly and then selling the second spent solvent stream for a net return of $0.75 million/year. By purchasing different raw materials, the solvent was ultimately eliminated from the process for a net return of $1.1 million per year in reduced treatment costs and sales revenues.
3. The byproduct stream was sent to a toll contractor, where it was converted back into one of the raw materials used in the manufacture of the herbicide. The net return was $0.25 million/year in reduced treatment and raw-material costs.

9.5.8 Waste Segregation and Conversion to Products in Dairy Production

Maola Milk and Ice Cream Company (Maola) is a multiproduct dairy producing several milk products (i.e., buttermilk, chocolate imitation milk drink), frozen desserts, juices, and fruit drinks. Before their pollution-prevention efforts proved successful, annual waste generation was 170,000 pounds of milk solids, which represented a potential annual value of $480,000. Most of the waste load from dairy processing consists of milk products that are either intentionally or inadvertently lost to the sewer system. To recover the lost value, Maola installed a recovery tank to recover product–water mixtures and to allow reuse of 90% of a rinse water in the process. A second tank was also installed to recover rinse material from the raw milk system.[4]

The two major uses for the milk solids identified by Maola were as an ice cream feedstock (high-value use) and as animal feed. After implementation of the changes, Maola realized savings of more than $350,000 per year, along with a reduction of 170,000 pounds/year in milk solids and greater than 2 million pounds/year of water.

9.5.9 Reintroduction of Waste Solids into a Polymer Process

At a polymer production facility, a solid intermediate from routine vessel washings was allowed to accumulate in a waste basin. Periodically, the basin was dredged and the solids were sent to a secure landfill. When the time approached to redredge the basin and landfill the recovered solids, a method was found to separate the chemical intermediate using leased equipment, and to reintroduce it back into the process. For a lease cost of $400,000, a net savings of $1,695,000 resulted from reduced intermediate costs and avoided landfill costs. A major barrier that had to be overcome before the project could be implemented was to gain customer acceptance for the use of the recovered intermediate in the manufacturing process.

9.5.10 Nonwoven Waste Fabric Turned into a High-Value Product

The waste from a DuPont nonwoven fabric facility was being sold at low prices to textile-waste customers, who then converted the waste into disposable wiping cloths of various sizes. The wiping-cloth business includes everything from paper towels to table linens, and is a multibillion dollar business. An analysis of the wiping-cloth business and the unique properties of the nonwoven material resulted in DuPont establishing a strong niche in the business. In fact, the business has grown sufficiently to require manufacture of material to be sold directly into the wiping-cloth market. The net result was the transformation of waste from a low-value waste into a higher-value product.

Literature Cited

1. U.S. Environmental Protection Agency. November 1993. *DuPont Chambers Works Waste Minimization Project.* EPA/600/R-93/203. Washington, D.C.: U.S. EPA, Office of Research and Development.
2. Gavaskar, A. R., R. F. Olfenbuttel, L. A. Hernon-Kenny, J. A. Jones, M. A. Salem, J. R. Becker, and J. E. Tabor. September 1993. *Onsite Solvent Recovery.* EPA/600/R-94/026. Cincinnati, OH: U.S. EPA, Risk Reduction Engineering Laboratory.
3. Parsons, A. B., K. J. Heater, and R. F. Olfenbuttel. May 1994. *Wash Solvent Reuse in Paint Production.* EPA/600/SR-94/063. Cincinnati, OH: U.S. EPA, Risk Reduction Engineering Laboratory.
4. Tillman, J. W. September 1991. *Achievements in Source Reduction and Recycling for Ten Industries in the United States.* EPA/600/2-91/051. Cincinnati, OH: U.S. EPA, Risk Reduction Engineering Laboratory.

CHAPTER 10

Reactor Design and Operation

10.1 Introduction

Waste generation in any chemical manufacturing process can almost always be traced back to the heart of the process—the reaction system. Most processes that involve chemical reactions also involve side reactions that produce byproducts, which often end up as wastes downstream. Thus, the design and operation of both new and existing reactors is a critical element in preventing pollution at the source. Pollution prevention through reactor design and operation can best be described as improving the reaction selectivity so that undesirable reactions that generate waste are minimized, while at the same time producing the desired products.[1]

Historically, the approach to chemical-reactor design has been to achieve the highest yield at the lowest cost. However, many reaction pathways that produce high yields also generate toxic byproducts or use environmentally unfriendly raw materials, solvents, and catalysts, which must be dealt with downstream at a significant cost. Often, the costs for waste management were not considered in the original design.

Today, consideration of alternative pathways that avoid or reduce the use of toxic chemicals or chemicals listed, for example, as organic hazardous air pollutants (HAPs) by the Clean Air Act amendments is vital to designing and operating the low-cost processes of the future. However, do not miss out on opportunities to minimize waste in existing equipment. Many times, relatively simple equipment changes will improve yield while also reducing waste.

In this chapter, we discuss pollution-prevention approaches to reactor design and operation. Numerous examples are woven into the narrative to illustrate how the pollution-prevention concepts can be applied.

10.2 Pollution-Prevention Strategies

The strategies discussed in this chapter are intended to stimulate thinking and challenge existing practices in reactor design and operation, though not all strategies will apply in every case. The challenge to the engineer is to identify for a particular application which opportunities will lead to optimum design and operating conditions from a technical, economic, and environmental perspective.

10.2.1 Consider the Impact of Solvent Selection

Choose regulatory-unlisted compounds over regulatory-listed ones. For example, heptane is unlisted, while hexane is listed as an organic hazardous air pollutant. Another advantage of heptane is that it is less volatile than hexane (98.4°C normal boiling point for heptane vs. 69°C for hexane).

Use raw materials or intermediates that are already part of the process chemistry in place of new solvents. For example, in a polymerization process to make an elastomer, an inhibitor solution was added at various points in the process to prevent uncontrolled polymerization of the reactive monomer intermediate used to make the elastomer. Originally, the inhibitor was added in a toluene/glycol ether solvent mixture. These solvents were later incinerated as a hazardous waste. The solution was to replace the existing solvents with the intermediate monomer, which had the same dissolution capacity as the toluene/glycol ether mixture.

Avoid dense nonaqueous phase liquids (DNAPLs), such as chlorinated organics. When spilled on the ground, DNAPLs quickly travel downward in the soil, become trapped in subsurface pockets, and are very difficult to remediate. It costs $50–$1000 per cubic yard to remediate DNAPL-contaminated soil.

Think about the feasibility of single solvents instead of solvent mixtures. Often, mixtures are harder to recover and reuse directly in the process because it is difficult to maintain the right blend during recovery.

Consider the ease of separating the solvent from the desired product and other feed materials. Try to avoid solvents that have a normal boiling point close to that of the product and raw materials or that form undesirable azeotropes. Also, volatile solvents increase air emissions and make recovery by condensation or adsorption more difficult.

For a series of reaction steps or reactors, consider minimizing the number of different solvents; "optimizing" the solvent for each step will lead to suboptimization of the overall train. After all, each additional solvent requires another separation step.

10.2.2 Improve the Quality of Raw Materials

Work with your suppliers or install dedicated equipment to remove impurities *before* they are introduced into the process. Often, suppliers already have the infrastructure in place to remove them. In addition, more accurately measure raw-material purity to minimize constituents available for side reactions and byproduct formation and to limit the quantity of these impurities that ends up in waste streams.

For example, a process for making an aromatic compound produced a hazardous-waste stream of heavy tars that was incinerated off site.[2] These heavy, viscous tars entrapped salable product, resulting in a significant yield loss (6.5 lb of salable product lost to waste/100 lb of salable product recovered). Two types of tar were formed in the reactor—acid tar due to acidity in one of the raw materials and thermal tar due to the high-temperature reaction. The ratio of acid tar to thermal tar in the waste stream could not be distinguished by laboratory analyses, which discouraged attempts to reduce waste. To combat acid-tar formation, a neutralizing agent was added to the reaction step, and the neutralizing agent became part of the waste stream.

The solution was for the supplier of the "acidic" raw material to institute better on-line pH control at its facility to keep residual acidity low. With this improvement in place, thermal-tar formation was then reduced by lowering the reactor operating temperature. The result was a 60% reduction in waste, which equated to a net present value of over $6 million (10 years, 12% discount rate).

10.2.3 Redesign the Reactor

Consider a fluidized catalyst bed over a fixed catalyst bed where hot spots lead to unwanted byproducts. In the oxychlorination process to manufacture ethylene dichloride (EDC), an intermediate in the production of vinyl chloride monomer, ethylene reacts with dry HCl and either air or oxygen to produce EDC and water.[3] The EDC is then cracked or pyrolyzed to form vinyl chloride and HCl. The oxychlorination reaction is carried out in the vapor phase in either a fixed-bed or fluidized-bed reactor containing a copper chloride catalyst. Good temperature control of this highly exothermic reactor is critical to minimizing formation of unwanted byproducts, especially at temperatures above 325°C. The fluidized-bed reactors are able to run at 220–235°C, while the fixed-bed reactors must run at 230-300°C. The excellent temperature control of the fluidized-bed reactor (less than 10°C variability across the bed), combined with its lower operating temperature and increased yield, has led to its commercial development over older fixed-bed designs. The absence of hot and cold spots in the fluidized bed is another advantage. The trend has also been to use oxygen instead of air to reduce vent-gas volume by over 95% and to lower capital costs for new system designs.

For batch processes, match heat removal and reactant addition rates to avoid temperature spikes. In highly exothermic batch reactions, the reactor temperature profile may look as it does in Figure 10-1. Often, this type of temperature spike encourages formation of unwanted byproducts, such as tars and other degradation products. A possible solution is to design a semibatch reactor where all reactants except one are premixed and added as a single batch to the reactor. The remaining reactant is then added continuously to the reactor with its feed rate varied to maintain a preset temperature in the reactor.

10.2.4 For Equilibrium Reactions, Recycle Byproducts

If byproducts are recycled to the reactor for equilibrium reactions, the byproducts will decompose at the same rate that they are formed. A key decision in the development of new processes is whether to recover and recycle or to remove from the process byproducts formed in secondary reversible reactions of the type:[4]

$$\text{A} + \text{B} \rightarrow \text{Product} \quad (1)$$

$$\text{A} + \text{Product} \leftrightarrow \text{Byproduct} \quad (2)$$

Reaction 2 is reversible, but Reaction 1 is not. If byproduct is removed from the process, a new exit stream is added to the flowsheet, whereas if it is recycled, one is not. Removal of byproducts increases raw-material costs because of the loss in selectivity to the desired product, and waste-disposal costs if the byproduct must be sent to treatment or disposal.

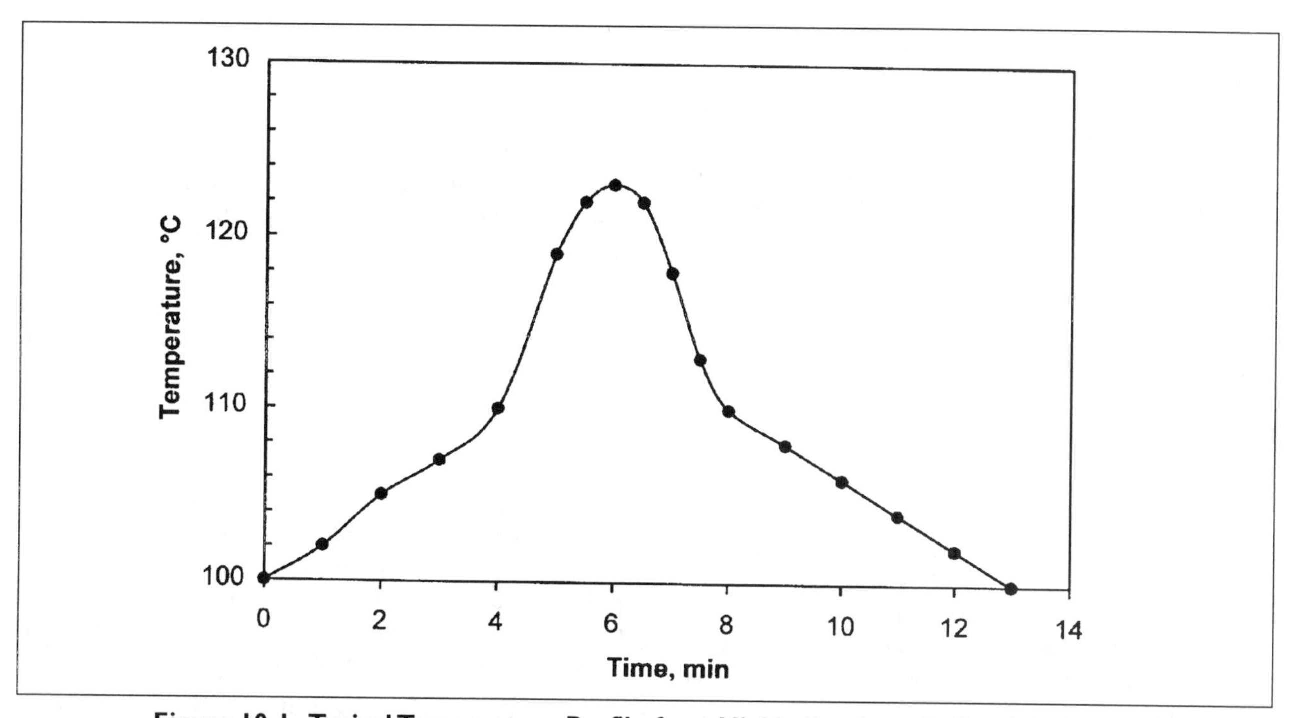

Figure 10-1. Typical Temperature Profile for a Highly Exothermic Batch Reaction

Conversely, if byproduct is recycled and allowed to reach its equilibrium level (that is, byproduct decomposes at the same rate that it is formed), then recycle costs increase, but there is no selectivity loss. Thus, recycle of byproduct leads to a cleaner process, and may be the lowest cost option overall.

An example is the steam reforming of natural gas to make synthesis gas ($CO + H_2$) for methanol production.[5-6] The key gas-phase reactions are the reforming reaction;

$$CH_4 + H_2O \leftrightarrow CO + 3H_2 \qquad (3)$$

and the water-gas shift reaction;

$$CO + H_2O \leftrightarrow CO_2 + H_2 \qquad (4)$$

Under typical reforming conditions (Reaction 3), about one-third of the CO produced shifts to CO_2 by the water-gas shift reaction (Reaction 4). As a result, the synthesis gas has an H_2:CO ratio of about 5:1. However, a lower H_2:CO ratio is needed. This is accomplished by recovering and recycling CO_2 back to the reactor to drive the equilibrium of the water-gas shift reaction (Reaction 4) to the left. This yields a synthesis gas product with an H_2:CO ratio as low as 3:1. If an even lower H_2:CO ratio is needed, additional CO_2 is imported to the reactor from outside the process.

10.2.5 For Sequential Reactions, Remove Product as it Forms

In sequential reactions, the desired product may be an intermediate in the reaction sequence. An example of this type of reaction is

$$A + B \leftrightarrow C \qquad (5)$$

$$C + B \leftrightarrow \text{Product} \qquad (6)$$

$$\text{Product} + B \leftrightarrow \text{Byproduct} \qquad (7)$$

From a pollution-prevention perspective, the key is to design the reactor to minimize the formation of byproduct. One way to do this is to remove product as it is formed to keep its concentration in the reaction mass low. This favors Reaction 6 over Reaction 7.

If Reactions 5 and 6 are fast relative to Reaction 7, mixing can also be very important. In this case, feed two moles of B per mole of A and mix well; then, no excess B would be available to react with product to form byproduct.

In the manufacture of a fluorochemical product, the following sequential reactions take place:

$$A + HF \rightleftharpoons \text{Intermediate} + HCl \qquad (8)$$

$$\text{Intermediate} + HF \rightleftharpoons \text{Product} + HCl \qquad (9)$$

$$\text{Product} + HF \rightleftharpoons \text{Byproduct} + HCl \qquad (10)$$

In this case, the desired product is actually an intermediate labeled "Product" in the sequence of reactions shown. The more volatile byproduct represents a yield loss and is a greenhouse gas. In the original design, the

HCl, product, and byproduct were removed from the reaction mass using a simple reflux column with condenser located above the reactor. However, some of the product refluxed back to the reactor with condensed reactant A. As a result, the product concentration in the reactor favored formation of the undesirable byproduct (equilibrium stronger to the right in Reaction 10).

The solution was to add a stripping section with reboiler to the lower half of the existing column to reduce reflux of product from the column to the reactor, that is, to improve separation between product and the less volatile reactant A. The result was a 30% reduction in byproduct formation.

10.2.6 Match the Mixing to the Reaction

Consider adding or improving baffling, installing a higher speed motor on agitators, using a different mixing blade, installing multiple impellers, or using jet mixers, pump recirculation, or in-line motionless mixers. Feed mixing is most important when reaction rates are faster than mixing rates. For example, use in-line high-efficiency mixers for fast reactions.

The importance of mixing as a pollution-prevention tool in reactive systems is enormous.[7] Chemical processes almost always require bringing reactants into close contact with the help of some type of mixing device. For fast reactions or viscous fluids, mixing is often slow relative to the reaction rate. As a result, desired reactions are slowed down and stopped before reaching completion, undesirable reactions are enhanced, and product selectivity is lowered. For example, large stirred tanks often have characteristic mixing times on the order of minutes; therefore, they are poorly suited for systems with fast competing reactions, because the product distribution is strongly dependent on mixing and often requires that homogeneity be achieved rapidly. For an excellent overview on mixing in single-phase chemical reactions, see Bourne.[8]

Turbulent jet mixers are specially suited for reactions in which very fast mixing is required.[8] Mixing times of a few milliseconds on a small scale to a few seconds on a large scale are feasible. Examples include the precipitation of barium sulfate,[9] injecting a feed stream containing either a catalyst or monomer in polymerization reactors,[10] and oxidation of ethylene and ammonia.[11] On the other hand, they are not very practical if the viscosity is high or suspended solids are present.

A typical jet-mixer-in-tank arrangement is shown in Figure 10-2. A high-velocity jet is injected into the vessel, wherein the jet fluid entrains and mixes with the tank fluid. The correlation shown in Figure 10-2 will predict the blend time with a relative standard deviation of 12%.[12] For minimum power input, the aspect ratio of the vessel, R, is given by:

$$R = \frac{H}{T} = 0.707 \qquad (11)$$

In a number of cases, jet mixers can replace or be used instead of mechanical agitators. Jet mixers are often a lower-cost solution and avoid the fugitive emissions associated with agitator seal leaks.

For viscous materials, the best alternatives often involve either motionless mixers or extruders. Reactive mixing processes are also conducted in packed beds, tray columns, fluidized beds, centrifugal pumps, jet-tank configurations, and the like. The key is matching the type of mixing to the reaction chemistry and physical properties of the constituents.

For example, an alkaline oxidation process is used to manufacture an intermediate for a major health product.[7] The primary issues with this process are very fast, competitive, and consecutive reactions, optimizing the reactant mole ratio, and controlling the large exothermic heat of reaction. The chemistry is as follows:

$$RS + H_2O_2 \rightarrow RSO + H_2O \qquad (12)$$
$$RSO + H_2O_2 \rightarrow \mathbf{RSO_2} + H_2O \qquad (13)$$
$$RSO_2 + H_2O_2 \rightarrow RSO_3 + H_2O \qquad (14)$$
$$RSO_2 + RSO \rightarrow RSO_3 + RS \qquad (15)$$

where RSO_2 is the desired product (Reaction 13). Because the reaction time for the desired reaction is about one second, extremely rapid and effective mixing is essential to avoid localized high concentrations of H_2O_2 from causing formation of the undesired RSO_3 (Reaction 14).

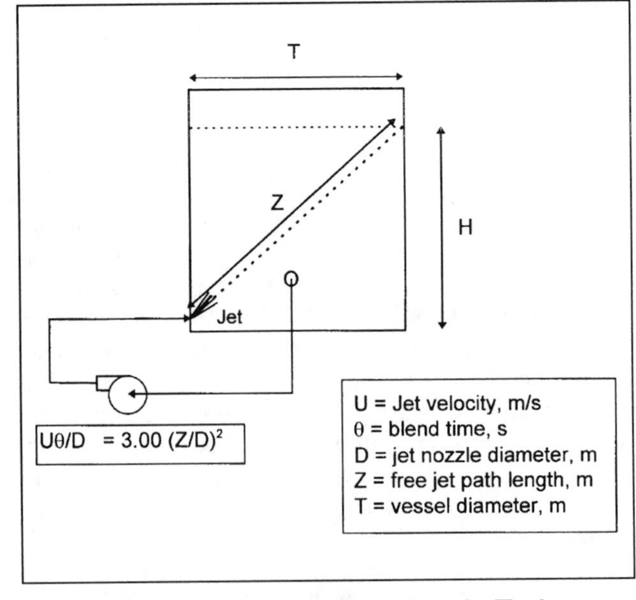

Figure 10-2. Typical Jet-Mixer-in-Tank Arrangement

Common methods, such as distillation or precipitation, for removing RSO_2 from the reactor to maximize yield were not feasible due to limitations in time, temperature, pH, and concentration for the primary reaction. The right technology in this case was an in-line mixer. The reactor feeds (except for H_2O_2) are preblended and then mixed with H_2O_2 in the in-line mixer for a residence time of about 1 second. The temperature is allowed to rise adiabatically to 95°C. In this way, formation of unwanted byproducts is minimized.

Alternatively, if stirred-tank reactors are available, a circulation loop with in-line mixer *and* heat exchanger offers a great deal of flexibility for reactions of this type (see Figure 10-3). This can improve not only mixing, but also temperature control. A high circulation rate also allows for dilution of the reactants to favor first-order reactions (e.g., Reaction 13) over second-order reactions (e.g., Reaction 15). This is important because second-order reaction steps are often the cause of tar formation.

Poor mixing can also result in poor heat transfer and hot or cold spots that lead to lower product yields. Technology has greatly improved over the years, so accomplishing pollution prevention through improved mixing is highly probable.[13,14]

10.2.7 Improve the Way Reactants Are Added to the Reactor

Improved dip-tube or sparger designs can make a difference. It may be better to stage the addition of a key reactant. Premixing of reactants might be the right approach, or simultaneous addition might be better. Avoid adding low-density reactants above the liquid surface in a batch reactor. Distribute feeds better to equalize residence time and temperature and minimize under- and over-reactions that lead to byproducts.

Chlorination reactions in the manufacture of intermediates for an insecticide and herbicide generated over 4000 gallons of spent caustic scrubber waste per reactor batch from scrubbing HCl and Cl_2. To reduce the waste load to the site incinerator, a plant team installed an improved bottom-nozzle dip-tube for adding chlorine and instituted a control strategy specifically designed to minimize off-gas flow from the reactor. Improved chlorine uptake (i.e., higher mass-transfer rate) coupled with better control of off-gas flow (i.e., increased residence time of chlorine in reactor) resulted in an 88% reduction in hazardous waste and a savings of $200,000 per year.

10.2.8 Modify Reactor Cooling or Heating

Heat transfer is an important issue in the design of most reactors, particularly for high-temperature, highly exothermic reactions. For example, with jacketed vessels or tubular reactors, it might be best for the cooling or heating medium flow to be cocurrent with respect to the reaction mass, countercurrent, or a combination of both.

Multitubular fixed-bed catalytic reactors are often used for highly exothermic reactions. Heat removal is accomplished by passing a coolant around the outside of the

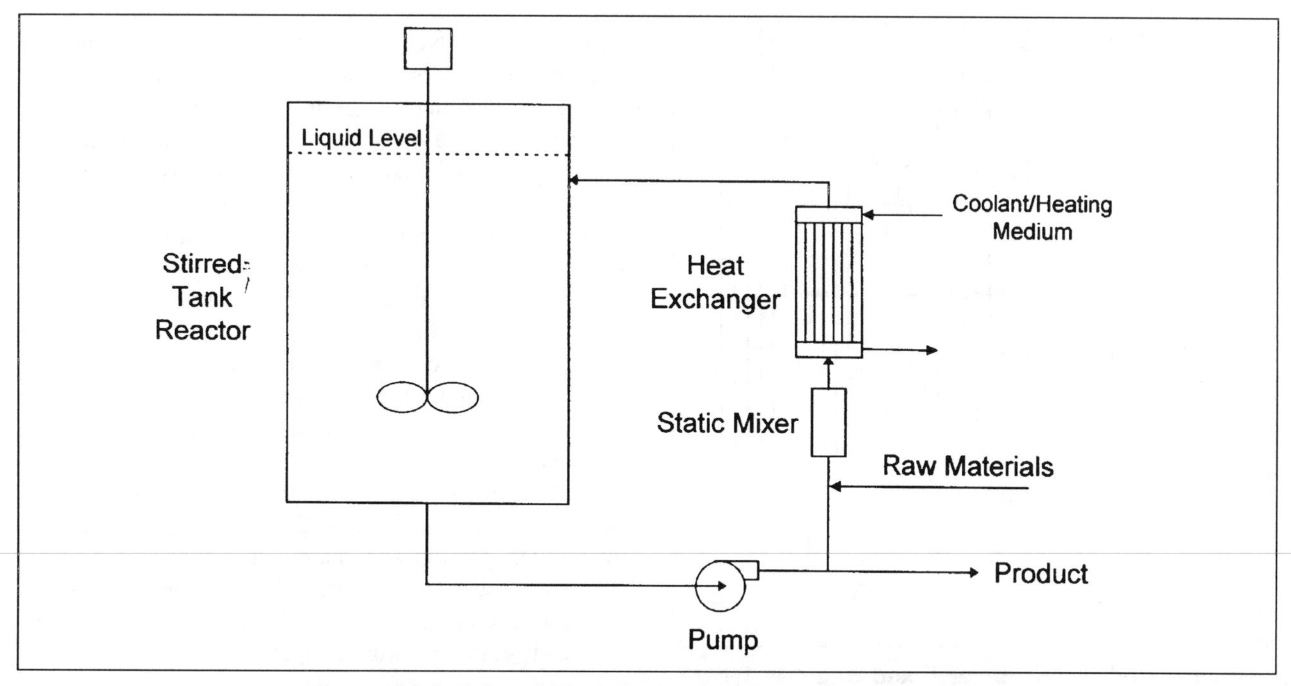

Figure 10-3. Stirred-Tank Reactor With In-line Mixer and Heat Exchanger for Fast, Competitive, and Consecutive Reactions

tubes. The challenge is to ensure good temperature control inside the tubes as the reaction proceeds from inlet to outlet. Often, the reactor shell is baffled to promote cross flow, and flow is usually either cocurrent or countercurrent. The limitations of this design from a pollution-prevention perspective are discussed in Filho and McGreavy.[15] In the inlet zone of the tubes, where the reactant concentrations and required heat removal are highest, feedback of enthalpy by countercurrent flow of coolant is undesirable, because it encourages secondary reactions, which decrease yield. Further down the tubes, however, where the reactant concentrations, and hence reaction rates, are lower, countercurrent flow to help increase the reaction rate is more desirable. In either case, the coolant temperature needs to be controlled to avoid secondary reactions. A reactor design that uses cocurrent flow in the initial reactor zone and countercurrent flow in the later stages of the reaction would be a better design, as shown in Figure 10-4.

In a recent paper, Arakawa et al. describe the benefits of a novel nonadiabatic reactor design that combines the concept of a welded-plate heat exchanger with a fixed, solid catalyst bed.[16] The catalyst is housed in alternating channels and a heat-transfer fluid flows in the complementary channels. Cocurrent flow, countercurrent flow, and crossflow of the reactants and heat-transfer medium are feasible. The nonadiabatic design also features plug flow (vs. backmixing in a fluidized-bed reactor), low-pressure drop (vs. relatively high-pressure drop in down-flow packed-bed and tubular reactors), and a controllable temperature profile. Typical adiabatic operation requires that all enthalpy of reaction enters the reactor as sensible heat in the feed for endothermic reactions, or exits the reactor as sensible heat in the products for exothermic reactions. With nonadiabatic technology, the feed temperature can be significantly reduced for endothermic reactions, thereby lowering the undesirable thermal degradation of feed materials and increasing the yield of the desired product. For exothermic reactions, the maximum allowable temperature is usually limiting. Using a nonadiabatic reactor, the maximum reactor temperature can be lowered, thus decreasing the rate of sequential reactions that convert product into byproducts.

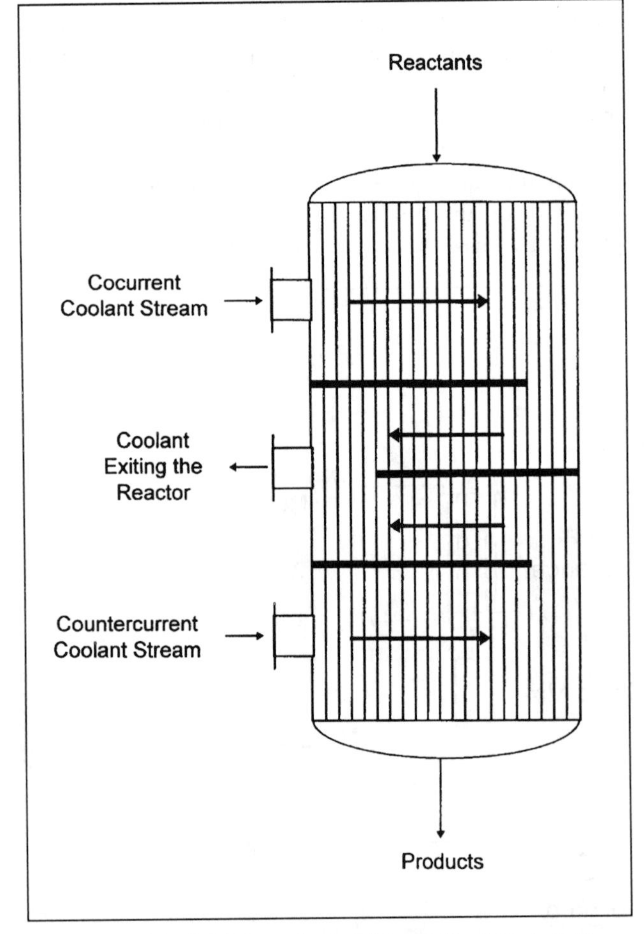

Figure 10-4. Multitubular Fixed-Bed Catalytic Reactor Design[15]

Consider whether a different diluent could be used as a heat sink, for example, steam or carbon dioxide instead of nitrogen or air. In partial oxidation reactions involving hydrocarbons, air has historically been used as the source of oxygen. Some CO_2 and H_2O are also produced during the oxidation, and economics drive a process designer to minimize the amount of CO_2 formed.[4] Some nitrogen oxides (NO_x) are also generated. One approach to eliminate NO_x formation is to eliminate nitrogen from the system by using oxygen instead of air. However, the nitrogen often serves as a heat sink for the exothermic reactions and moderates the temperature rise in the reactor. An alternative is to use pure oxygen or enriched air for the oxidation, but recycle CO_2 around the reactor, in which case CO_2 serves as the heat sink.

Steam can also be used as a heat sink in certain high-temperature reactions. Unlike N_2 and CO_2, steam is a condensable, which can be a desirable feature in downstream separation schemes. At the same time, steam use introduces a potential wastewater emission to the process.

In short, the goal is to minimize hot spots or unwanted temperature variations that lead to undesirable byproducts. Startup and shutdown should also be considered, because temperature will often vary significantly as the reaction mass is heated and cooled to and from the steady-state operating temperature.

10.2.9 Consider More Forgiving Regions of Operation

Considering more forgiving regions of operation improves process control and stability. For example, try to avoid chemistries that require pH to be controlled to within one pH unit to avoid significant byproduct formation.

Use pH buffers wisely. Try to choose buffers that do not significantly add to waste treatment downstream. For example, a carbon dioxide buffer may be preferred over an organic-acid buffer. The organic acid will often end up in a wastewater stream, increasing chemical oxygen demand (COD) load to the wastewater treatment facility. With a carbon dioxide buffer, it is possible to take advantage of CO_2 stripping to raise pH (if necessary), and hence reduce base consumption in the process.

10.2.10 Add, Change, or Optimize the Catalyst

Consider heterogeneous catalysts over homogeneous ones. Homogeneous catalysts often result in metal contamination of water and solid-waste streams, because the metal is not fixed to an inert, insoluble substrate.

Consider new catalyst technology. Catalysts have improved over the years, and processes built 20 to 50 years ago were often optimized to maximize conversion at the expense of selectivity to the desired product. In the past, the economic equation favored high conversion, because waste-management costs were much less than the costs for separation and recycle of the unconverted reactant. Nowadays, however, reaction pathways that sacrifice conversion for higher selectivity to the desired product are often more desirable economically as well as environmentally. Eliminating a byproduct eliminates a separation step downstream to purify the desired product and also eliminates the need for waste treatment of that byproduct.

For example, in an old process for making phosgene from the catalytic reaction of CO with Cl_2, a new catalyst minimized formation of carbon tetrachloride and methyl chloride, eliminating the need for a $1 million end-of-pipe treatment device.

10.2.11 Optimize Reactant Feed Ratios and Addition Sequence

Optimizing reactant ratios is often important in reducing excess constituents that become involved in side or byproduct reactions. In some cases, the excess reactant should be the limiting reactant. For example, in the alkaline oxidation process described in Reactions 12 through 15, a key to minimizing byproduct formation was to limit the amount of excess H_2O_2 that was available to react with RSO_2 by the undesirable Reaction 14.

Pay particular attention to the molar excess for minor ingredients, such as initiators, inhibitors, and catalysts. More is not always better. In some cases, modifying the sequence of raw-material addition will reduce the amount of or change the composition of intermediates and byproducts.

10.2.12 Know, Analyze, and Control Critical Parameters

Pollution prevention requires control of critical process parameters, such as pH, temperature, pressure, residence time, raw-material feed rates, and reaction rates. If the pH of the feed is a critical variable affecting byproduct formation, it needs to be measured and controlled. In some cases, high operating temperatures lead to unwanted degradation reactions, such as tar formation. Also, long residence times typical of large continuous stirred-tank reactors can favor byproduct formation.

One tool that can be used to help understand critical control parameters is the RC1 Reaction Calorimeter. The RC1 is a 2-L agitated batch/semibatch reactor used to study the reaction chemistry, quantify heat effects of reactions, and determine reaction kinetics. Controls include temperature, pressure, pH, and flow. Temperature ramps are also possible. Data output includes temperature, pressure, pH, mass, heat duty, and Fourier transform infrared (FTIR) peaks for compound identification.

In one case, the RC1 was used to study the batch reaction for making a key heterocycle for a specialty chemical product. By tracking heat duty, pH, temperature, mass, and FTIR peaks for the reaction, the researchers learned that the key to maximizing yield was to control pH at 11.5 over the entire batch cycle, rather than allowing pH to naturally drop with time. This also eliminated the need for a heating step in the batch cycle, and reduced the cycle time by 30%.

An important aspect of this device is the ability to characterize existing full-scale equipment (that is, to actually measure heat-transfer coefficients!), so that the intended full-scale process can be simulated at small scale. For example, in another business, the RC1 has been used to understand which reactions are heat-transfer limited by looking at heat duty as a function of time. The hope is to optimize the heat transfer to increase capacity with minimal investment. Most likely, pollution-prevention benefits will also be realized. (Using reaction calorimetry to scale up heat transfer is discussed in Landau and Blackmond.[17])

The RC1 is a window to the reaction as it is happening. This can only benefit pollution-prevention programs.

10.2.13 Routinely Calibrate Instrumentation

The goal is to avoid faulty set points and poor control of the reaction process. For example, in a process to make an agricultural product, a batch reactor produced a water slurry containing the desired product. The water slurry was filtered to capture the product solids, followed by drying and packaging. The aqueous filtrate from the filter was sent to wastewater treatment. The concentration of the desired product in the filtrate increased dramatically if the reactor pH rose much above 8.0. This is because the product was a weak organic acid, HA, with a pK_a of 7.7 (the pK_a is the pH where [HA (aq)] = [A^-]). At pH 8 or above, the deprotonated form of the weak organic acid, A^-, dominated and increased product solubility. In fact, in one instance, the reactor pH was high enough that the majority of the product solids were solubilized and passed through the filter cloth with the filtrate. These pH excursions have been reduced substantially through better attention to pH control in the reactor.

10.2.14 Other Strategies

Consider conversion from a batch process to a continuous process to avoid the wastes associated with startup, shutdown, and equipment cleaning, which are inherent with batch operations. However, this may not be feasible for multiproduct plants that produce small quantities of product over short campaigns.

Provide a separate reactor for recycle streams, such as a back-reactor to convert off-spec product back to its raw materials.

Additional case studies reflecting some more of the pollution prevention concepts discussed in this chapter are available.[2]

Literature Cited

1. Hopper, J. R. 1995. "Pollution Prevention Through Reactor Design." *Industrial Pollution Prevention Handbook*, H. M. Freeman, Ed. 343–360. NY: McGraw-Hill.
2. U.S. Environmental Protection Agency. November 1993. *DuPont Chambers Works Waste Minimization Project*. EPA/600/R-93/203. Washington, D.C.: U.S. EPA, Office of Research and Development.
3. Randall, P. M. November 1994. "Pollution Prevention Strategies for the Minimizing of Industrial Wastes in the VCM-PVC Industry," *Environmental Progress*. 13(4): 269–277.
4. Douglas, J. M. January 1992. "Process Synthesis for Waste Minimization." *Industrial & Engineering Chemistry Research,* 31(1): 238–243.
5. Dingler, J. E., S. Nirula, and W. Sedriks. February 1983. "Costs of Synthesis Gases and Methanol: Part I." *Process Economics Program Report No. 148*: 103. Menlo Park, CA: SRI International.
6. Dingler, J. E., S. Nirula, and W. Sedriks. February 1983. "Costs of Synthesis Gases and Methanol: Part II." *Process Economics Program Report No. 148:* 4–15. Menlo Park, CA: SRI International.
7. Muzzio, F. J., and E. L. Paul. 1995. "Mixing as a Tool for Pollution Prevention in Reactive Systems." *Industrial Pollution Prevention Handbook,* H. M. Freeman, Ed. 455–466. NY: McGraw-Hill.
8. Bourne, J. R. 1985. "Mixing in Single-Phase Chemical Reactors." *Mixing in the Process Industries*, N. Harnby et. al., Eds. 170–184. London: Butterworths.
9. Van Leeuwen, M. L. J., S. L. Bruinsma, and G. M. Rosmalen. June 1996. "Influence of Mixing on the Product Quality in Precipitation." *Chemical Engineering Science*. 51(11): 2595–2600.
10. Forney, L. J., N. Nafia, and H. X. Vo. November 1996. "Optimum Jet Mixing in a Tubular Reactor." *AIChE Journal*. 42(11): 3113–3122.
11. Ajinkya, M. B. 1983. "Mixing of Gases." *Handbook of Fluids in Motion*, Chap. 9, N. P. Cheremisinoff and R. Gupta, Eds. Ann Arbor, MI: Ann Arbor Science.
12. Grenville, R. K., and J. N. Tilton. April 1996. "A New Theory Improves the Correlation of Blend Time Data From Turbulent Jet Mixed Vessels." *Transactions of the Institute of Chemical Engineering,* 74(A): 390–396.
13. Bakker, A., and L. E. Gates. December 1995. "Properly Choose Mechanical Agitators for Viscous Liquids." *Chemical Engineering Progress*. 91(12): 25–34.
14. Myers, K. J., A. Bakker, and D. R. Gates. June 1997. "Avoid Agitation by Selecting Static Mixers." *Chemical Engineering Progress*. 93(6): 28–38.
15. Filho, M. R., and C. McGreavy. 1993. "Multitubular Reactor Design for Waste Minimization." *Proceedings of the Symposium on Effluent Treatment and Waste Minimization*, Symposium Series No. 132: 247–259. Leeds, U.K.: Institution of Chemical Engineers.
16. Arakawa, S. T., R. C. Mulvaney, D. E. Felch, J. A. Petri, K. Vandenbussche, and H. W. Dandekar. 1998. "Increase Productivity with Novel Reactor Design." *Hydrocarbon Processing*. 77(3): 93–100.
17. Landau, R. N., and D. G. Blackmond. November 1994. "Scale Up Heat Transfer Based on Reaction Calorimetry." *Chemical Engineering Progress*. 90(11): 43–48.

CHAPTER 11

Use of Water as a Solvent and Heat Transfer Fluid

11.1 Introduction

Water is critical to life and, until this century, has been considered abundant and safe. However, because of the increased world population and the development of synthetic materials that are not readily biodegradable, water is becoming a much more valuable commodity. In other words, it is becoming scarce and is no longer free. The two properties that make water important to life—a universal solvent and good heat-transfer fluid—also make water important to the manufacturing industry.

Even though water is not a contaminant, it is the major component of most wastewater streams (normally greater than 98% by weight). Usually, only a small amount of this waste water is water of reaction (i.e., water made by the chemical reactions). In addition, relatively small amounts of water usually leave as part of the salable product or as a vapor in vent-gas streams from tanks, reactors, and scrubbers. The majority of the contaminated water leaving a chemical process is introduced to the process as a diluent, solvent, or heat-transfer fluid. Pollution-prevention efforts, then, must ultimately focus on the source, use, contamination, and disposal of water in a chemical process. For this reason, water use and reuse is now being studied by many industry, government, academic, and environmental groups.[1–4]

Two key questions that must be answered as part of any pollution-prevention program are: "Why use water?" and, if water is necessary, "How can water use be optimized?" In almost all cases, pollution-prevention programs require that the process area or site develop a water mass balance for the process. It has been somewhat eye-opening to the authors to discover how few plant sites have developed even rough water balances for their manufacturing areas. If the sources and amount of water usage within a process area cannot be identified, pollution-prevention efforts will ultimately be stymied by this lack of knowledge. The "Collect Data" step in the Assessment Phase of any pollution-prevention program, then, should include the development of a water balance for the process.

In this chapter we discuss pollution-prevention approaches to optimize the use of water as a solvent and heat-transfer fluid. These approaches address water conservation, as well as recycle and reuse.

11.2 The Incentive for Source Reduction of Water Use

Industrial users of fresh water are being driven to adopt water conservation, recycle, and reuse practices for a number of reasons. These include pressures for cost reduction, shrinking water supplies (seasonal or year-round), regulatory and corporate movement toward zero emissions discharge processes, and the desire to enhance community relations.

While the last three issues are self-explanatory, the idea of cost reduction through water conservation, recycle, and reuse merits a brief discussion. The cost for treating fresh water has been published as $1 to $2 per 1000 gallons. Although there are some savings associated with the avoided cost of water treatment, a much greater savings is possible by avoiding the cost of building and operating wastewater treatment facilities. This statement is true not only for new site installations, but also for existing sites that are currently replacing or planning to replace or expand wastewater treatment facilities. Table 11-1 summarizes the *minimum incremental* investment and operating cost for building and operating an aerobic, deep-tank, activated-sludge biological wastewater treatment system.

If the plant were to reduce its wastewater...	New capital investment would decrease by...	Annual cash operating cost would decrease by...
flow	$3,000 per gpm	$300 per gpm
organic loading	$6,000 per pound of organic per hour*	$2,000 per pound of organic per hour*

* Assumes 1 lb biochemical oxygen demand (BOD) per lb organic.

Table 11-1. Minimum Incremental Investment and Operating Cost for Aerobic, Deep-Tank, Activated-Sludge Biological Treatment Systems

Minimum incremental investment savings are based on 1997 dollars, while incremental annual cash operating-cost savings assume 1999 dollars.

Some fraction of the wastewater produced in a chemical manufacturing process is often not amenable to biological treatment. This is because certain compounds in the wastewater (e.g., certain classes of chlorinated organics, metals, pesticides, etc.) are toxic to the microorganisms or are biorefractory, or the wastewater is too concentrated in organics, ammonia, or inorganic salts. As a result, the process area must either segregate and pretreat the problem wastewater streams before biological treatment using nonbiological treatment technologies such as stripping, carbon adsorption, ion exchange, chemical oxidation, or, in some instances, posttreat the entire wastewater using technologies such as carbon adsorption, sand filtration, and alkaline precipitation. Table 11-2 summarizes the *minimum* reduction in investment and cash operating cost that can be expected by avoiding installation of nonbiological pre- and posttreatment technologies. In certain regions of flow, these incremental investment and cash operating costs will be much higher. For nonbiological technologies, the impact of contaminant loading varies substantially from one technology to another. For example, carbon adsorption and chemical oxidation technologies are much more sensitive to organic loading than are air and steam stripping.

In summary, the total incentive for reducing water use within a process will be the sum of (1) the avoided cost of fresh water treatment before its use as a solvent or heat-transfer fluid, and (2) the avoided investment and operating cost associated with the installation and operation of both biological and nonbiological treatment facilities to decontaminate the resulting wastewater.

11.3 Water Reuse

The first step in reducing water usage within a process is to consider water reuse. Water reuse involves two key steps: first, matching water quality with the criteria for its next use, and, second, comparing the cost of attaining a specified purity with the sum of the costs for freshwater treatment and wastewater treatment and disposal.

If the plant were to reduce its wastewater...	New capital investment* would decrease by...	Annual cash operating cost** would decrease by...
flow	$1,100 per gpm	$330 per gpm
organic loading	highly sensitive to technology	highly sensitive to technology

* 1997 U.S. dollars.
** 1999 U.S. dollars.

Table 11-2. Minimum Incremental Investment and Cash Operating Costs for Nonbiological Treatment Technologies

When choosing a technology to improve the quality of water for reuse, the following two aspects should be considered:[2]

- The basic chemical aspects of organic, inorganic, and biological constituents
- The phase state of the compounds (e.g., dissolved vs. suspended solids)

For example, are certain constituents present, such as metals, that will form complexes with key organic chemicals in the process or precipitate on vessel surfaces when the pH is raised? If so, ion exchange or alkaline precipitation might be considered to remove calcium and other heavy metals. If volatile organic compounds are the problem, can they be air- or steam-stripped before the water is reused?

Five basic contaminant categories represent key quality issues associated with water use and reuse. They are:

1. *Inorganics*, such as elemental ions, transition-metal complexes, and various oxides of metalloids and nonmetals.
2. *Organics*, such as petroleum hydrocarbons, alcohols, and aldehydes, and noncompound-specific measurements, such as chemical oxygen demand (COD), biochemical oxygen demand (BOD), and total organic carbon (TOC).
3. *Dissolved components* that are smaller than colloidal particles (less than about 0.1 micron in size) and are dispersed as individual molecules or ions in water.
4. *Suspended solids* that are solid matter present in particle sizes above about 0.1 micron and that can be filtered from solution.
5. *Biological constituents,* including bacteria, and fungi, and possibly pathogenic organisms.

In addition, pH (which is a measure of the hydrogen ion activity in an aqueous stream) represents another key quality issue associated with water use and reuse.

Table 11-3 shows which water and wastewater treatment technologies can be used to address one or more of the five basic contaminant categories previously listed. Byers et al.[2] briefly describe each of the technologies shown in Table 11-3. Each technology description includes a process schematic, along with a discussion of technology applicability, stream qualities, and current

	Contaminants				
Technology	Inorganic	Organic	Dissolved	Suspended	Biological
Biological oxidation and biotreatment	✓	✓	✓	✓	✓
Carbon treatment	✓	✓	✓		
Centrifuge separation				✓	
Chemical oxidation	✓	✓	✓		✓
Crystallization	✓	✓	✓		
Electrodialysis	✓		✓		
Evaporation	✓	✓	✓		
Filtration	✓	✓		✓	✓
Flotation	✓	✓		✓	
Gravity separation or settling	✓	✓		✓	✓
Ion exchange	✓	✓	✓		
Membrane separation	✓	✓	✓	✓	✓
Precipitation	✓	✓	✓	✓	
Solidification or stabilization	✓	✓	✓	✓	
Solvent extraction	✓	✓	✓		
Stripping	✓	✓	✓		
Thermal treatment (drying, incineration, other)		✓	✓	✓	✓

Table 11-3. Feasible Water Treatment Technologies for Each of the Five Basic Contaminant Categories

applications. Optimum performance for most of the technologies shown in Table 11-3 requires either separate pretreatment to adjust pH or improve waste-stream quality, or the addition of chemicals to affect pH within the treatment process itself.

11.4 The Pollution-Prevention Continuum

The pollution-prevention continuum, depicted in Figure 11-1, shows the relative merits of options that are available to process, project, and environmental engineers to eliminate or reduce the use of water as a solvent. Figure 11-2, on the other hand, shows the hierarchy of pollution-prevention options that are available to eliminate or reduce the frequency of water use as a heat-transfer fluid. At a minimum, *complete a water balance* for the manufacturing process! Do not be surprised if the water balance leads to far more insight and, as a result, better pollution-prevention ideas. The decision of how far to move toward a zero waste and emissions design, of course, will depend on a number of factors, including corporate and business environmental goals, economics, and applicable regulations.

11.5 Pollution-Prevention Strategies

The first and most important step in water conservation is to view water in the same way as any other raw material—as a valuable process ingredient. By applying the principles of mass and energy conservation to water, the major uses and misuses of water can be identified (see the case history on "Water Monitoring"). The second step is to identify the water-purity specifications required for each point where water is used in the process. Once the required purity specifications are known, the third step is to consider multiple uses of the same water within the process area or plant boundaries. Two examples are described below in the case histories "Water Reuse as a Solvent" and "Water Reuse for Energy."

Treating water as a valuable process ingredient, coupled with matching water-purity specifications to its intended use, permit a process area to identify opportunities for water conservation, recycle, and reuse by

- comparing design conditions to actual operating conditions
- replacing higher-quality water with lower-quality water

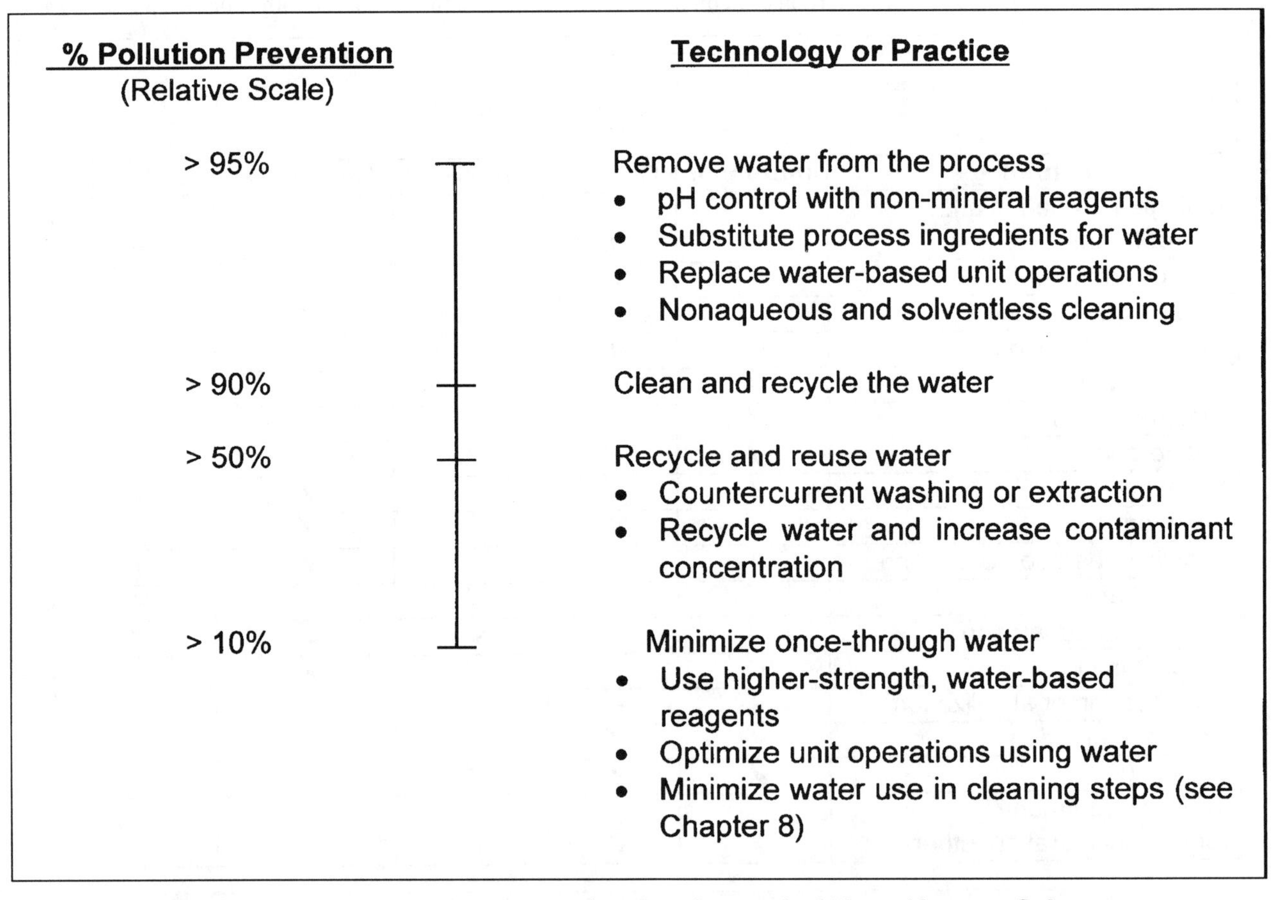

Figure 11-1. The Pollution-Prevention Continuum for Water Use as a Solvent

- analyzing process operations for changes that reduce water consumption
- improving housekeeping and eliminating leaks

Finally, consider these two questions: "Why allow water to contact process materials?" and "Why use water in the first place?" Several of the case histories below address these two questions.

11.6 Water Pinch Analysis

Pinch analysis is a familiar heat-integration technique to construct plots and perform simple calculations that yields powerful insights into heat flows through processes. The technique is based on rigorous thermodynamic principles, and is widely used to determine the scope for energy savings in industrial operations and to define possible process changes to reduce intrinsic energy consumption.

By taking advantage of certain parallels between the principles of heat transfer and mass transfer, a similar analysis can be extended to address mass-transfer problems associated with water. A manufacturing process has both input water streams and output water streams. The sum of the mass flows for the input streams defines the water demand, while the sum of the mass flows for the output streams defines the water sources. By plotting water purity (contaminant concentration) versus water flow (the mass of water required for different parts of the process), the minimum water needed (both mass and purity) or water pinch can be determined. At the pinch, mass-transfer driving forces are at a minimum or concentrations have moved to limiting values set by corrosion limitations, and so on.[3] Figure 11-3 shows a typical composite concentration curve and the resulting minimum water-supply line for the process water profiles of interest. In addition, this plot defines the optimum path for water to move through the process to attain maximum utilization.[3, 5]

11.7 Case Histories: Water as a Solvent

Four case histories follow that discuss pollution-prevention alternatives to the use of water as a solvent.

11.7.1 Water Reuse as a Solvent

The Kinnear Dorr/Wayne Dalton Corporation, located in Centralia, Washington State, is a manufacturer of wooden parts for overhead garage doors.[6] The primary waste stream associated with the manufacture of wood parts is wastewater containing glue washdown wastes. Changes in water regulations eliminated the option of spraying the water on a local landfill. The company identified that a new pretreatment system would require an annual operat-

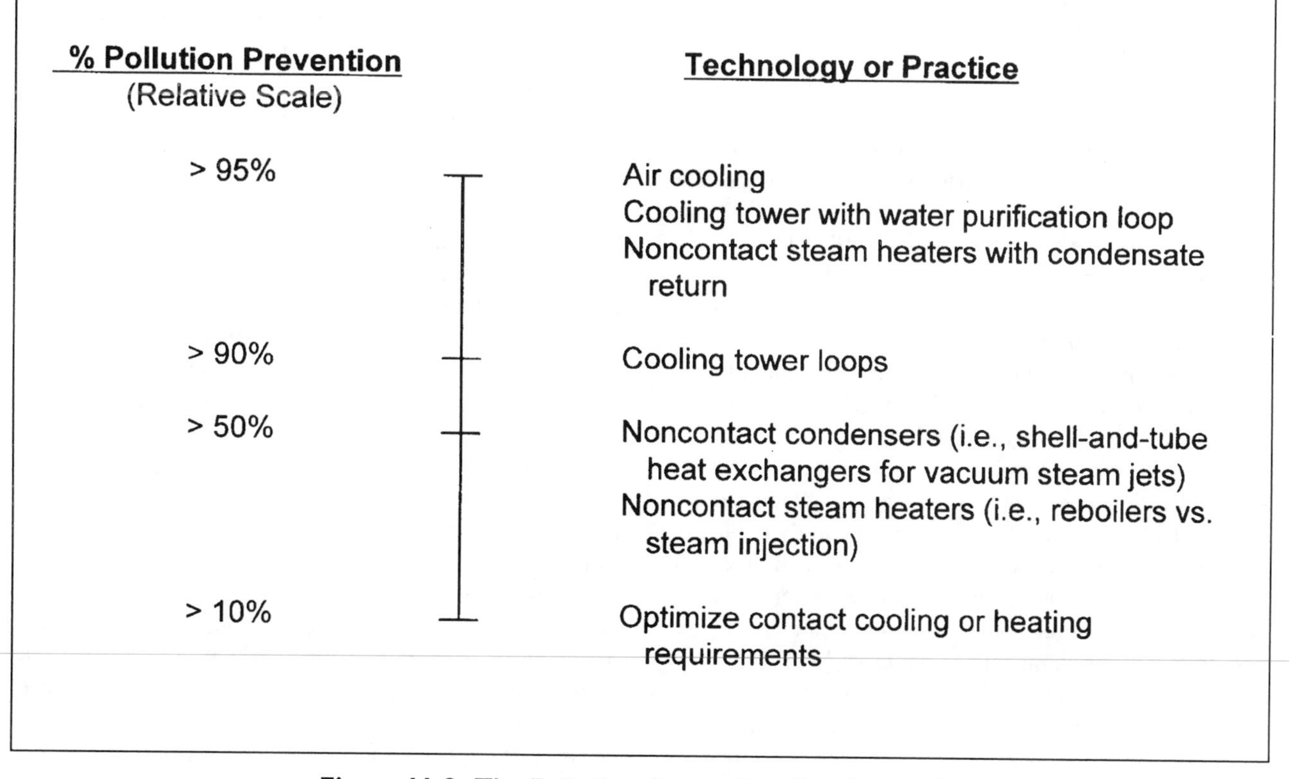

Figure 11-2. The Pollution-Prevention Continuum for Water Use as a Heat-Transfer Fluid

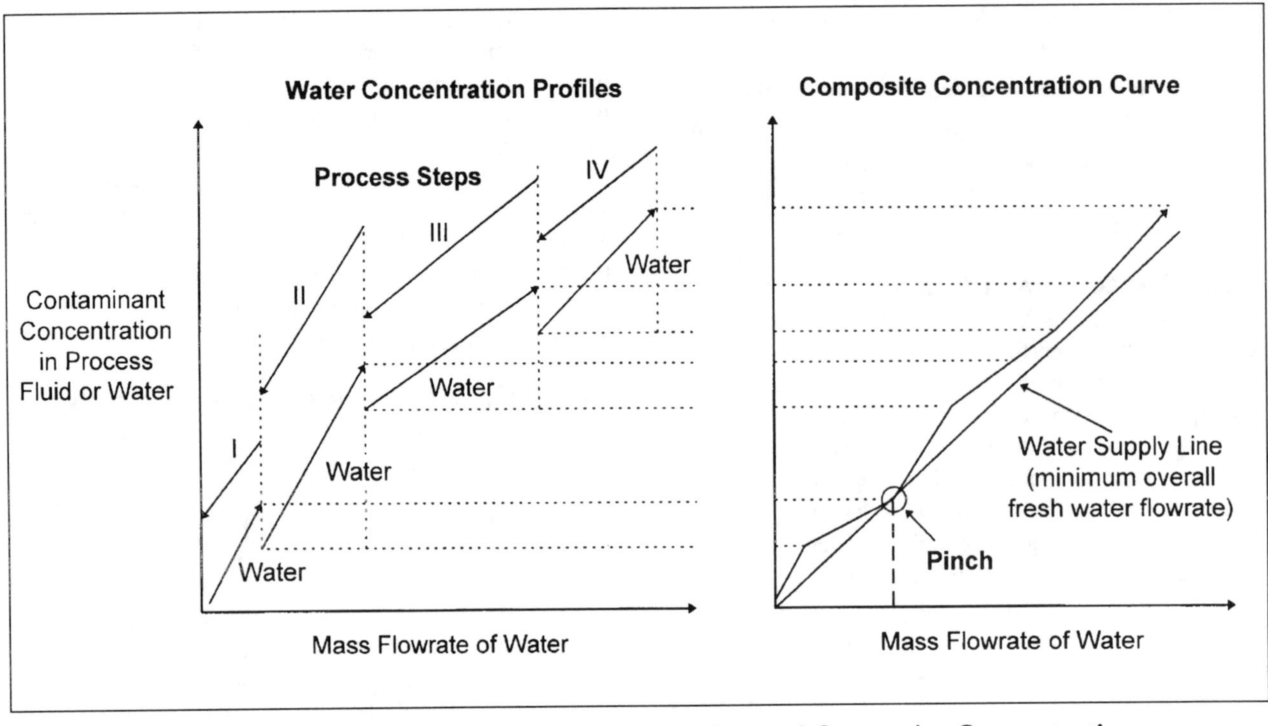

Figure 11-3. Typical Water Concentration Profiles and Composite Concentration Curve for Water Pinch Analysis

ing cost of $16,600. In response, the operators pursued the option of reusing the wash water to prepare the liquid glue formulations from dry glue instead of purchasing liquid glue formulations directly. The new system involved two barrels, pumps, and a fiberglass settling tank for a total capital cost of $1500. The amount of wash water was further reduced by lining the mixing barrels with plastic bags. The actions taken by the employees on a small water stream resulted in an internal rate of return (IRR) of more than 1000% on the $1500 capital investment.

11.7.2 pH Control

Several specialty alcohols produced at the DuPont Chambers Works site contain residual acidity that must be removed before the product can be sold.[7] To remove the acidity, a batch of alcohol crude enters a wash kettle and is mixed with water, chemical scavenging agents, and isopropyl alcohol. The mixture is agitated and then allowed to settle. The product layer is washed a second time with water, scavengers, and isopropyl alcohol. After agitation, settling, and separation, the specialty alcohol is filtered and drummed for shipment as final product.

An analysis of the aqueous layer showed that the water contained isopropyl alcohol, product, and inorganic chemicals. Because the water-wash step was designed to remove residual acidity, other technologies were assessed to replace the water-based system. A chemical acid neutralization step was implemented resulting in a 100% reduction in waste generation. For only $40,000 investment, the cost savings associated with higher yields, lower wastewater treatment costs, and elimination of replacement costs for the wash solutions resulted in a $272,000 net present value.

11.7.3 Water Scrubbing

The manufacture of a chemical intermediate requires a separation process that involves water scrubbing of a reactor off-gas, followed by extraction of the reactant and product into benzene and high-purity distillation of the reactant and product. The benzene is recovered for recycle, and the aqueous raffinate stream from the extractor is stripped of benzene before discharge to biological treatment. An analysis of the properties of the materials in the system indicated that the product solidifies at 60°C, while the reactant solidifies at –20°C. A pollution-prevention brainstorming session identified that the water-scrubbing and benzene-extraction steps could be replaced with a heat exchanger (operating at 65°C to condense most of the product from the reactor off-gas), followed by a spray column (using a recirculation loop of chilled liquid reactant at –10°C) to condense the reactant and any remaining product from the off-gas.[8] Removing the water scrubber and benzene extraction system resulted in a greater than $5.5 million reduction in capital investment for the new facility.

11.7.4 Nonaqueous Cleaning

In a sold-out market situation, a DuPont intermediates process was operating at 56% of its peak capacity. The major cause of the rate limitation was traced to poor decanter operation. The decanter recovered a catalyst, and its poor operation was caused by fouling from catalyst solids. Returning the process to high utility required a 20-day shutdown. During the shutdown, the vessel was pumped out and cleaned by water washing. The solids and hydrolyzed catalyst were then drummed and incinerated.

An analysis of the process and its ingredients indicated that the decanter could be bypassed and the process run at a reduced rate while the decanter was cleaned. An organic-process ingredient was used to clean the decanter instead, enabling recovery of the catalyst (value of \$200,000 per year). The use of the organic process ingredient cut the cleaning time in half, and that, along with continued running of the process, eliminated the need to buy the intermediate on the open market. The results were a 100% elimination of a hazardous waste (125,000 gallons per year) and a cash flow savings of \$3,800,000 for each avoided shutdown.

11.8 Case Histories: Water as a Heat-Transfer Fluid

11.8.1 Noncontact Cooling

The production of polyester polymers requires high vacuums (less than 2 mm Hg absolute). Multistage steam jets are used to attain these high vacuum levels, normally with direct-contact, water-cooled condensers between each jet stage. For a 4000-lb/h polymer production facility, approximately 165 gallons per minute (gpm) of water is required to condense the steam in the contact condensers. The water was originally once-through ground or river water, passing through the condensers to the wastewater treatment facility. Water shortages forced the installation of cooling towers to cool the water, and the resulting cooling tower blowdown reduced the makeup water flow to 6 gpm (a 96% reduction). A further reduction of water usage can be realized by condensing the steam with shell-and-tube heat exchangers, thus reducing the makeup water flow to less than 2 gpm (an overall 99% reduction in water consumption). For an average cost of water of \$3.15 per 1000 gallons (freshwater plus wastewater treatment costs), the costs for water use were reduced from \$260,000/yr to \$10,000/yr, to as little as \$2000/yr if shell-and-tube heat exchangers were implemented.

11.8.2 Water Monitoring

The J.W. Lees & Co. Greengate Brewery in England was faced with ever-increasing purchased water and municipal sewer costs. A comprehensive survey of the water distribution system within the brewery was undertaken to identify appropriate locations for the installation of additional water flowmeters.[4] The flowmeters recorded that there was significant water usage when the plant was shut down, that the pattern of water use did not correspond to production activity, and that water usage was consistently above that expected for certain plant locations.

For example, it was discovered that solenoid valves on the emergency water supply to a refrigeration plant had failed. In other areas, improved operational techniques reduced water usage to approximately 25% of its original value. These operational improvements led to savings of more than \$50,000 each year in purchased water and sewer costs with no capital expenditures.

11.8.3 Water Reuse for Energy

The most significant waste stream generated at the Amital Spinning Corporation is 320,000 gallons of water used to process 12 dye batches of yarn each day.[6] Water usage and disposal cost the company over \$26,000 per month. Amital purchased three 5,000-gallon tanks at a capital cost of \$13,000 to store hot noncontact cooling water, which was then used in the color kitchen for the preparation of dye liquors. The use of this hot water allows the dye liquors to be prepared at high temperatures; therefore, steam requirements during dyeing are reduced, process water is recovered, and the remaining chemicals are recycled. Not only did they realize a savings in energy and chemical use, but they also realized a reduced cycle time for heating the dye bath. The changes resulted in a 60% reduction in water use and waste generation, a \$13,000 per month cost savings (greater than 1200% IRR), and a production increase from 12 to 20 yarn dye batches per day.

Literature Cited

1. Breske, T. C. 1997. "Industrial Water Conservation, Recycle and Reuse: A Literature Survey." Paper read at The NACE International Annual Conference and Corrosion Show, March 9–14, New Orleans, Paper No. 608, National Association of Corrosion Engineers.
2. Byers, W., W. W. Doerr, R. Krishnan, and D. Peters. 1995. *How to Implement Industrial Water Reuse, a Systematic Approach.* Center for Waste Reduction Technologies. New York: American Institute of Chemical Engineers.
3. Smith, R. 1995. "Wastewater Minimization." *Waste Minimization Through Process Design*, Alan P. Rossiter, Ed. 93–108. New York: McGraw-Hill.
4. Terrell, R., and M. Holmes. 1994. "Is Zero Aqueous Discharge a Practical Option?" *Water Use and Reuse*, David Newton and George Solt, Eds. 1–22. Rugby, U.K.: Institution of Chemical Engineers.
5. Linnhoff, B. August 1994. "Use Pinch Analysis to Knock Down Capital Costs and Emissions." *Chemical Engineering Progress.* 90(8): 32–57.
6. U.S. Environmental Protection Agency. September 1991. *Achievements in Source Reduction and Recycling for Ten Industries in the United States.* EPA/600/2-91/051. Cincinnati, Ohio: U.S. EPA, Risk Reduction Engineering Laboratory.
7. U.S. Environmental Protection Agency. November 1993. *DuPont Chambers Works Waste Minimization Project.* EPA/600/R-93/203. Washington, D.C.: U.S. EPA, Office of Research and Development.
8. Grant, J. J., III, and K. L. Mulholland. June 1993. "Waste Minimization in Process Development." Paper read at the AIChE 1993 Summer National Meeting, Paper 36B. American Institute of Chemical Engineers, New York.

CHAPTER 12

Organic Solvents

12.1 Introduction

Organic solvents are used to dissolve materials for easier processing, to extract materials from a mixture, to create or coat films, and to provide a medium for liquid-phase reactions. Because a large fraction of organic compounds used as solvents fall into one or more of the following categories—flammable, carcinogenic, toxic, volatile, ozone-depleting, contributor to global warming—solvent losses must be controlled and purges safely managed. In this chapter we address approaches that process, project, and environmental engineers can use to reduce or eliminate solvent emissions at the source—a cash-generating opportunity!

In the past, the major drivers for organic-solvent selection were ease of dissolution of the solute (i.e., the material being dissolved), compatibility (i.e., reactivity and affinity), safety (i.e., flammability and toxicity), and cost. However, regulatory restrictions have now become just as important. A looming concern is the environmental fate of waste solvents, especially if destruction forms persistent, toxic, and bioaccumulative byproducts.

12.2 Cost as a Driver for Pollution Prevention

Some of the costs associated with organic solvents may be hard to quantify. In addition to the purchase cost for fresh makeup solvent, there are waste-solvent handling and disposal costs, added costs and investment to ensure safe operation, and environmental permitting costs. Much of the chemical industry buys business-interruption insurance, which is higher if flammable liquids are present. The 3M Company has estimated that solvent purchase costs represent only 20% of the total cost associated with flammable solvents.[1] Additional costs included allocated investment for electrical safety upgrades, additional permitting costs, the cost of environmental resources, and a prorated cost for loss of business based on actual experience with fires. As a result, pollution prevention of organic solvents means increased cash flow for the business.

12.3 Regulations

Gas-phase emissions of organic solvents (often called hazardous air pollutants, or HAPs) have come under increasing regulation in the past 10 years. In the United States, compliance with these regulations is implemented through comprehensive Clean Air Act (CAA) Title V operating permit programs administered by the states. The need to control gaseous emissions will depend on a number of factors, including whether the plant is located in an ozone attainment or nonattainment area; is a major or minor source; is a new or existing source; and whether the vent contains one or more HAPs. Regulations and compliance options will vary from state to state. When controls are mandated, the trend is to require greater than 95% control efficiency. For example, the 1990 CAA Amendments require many new and existing major sources of HAPs to achieve emissions limits reflecting the application of maximum achievable control technology (MACT). MACT requires control efficiencies on the order of 98% for process vents and transfer operations and 95% for solvent storage tanks that are subject to control. However, there are always exceptions.

A good number of solvents (particularly aromatic, halogenated, and nitrogen-bearing organic compounds) that end up in process wastewater become priority pollutants under the Clean Water Act. These compounds must be controlled to extremely low levels in the treated wastewater discharge—typically less than hundreds of micrograms per liter (μg/L).

In addition, solid, liquid, semi-solid, or containerized gas waste streams containing organic solvents will be classified as a "hazardous waste" under the Resource Conservation and Recovery Act (RCRA) if they:

- Exhibit one or more of the four EPA-defined characteristics: corrosiveness, ignitability, reactivity, and toxicity
- Are specifically listed in the regulations, that is, F-, K-, P-, and U-listed wastes
- Are declared hazardous by the generator based on its knowledge of the waste

Many organic-solvent wastes will exhibit the characteristic of ignitability (flash point below 140°F) or toxicity (by failing the toxicity characteristic leaching procedure (TCLP)). The objective of the TCLP is to determine whether the toxic constituents in a solid-waste sample will leach into groundwater if the waste is placed in a solid-waste landfill. Table 12-1 shows the compounds that are included on the TCLP list. In addition, many spent halogenated and nonhalogenated solvents from "nonspecific" sources are listed as F-wastes under RCRA. In general, there is an incentive to select solvents that have a flash point above 140°F and that are not on the TCLP list. Table 12-2 shows typical solvents that have flash points above and below 140°F.

Contact your business environmental specialist(s) for more details on the applicability of the CAA, Clean Water Act, and RCRA regulations to particular solvents or solvent blends.

12.4 The Pollution-Prevention Continuum

The pollution-prevention continuum depicted in Figure 12-1 shows the relative merits of options that are available to process, project, and environmental engineers to minimize or eliminate organic solvent emissions. The decision of how far to move toward a zero waste and emissions design will depend on a number of factors, including corporate and business environmental goals; economics; and applicable regulations. Figure 12-2 shows an organic-solvent selection hierarchy that can be used to help roughly guide solvent selection.

12.5 Source Reduction of Solvents

The highest level of source reduction for solvents is to develop a solventless process. Examples of solventless processes include powder-coating technology to replace solvent-based paints, melt spinning in place of solution spinning, solid-phase reactions instead of homogeneous-solution reactions, and new process chemistries that eliminate the need for solvents as carriers for solids and as wash liquids.

The next step down in the hierarchy is to identify a process intermediate, product, or feed material that can be used as a solvent. For example, can intermediate monomers be used as a solvent to add catalysts, initiators, and inhibitors to polymerization reactors? Can chilled, low-volatility reactants be used to scrub off-gas streams to recover valuable reactants and products, instead of introducing a new organic solvent?

The third step down in the hierarchy is to use water as

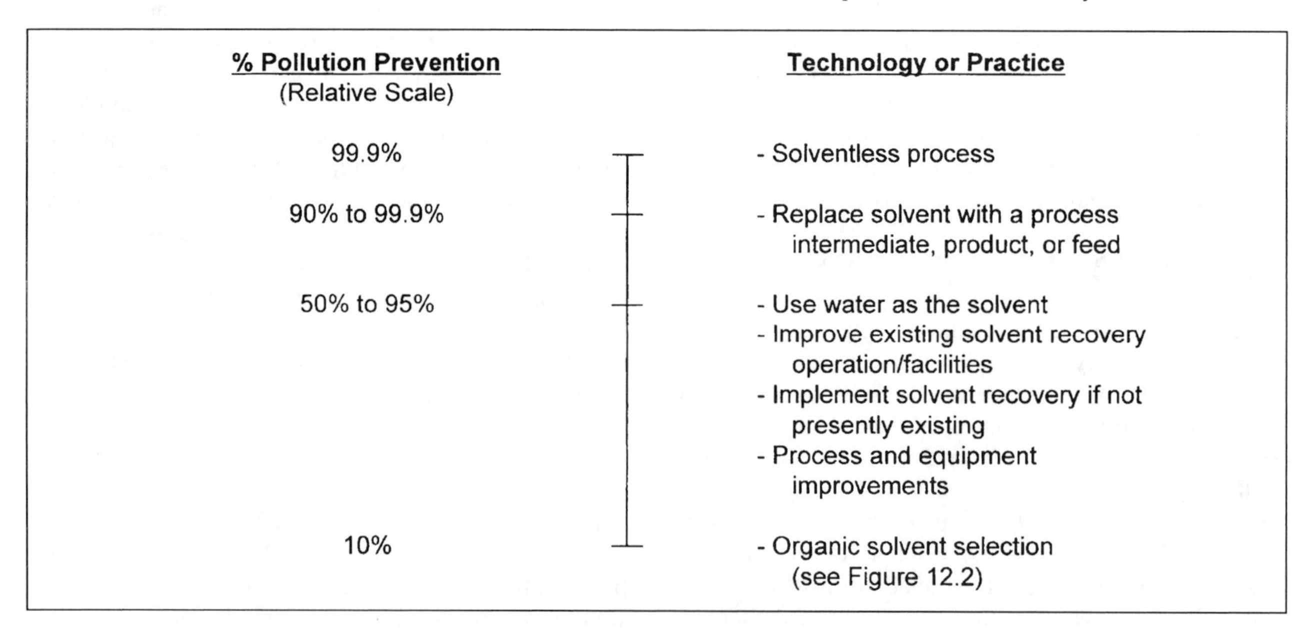

Figure 12-1. The Pollution Prevention Continuum For Organic Solvents

Compound	ID No.	Regulatory level in TCLP extract (mg/L)
Arsenic	(D004)	5.0
Barium	(D005)	100.0
Benzene	(D018)	0.5
Cadmium	(D006)	1.0
Carbon tetrachloride	(D019)	0.5
Chlordane	(D020)	0.03
Chlorobenzene	(D021)	100.0
Chloroform	(D022)	6.0
Chromium	(D007)	5.0
o-Cresol*	(D023)	200.0
m-Cresol*	(D024)	200.0
p-Cresol*	(D025)	200.0
2,4-D	(D016)	10.0
1,4-Dichlorobenzene	(D027)	7.5
1,2-Dichloroethane	(D028)	0.5
1,1-Dichlororethylene	(D029)	0.7
2,4-Dinitrotoluene	(D030)	0.13#
Endrin	(D012)	0.02
Heptachlor (and its hydroxide)	(D031)	0.008
Hexachlorobenzene	(D032)	0.13#
Hexachloro-1,3-butadiene	(D033)	0.5
Hexachloroethane	(D034)	3.0
Lead	(D008)	5.0
Lindane	(D013)	0.4
Mercury	(D009)	0.2
Methoxychlor	(D014)	10.0
Methyl ethyl ketone	(D035)	200.0
Nitrobenzene	(D036)	2.0
Pentachlorophenol	(D037)	100.0
Pyridine	(D038)	5.0#
Selenium	(D010)	1.0
Silver	(D011)	5.0
Tetrachloroethylene	(D039)	0.7
Toxaphene	(D015)	0.5
Trichloroethylene	(D040)	0.5
2,4,5-Trichlorophenol	(D041)	400.0
2,4,6-Trichlorophenol	(D042)	2.0
2,4,5-TP (Silvex)	(D017)	1.0
Vinyl Chloride	(D043)	0.2

* If o-, m-, and p-cresol concentrations cannot be differentiated, the total cresol concentration is used. The regulatory level for total cresol (D026) is 200 mg/L.

\# Quantitation limit.

Table 12-1. Compounds on the TCLP List

Compound	Boiling Point (°F)	Flash Point (°F)
Flash Point < 140°F		
Acetone	133	0
Benzene	176	12
Cyclohexane	179	-4
n-Decane	345	115
Ethyl acetate	24	24
Ethyl alcohol	173	55
Ethyl ether	95	-49
d-Limonene	351	113
Methyl acetate	14	14
Methyl alcohol	147	52
Methyl ethyl ketone	21	21
n-Pentane	97	-40
Propyl alcohol	207	77
Toluene	231	40
o-Xylene	292	90
Turpentine	300	95
Flash Point > 140°F		
Aniline	363	158
Benzyl chloride	355	153
Cyclohexanol	321	154
Dibutyl sebacate	660	352
Diethyl malonate	390	199
Diethyl oxalate	365	151
Diethyl succinate	423	194
Diethyl sulfate	409	158
Dimethyl sulfate	372	181
Dodecane	421	165
Ethylene glycol	387	232
1-Hexanol	315	145
1-Methylcyclohexanol	334	149
N-methyl-2-pyrrolidone	400	187
Nitrobenzene	411	190
Tetradecane	488	212
Trimethyl phosphate	387	194

Table 12-2. Flash Points for Typical Organic Solvents

Nonpolar Organics	Polar Organics
Aliphatics (Flash Point above 140°F)	Alcohols
Aliphatics (Flash Point below 140°F)	Organic Acids
Aromatics	Other Oxygenates
Halohydrocarbons	Nitrogen-bearing Organics
1. HCFCs/HFCs	Halogenated Organics
2. Chlorocarbons	Chlorinated Organics
3. CFCs	
Listed or Acutely Toxic Compounds	

Figure 12-2. Organic Solvent Selection Hierarchy (in order of decreasing desirability)

the solvent. Water itself is not a pollutant; however, materials dissolved in the water will often require treatment. The waste-reduction potential is then highly dependent on contaminant solubility in the water.

Solvent-recovery operations exist in most plants that use large volumes of solvent; however, many batch processes producing high-value products do not consider solvent recovery to be worth the investment. In today's economic and environmental climate, however, all solvents should be recovered at the highest efficiency, whenever possible.

Organic solvent losses can often be reduced by process and equipment modifications. Examples include closed-loop addition of ingredients to batch processes, welded instead of gasketed flanges, and improved seals on vessel lids, agitators, pumps, and valves.

Finally, the trend in industrial solvent selection is toward environmentally friendly chemicals; that is, toward water-based and hybrid solvents that are non-ozone-depleting, noncarcinogenic, low vapor pressure, biodegradable, nonpersistent, nonbioaccumulative, and not on a United States EPA regulatory list. Particularly in cleaning applications, preference is given to solvents with flash points greater than 200°F. The move appears to be away from halogenated and aromatic compounds and toward solvents such as *n*-methyl-2-pyrrolidone, dibasic esters, low-vapor-pressure terpenes (e.g., *d*-limonene), and aliphatic hydrocarbons.

Another environmental concern is a solvent classified as a dense (more dense than water), non-aqueous phase (not soluble in water) liquid (DNAPL). When spilled on the ground, DNAPLs quickly travel downward in the soil, become trapped in subsurface pockets, and are then very difficult to remediate. Experience shows that it costs $50–$1000 per cubic yard to remediate DNAPL-contaminated soil.

12.6 Solvent Selection

In many applications, conventional organic solvents are still highly desirable because of familiarity, low cost, ease of handling, and ease of disposal. The selection of a solvent involves a careful balance of the optimum physical and chemical properties, workplace safety (e.g., toxicity and flammability), cost, and environmental regulations. Solvent-selection systems described in the "Solvent Selection Databases" section below are currently limited in scope; however, the development of more sophisticated selection software is in progress. To learn more about the development status of these software systems, contact the developers of the software directly.

Because there are no exact drop-in replacements, and until more rigorous solvent-selection software programs are available, the following series of steps can be used to select a solvent.[2] In short, there is nothing better than a well-rounded, environmentally conscious organic chemist when it comes to solvent selection.

12.6.1 Step 1: Identify Important Chemical and Physical Solvent Properties

The first step is to identify the physical and chemical properties that the solvent must possess for a given application. If a solvent is already in use, use its properties as the initial list of desirable and undesirable properties.

Performance Properties. A major use for solvents is to facilitate separation via liquid–liquid extraction, azeotropic distillation, crystallization, and absorption. A solvent's dissolution capacity, selectivity for the solute, and distribution between the vapor and liquid phases are important properties that can all be related to a solvent's activity coefficients.

These same performance properties are also applicable to many cleaning solvents, particularly a high capacity for the soil (solute) being removed. On the other hand, cleaning systems that use surfactants want a solvent that has a very low capacity for the soil. In these systems, the surfactant is attracted to the surface being cleaned, rather than to the soil, thus displacing the soil from the surface. Due to the low attraction between the soil and solvent, an easy separation results.

Distillation is a key technology for solvent recovery. In a separation process where the chosen solvent must be easily distilled from the extracted solute, the relative volatility should be high to ensure easy separation, and the enthalpy of vaporization should be low to minimize energy consumption.

Desired performance properties for solvents can be obtained from solvent manufacturers, the open literature, reference handbooks, such as the *Merck Index*,[3] *Perry's Chemical Engineers' Handbook*,[4] and Reid, Prausnitz, and Poling's *The Properties of Gases and Liquids*,[5] and the Design Institute for Physical Properties Research (DIPPR) physical properties database,[6] which is available through the American Institute of Chemical Engineers.

Flash points, explosion limits, toxicity, and so on, can be obtained from Material Safety Data Sheets (MSDS) for each solvent.

Reactivity Properties. The majority of applications require solvents to be stable and unreactive, which includes stability in the presence of air and water. Reactivity in the environment, or lack thereof, must also be considered. An example is chlorinated solvents that are unreactive until they reach the upper atmosphere, where they catalyze ozone degradation.

Cost and Other Considerations. The total cost of a solvent includes the sum of purchase, recovery, permitting, and disposal costs. Other application-specific properties that might need to be considered are density, viscosity, surface tension, enthalpy of combustion, freezing point, purity, and market availability.

12.6.2 Step 2: Identify Constraints on Important Solvent Properties

Environmental, Health, and Safety Constraints. Even though equipment design, operating procedures, and operator training can address environmental, health, and safety constraints, the chemical solvents themselves must be inherently harmless with respect to the following properties: biochemical oxygen demand (BOD), global-warming potential, ozone-degradation potential, vapor pressure, toxicity, explosion limits, and flash point.

For example, when choosing a solvent, consideration must be given to the degree of hazard posed by flammable and combustible liquids. The Occupational Safety and Health Administration (OSHA) defines a flammable liquid (Class I) as a liquid having a flash point below 100°F and having a vapor pressure not exceeding 40 psia at 100°F. A combustible liquid is a liquid with a flash point at or above 100°F. Combustible liquids are further subdivided as Class II (100°F ≤ flash point < 140°F), Class IIIA (140°F ≤ flash point < 200°F), and Class IIIB (flash point ≥ 200°F).

Liquids having flash points at or above 100°F may be treated as nonhazardous, unless they are heated to temperatures at or above their flash points. (This is different from the characteristic "hazardous waste" classification under RCRA due to ignitability (flash point below 140°F) of a solvent waste as described in Section 12.3.) Hazardous locations require special precautions, such as explosion-proof equipment and wiring. Such installations average two to three times the cost of similar installations in nonhazardous (e.g., general-purpose) locations.

All else being equal, there is a strong incentive to (1) choose a solvent with a flash point at or above 140°F, and (2) for liquids with flash points at or above 140°F, to choose the liquid with the highest flash point.

Regulatory Constraints. The presence or absence of a solvent on various regulatory lists, for example, the U.S. EPA's EPCRA 313 list, can either restrict or increase the cost of its use. Section 12.7, "Solvent Selection Databases," describes resources that can be used to identify which regulations will impact the use of a specific solvent.

12.6.3 Step 3: Compile Data for all Properties and Rank Solvents Satisfying the Target Constraints

Once the physical and chemical properties are determined for each candidate solvent, they are compared against the constraints listed in Step 2; only those satisfying the constraints are considered further. If all candidates are rejected, however, the constraints will then need to be reviewed to determine if they are too restrictive or conflict with one another.

When several candidates satisfy the constraints, they must be ranked. If cost is a key criterion, be sure to include not only purchase price but also the associated recovery, permit, safety, and disposal costs.

12.6.4 Step 4: Evaluate the Top Solvent Candidates Using Simulation

Chemical process simulators, such as Aspen Technology's ASPEN PLUS (Cambridge, MA), Simulation Sciences' PRO/II (Brea, CA), and Hyprotech, Ltd.'s HYSIM (Calgary, Alberta, Canada), can be used to analyze the performance of the top-ranked organic solvent(s). However, other "simulations" of solvent storage, transportation, regulation, disposal, and so forth, must also be considered. In many cases, bench-scale or field testing will be required.

12.7 Solvent-Selection Databases

The U.S. EPA has created a free, public, integrated environmental information system for the Internet called Enviroene. Accessible from within Enviroene is a Solvent Substitution Data System (SSDS) that contains six solvent substitution databases. These include:

- Solvent Alternatives Guide (SAGE)
- Hazardous Solvent Substitution Data System (HSSDS)
- Integrated Solvent Substitution Data System (ISSDS)
- Department of Defense Ozone Depleting Chemical/Substance Information
- Solvents Database (SOLV-DB)
- Solvent Handbook Database System (SHDS)
- Materials Compatibility Database (MATCOMPAT)

SAGE assists in the selection of nonpolluting industrial surface-cleaning alternatives. It is a PC-based logic-tree system that evaluates the user's present operating scenario and then identifies possible surface-cleaning alternative solvent chemistries and processes that best suit the defined operating and material requirements. Subsequent versions will incorporate additional surface-cleaning requirements such as paint stripping, electronic manufacturing, machinery, and printing equipment cleaning.

The HSSDS is an on-line, comprehensive system of information on alternatives to hazardous solvents and related subjects. The HSSDS allows scanning of technical product data sheets provided by manufacturers, provides a link to vendor-supplied material safety data sheets for each product, and allows the user to perform keyword and topic searches.

NCMS databases contain information on environmental fate, health and safety data, regulatory status, chemical/physical properties, and suppliers for over 320 pure solvents and trade name mixtures. The emphasis is on alternatives to ozone-depleting chemicals used in manufacturing processes such as degreasing, painting, paint removal, circuit board assembly and cleaning, parts cleaning, and maintenance. The database is distributed on floppy disks.

In addition, the EPA and the University of Cincinnati are currently developing a new solvent substitution database called Program for Assisting the Replacement of Industrial Solvents (PARIS). PARIS will match the attributes of a multicomponent solvent mixture with a "designer" mixture containing components of lower toxicity. The program will most likely be licensed to a third party when available.

The Coating Alternatives Guide (CAGE), a PC-based system, is designed to allow users of coatings to easily find technically innovative, cost-effective, and low-polluting coatings for specific applications. CAGE currently addresses alternative, low-emitting coatings for metal and plastic-part coating operations. Future versions of CAGE will include alternative coatings for wood and other substrates.[7]

12.8 Case Histories

12.8.1 Solventless Process

At a DuPont facility, a higher alcohol was washed with an aqueous solution containing inorganic chemical scavengers to remove residual acidity. Isopropyl alcohol was added to the solution to assist separation of the alcohol product from the wash water. The water wash and the isopropyl alcohol addition steps were eliminated by neutralizing the residual acidity with a chemical agent added directly to the higher alcohol product instead. This process change reduced the aqueous waste load by 100%, and realized an internal rate of return (IRR) of 93%.[8]

Garden Way, a manufacturer of outdoor power equipment, required that the paints and finishes endure the rigors of outdoor use. For this reason, the paints had a lead content exceeding 10,000 ppm, which was above the regulatory limit of 5 ppm. The manufacturer switched to powder paint technology, which not only formed durable nonlead coats, but also reduced hazardous waste and volatile organic compound (VOC) emissions by 95%.[9]

12.8.2 Replace Solvent with a Process Intermediate, Product, or Feed

At a DuPont site, spent solvents used in the manufacture of intermediate monomers were incinerated as a hazardous waste. Alternative nonhazardous solvents were considered and rejected. However, the intermediate monomers were found to have the dissolution capacity of the original solvents and could replace them. By utilizing existing equipment, realizing savings in ingredients recovery, and reducing operating and incineration costs, the project achieved a 33% IRR and a 100% reduction in the use of the original solvents.

12.8.3 Water as a Solvent

At a DuPont facility, a surfactant (a waxy solid in its pure state) was sold as a solution in CFC-113. Because CFC-113 was found to contribute to ozone depletion in the upper atmosphere, the customer demanded a non-CFC solvent. A study found that water could be used in place of the original solvent with no new investment. By switching to a water solvent, 100% of the CFC emissions were eliminated and the project realized an IRR of 13%.[8]

At a DuPont facility, tars generated in a purification process were blended with a chlorinated solvent, and then sent off-site for incineration. The chlorinated solvent was eliminated by generating a water dispersion instead. The tar is filtered from the water and then burned for its fuel value. This process change resulted in a 100% reduction in the use of a chlorinated solvent and elimination of off-site incineration—2 million lb/yr of hazardous waste incinerated at a cost of $1.5 million per year.[8]

Amko Plastics, a decorative printer of packaging for consumers, retail stores, and industry, switched from solvent-based inks to water-based inks. The switch required a number of process and equipment modifications, which required additional new investment and operating costs. However, the company did reduce its VOC emissions by more than 85%, and has developed a recovery system for the ink-press wastes that eliminates the costly disposal of a hazardous waste. The payback for the recovery system was only one year.[9]

12.8.4 Improve Solvent Recovery

Equipment startups and shutdowns are frequent sources of waste. When process equipment starts up, there is usually a line-out period before the equipment operates as designed. At a DuPont site, the waste from a methanol/water distillation column was minimized by recirculating water back to the column until it reached its operating temperature. The project required modest investment in a tank and pump, but realized an IRR of 103% through solvent savings and reduced waste treatment costs.[8]

12.8.5 Implement Solvent Recovery

At a DuPont site, a mixture of solvents—tetrahydrofuran, toluene, and water—was disposed of as a hazardous waste. Two distillation columns were installed to separate the mixture into its pure components for recycle. Recovery and reuse of the solvents reduced raw-material and waste-disposal costs, with savings totaling approximately $6 million annually or $1/pound recovered solvent.[8]

In the pharmaceutical industry, waste disposal can constitute an appreciable fraction of the cost of bulk drug manufacture. Several examples of how solvent usage has been reduced are summarized below:[10]

- A process step requiring two solvents—methyl isobutyl ketone and acetonitrile—was modified to use only acetonitrile, which could then be recycled back to the process.
- A synthesis route was modified to eliminate use of methylene chloride, and then to recover and reuse 85% of the solvents.
- Toluene was replaced with isopropyl acetate, while also implementing a scheme to recover 90% of the isopropyl acetate solvent.
- Pharmaceutical granulation and coating operations were modified to eliminate using solvents such as chloroform, acetone, and methanol.
- Solvent recovery by distillation had been considered and rejected as possibly leading to impurity accu-

mulation. However, distillation was later implemented after disposal costs became excessive. The irony is that the quality of the recovered solvent was higher and less variable than the purchased solvent.

12.8.6 Process and Equipment Improvements

At a DuPont film-coating facility, a chlorocarbon had been the manufacturing and cleaning solvent of choice. Required federal emissions reporting showed a high rate of emissions, thus prompting a careful examination of the solution preparation procedures. The major source (70%) of emissions was from process tank lid and seal leaks. Upgrading the lid and shaft seals, along with procedural and other process changes, reduced the emissions by more than 90%. The upgrades required minimal investment. The result was $500,000 per year in additional revenue to the business from reduced solvent purchases.

12.8.7 Organic Solvent Selection

At a DuPont site, the manufacture of an automotive part involved degreasing, spray painting, and final cleaning. A new degreasing operation replaced the CFC-113 solvent and involved wiping the parts with a mixture of water and a biodegradable detergent. The spray painting and final cleaning operations were improved through minor equipment changes and better tracking of paint usage. The project required essentially no investment, reduced waste generation by 68%, and achieved significant ingredients savings of more than $500,000 per year.

Literature Cited

1. Maves, F. L. 1995. "Total Known Cost of Solvents." Paper read at 210th ACS National Meeting, August 20–24, Chicago, Ill.
2. Joback, K. G. 1994. "Solvent Substitution for Pollution Prevention. Pollution Prevention via Process and Product Modifications." *AIChE Symposium Series*, M. M. El-Halwagi and D. P. Petrides, Eds. 90(303): 98–103.
3. Budavari, S. 1996. *The Merck Index: An Encyclopedia of Chemicals, Drugs, and Biologicals,* 12th ed. Whitehouse Station, N.J.: Merck & Co.
4. Liley, P. E., R. C. Reid, and E. Buck. 1984. "Physical and Chemical Data." *Perry's Chemical Engineers' Handbook*, 6th ed., R. H. Perry and D. W. Green, eds. 3-1–3-291. New York: McGraw-Hill.
5. Reid, R. C., J. M. Prausnitz, and B. E. Poling. 1987. *The Properties of Gases and Liquids*, 4th ed. New York: McGraw-Hill.
6. Design Institute for Physical Property Data. 1996. *DIPPR Data Compilation of Pure Compound Properties*. New York: American Institute of Chemical Engineers.
7. Cornstubble, D. R., J. N. Baskir, and M. Kosusko. 1996. "A Personal Computer Guide for Selecting Alternative Coatings for Metal Parts and Products Painting." *Proceedings of the 1996 A&WMA 89th Annual Meeting and Exhibition*, pp 1–13. Nashville, Tenn.: Air & Waste Management Association.
8. U.S. Environmental Protection Agency. November 1993. *DuPont Chambers Works Waste Minimization Project.* EPA/600/R-93/203. Washington, D.C.: U.S. EPA, Office of Research and Development.
9. Tillman, J. W. September 1991. *Achievements in Source Reduction and Recycling for Ten Industries in the United States.* EPA/600/2-91/051. Cincinnati, Ohio: U.S. EPA, Risk Reduction Engineering Laboratory.
10. Venkataramani, E. S., F. Vaidya, W. Olsen, and S. C. Wittmer. November 1992. "Create Drugs: Case Histories of One." *Chemtech.* 674–679.

CHAPTER 13

pH Control as a Pollution-Prevention Tool

13.1 Introduction

It is not uncommon for the absence of pH or poor pH control to be at the heart of an environmental problem. The pH of an aqueous system can affect byproduct formation in chemical reactions, excess alkali consumption and inorganic salt formation in acid gas (e.g., HCl, HBr) scrubbers, the solubility of organic acids and bases and metals in wastewater streams, and air emissions of volatile organics from wastewater treatment systems. For this reason, engineers need to understand the aqueous chemistry of the system at hand, and how it impacts waste generation and emissions.

In this chapter we discuss pH control as a pollution-prevention tool. The discussion will not focus on the "how to" of pH control, but on situations where pH control, or lack thereof, can significantly affect waste generation and emissions.

13.2 Pollution-Prevention Strategies

The pH of an aqueous system is a measure of the activity of the hydrogen ion in solution, that is, $\{H^+\}$, where activity is the product of concentration ($[H^+]$ in moles/kg H_2O) and the aqueous-phase activity coefficient, γ_{H+}. From a pollution-prevention perspective, the goal is most often to identify a forgiving pH region of operation that minimizes byproduct formation in reactors, the solubility or volatility of organic compounds in aqueous-waste streams, the solubility of metals, and the consumption of mineral acids and bases to control pH.

13.2.1 Understand Whether Your Plant Adds Acid or Base to Control pH

Look for the opportunity to reduce pH control chemical demand. For example, the cost to neutralize hydrochloric acid is about $200/ton acid. Question all mineral-acid and -base additions to the manufacturing process. Consider the use of waste acid or base from another part of the plant or *an acid or base that is generated as a byproduct within the process.* This strategy saves money twice by reducing ingredient cost and the amount of base/acid ultimately used downstream for neutralization.

In a pollution-prevention brainstorming session for a process that makes an agricultural product, a concept was developed to convert Br_2, a waste byproduct of the reaction chemistry, to its acid form, HBr. The recovered HBr would then be recycled back to the same reactor in place of the mineral acid, H_2SO_4, which was being added as a new constituent to the process. The benefit of not adding H_2SO_4 would be to significantly reduce caustic consumption for acid neutralization (2 moles NaOH required per mole H_2SO_4).

In some instances, it may be possible to take advantage of CO_2 stripping to reduce base consumption in a process. For example, the first stage of a two-stage organic oxidation process can be satisfactorily operated between pH 6 and 8.3. The aqueous feed stream to the process contains waste mineral acid and a moderate concentration of weak organic acids. In the first-stage oxidation reactor, organic acids are partially degraded to CO_2 and H_2O. In the second oxidation stage, which must operate at pH 8.3, further destruction of the organic acids occurs. Two options are available: (1) operate the first-stage reactor at pH 8.3 with all the base being added to the first-stage reactor, and (2) split base addition between the first-stage reactor and a separate downstream pH adjustment step; that is, operate the first-stage reactor at pH 6, then adjust pH to 8.3 with base in a separate step before the second oxidation stage. Which option results in less overall base consumption?

Figure 13.1 shows *total required* NaOH consumption and CO_2 stripping rate in the first-stage reactor as a function of first-stage reactor pH for a typical wastewater feed to the process. Notice that total required NaOH consumption is approximately 40% less (25 lb-mole/h vs. 40 lb-mole/h) if the first-stage reactor is run at pH 6 (which maximizes CO_2 stripping), followed by pH adjustment to 8.3 in a separate step. In effect, the stripping of CO_2 acid gas acts as a "base" in the process.

13.2.2 Avoid Excessive Carbon Dioxide Scrubbing in Caustic Scrubbers

Scrubbers can often be operated at pH 8 or less and still remove chlorine, bromine, sulfur dioxide, and hydrochloric acid. Operating at higher pH wastes caustic by scrubbing carbon dioxide in addition to the target acid gases. For example, at pH 10 and typical carbon dioxide partial pressures found in oxidizer off-gases, the incremental caustic consumption for scrubbing carbon dioxide is \$100–\$700/yr per standard cubic foot per minute (scfm).

Consider a waste gas containing 1000 parts per million by volume (ppmv) HCl, carbon dioxide, nitrogen, and oxygen feeding a caustic scrubber. Figure 13-2 shows the moles of NaOH consumed per mole of HCl acid gas as a function of pH and CO_2 partial pressure. Much above pH 6, caustic consumption rises dramatically, as additional NaOH is needed to convert $CO_{2(aq)}$ (plus some $H_2CO_{3(aq)}$) to bicarbonate ion (HCO_3^-) and, ultimately, to carbonate ion (CO_3^{2-}) at a pH greater than 9 to 10. In this example, the HCl partial pressure in the scrubber off-gas is low enough at pH 4 that operating at higher pH consumes additional caustic only to scrub carbon dioxide. The goal, then, is to operate caustic scrubbers at a pH that maximizes acid-gas removal, while minimizing CO_2 removal.

13.2.3 Know the pK_a and pK_b of Weak Organic Acids and Bases in Your Process and Their Impact on Waste Emissions.

The solubility and volatility of weak organic acids and bases in water changes with pH. First, consider the solubility in water of a solid, salable product that is a weak organic acid. The objective is to maximize recovery of the product to minimize yield loss and emissions, or, in other words, to minimize the solubility of the salable product in water. The equilibrium reactions for the dissolution of the weak organic acid are given by[1]

$$HA_{(s)} = HA_{(aq)} \quad (1)$$

$$HA_{(aq)} = A^- + H^+ \quad (2)$$

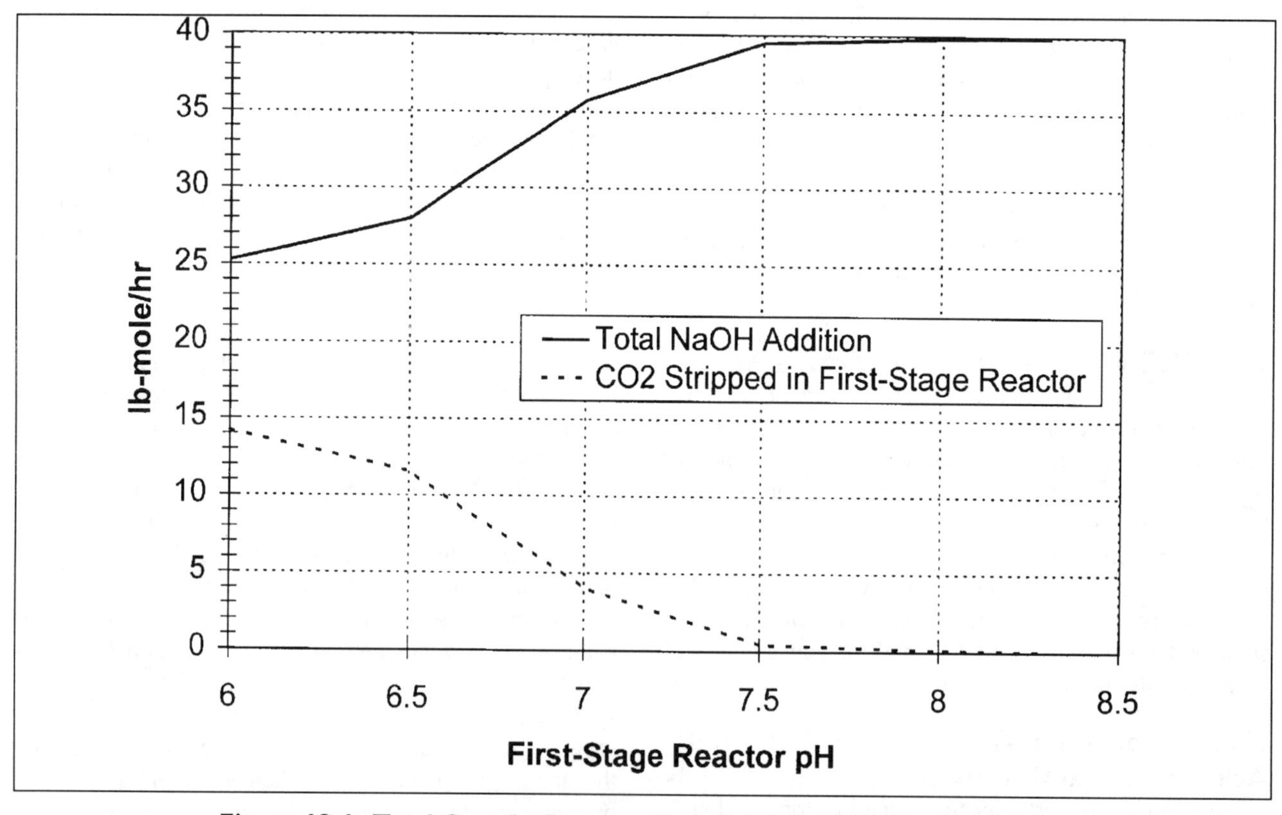

Figure 13-1. Total Caustic Consumption and CO_2 Stripping Rate from First-Stage Oxidation Reactor as a Function of First-Stage Reactor pH

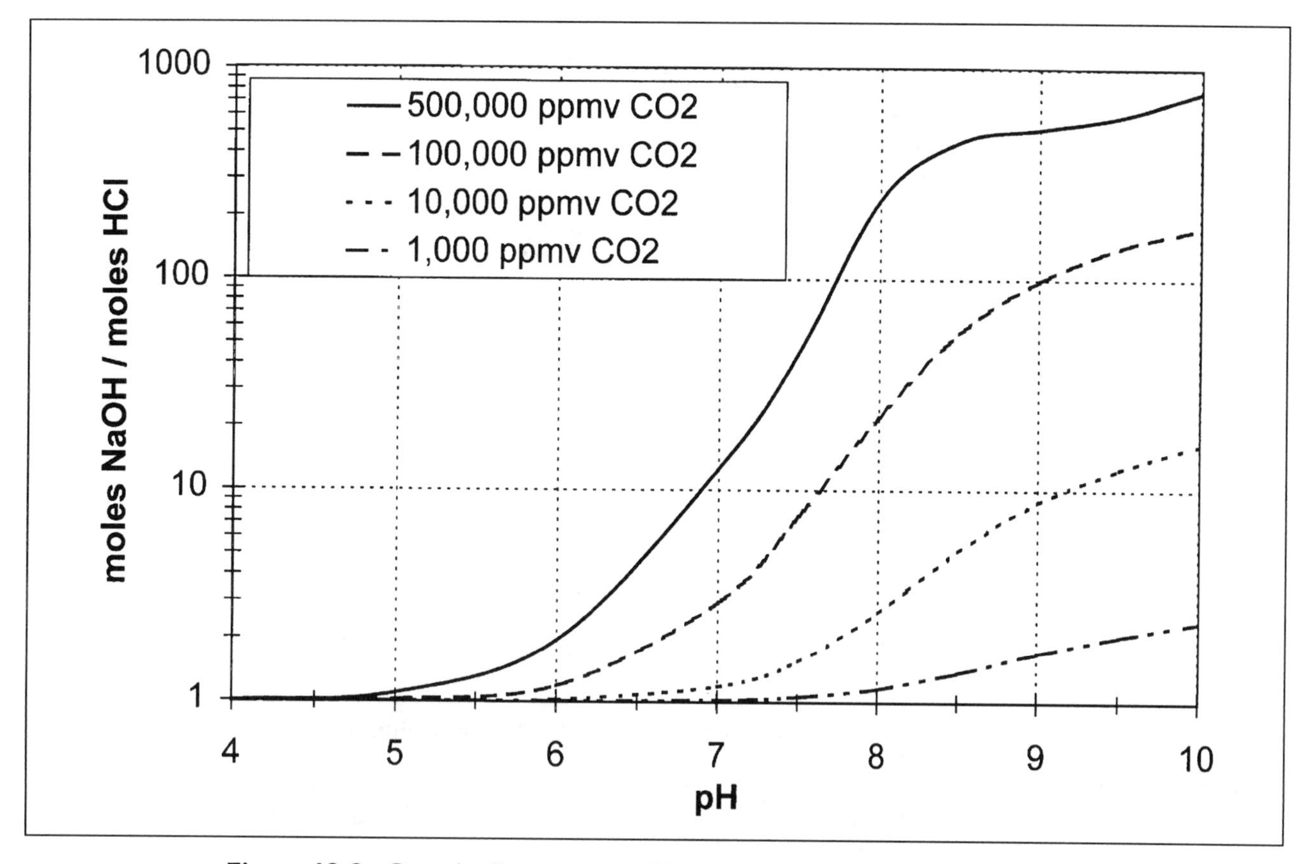

Figure 13-2. Caustic Consumption Versus pH and CO_2 Partial Pressure for an Acid Gas Scrubber Treating 1,000 ppmv HCl at 30°C and 1 atm

It is evident from these equilibrium reactions that the concentration of $HA_{(aq)}$, and hence the "solubility" of $HA_{(s)}$, are both affected by the hydrogen-ion activity (which is given by the pH of the system). Reducing the hydrogen-ion activity (that is, raising the pH) shifts the equilibrium of both Equations 1 and 2 to the right. As a result, the "solubility" or total concentration of HA in solution (that is, the sum of $[HA_{(aq)}]$ and $[A^-]$) increases with increasing pH.

The pH where $[HA_{(aq)}] = [A^-]$ is given by the pK_a for the organic acid, where K_a is the equilibrium dissociation constant for the reaction: K_a and pK_a are given by

$$K_a = \frac{\{A^-\}\{H^+\}}{\{HA_{(aq)}\}} \tag{3}$$

$$pK_a = -\log(K_a) \tag{4}$$

One-half pH unit below the pK_a, $[HA_{(aq)}] >> [A^-]$, while 0.5 pH units above the pK_a, $[A^-] >> [HA_{(aq)}]$. Minimizing the solubility of $HA_{(s)}$, then, requires operating *at least* 0.5 pH units *below the pK_a* for the weak organic acid. Figure 13-3 shows the solubility of a four-carbon dicarboxylic acid as a function of pH. Note how the solubility of the solid increases substantially at a pH above the first pK_a for the acid, pK_{a1}. From a pollution-prevention perspective, operating at pH less than 3.5 would minimize the organic-acid solubility in water.

Similarly, for weak organic bases, solubility in water will be minimized at a pH at least 0.5 pH units *greater than (14 - pK_b)*. K_b is the equilibrium constant for the reaction

$$B_{(aq)} + H_2O = HB^+ + OH^- \tag{5}$$

where

$$K_b = \frac{\{HB^+\}\{OH^-\}}{\{B_{(aq)}\}} \tag{6}$$

$$pK_b = -\log(K_b) \tag{7}$$

At a pH less than (14 - pK_b), the protonated form of the base (HB^+) will dominate, increasing solubility substantially. The most common organic bases are the amines.

In a process for making an agricultural product, a batch reactor produces a water slurry containing the desired product. The water slurry is filtered to capture the product

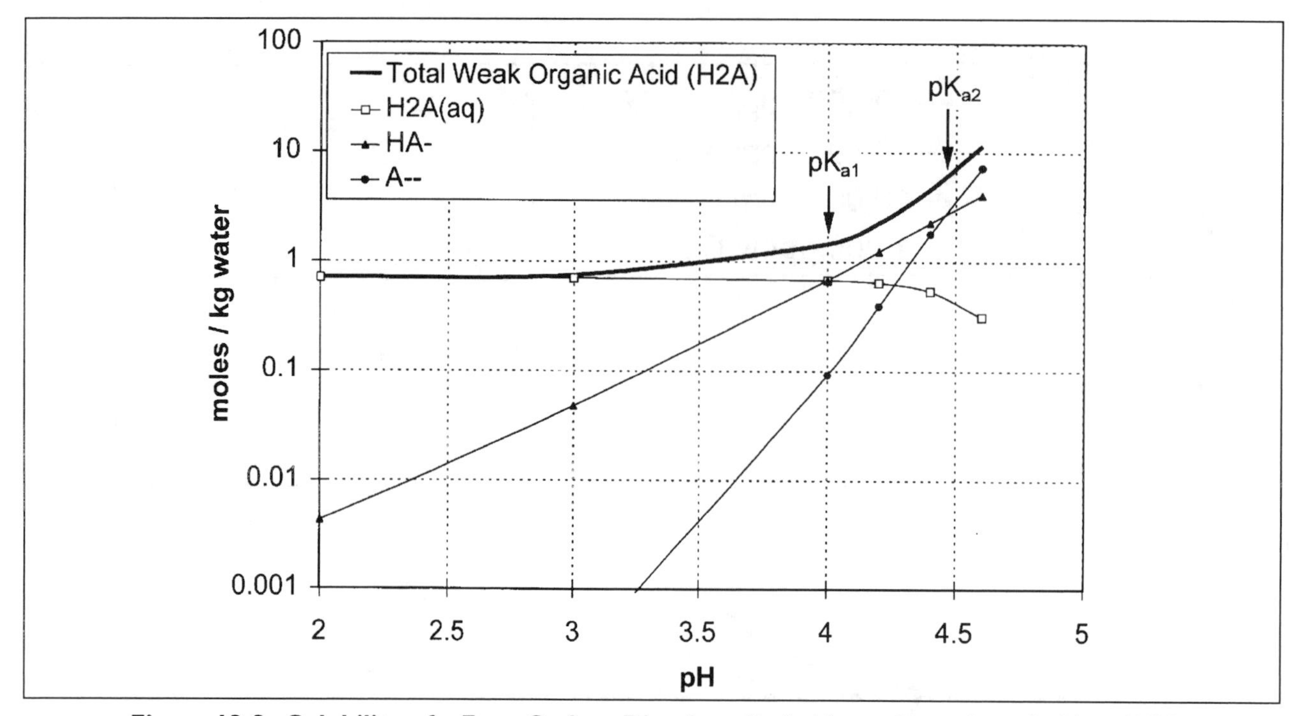

Figure 13-3. Solubility of a Four-Carbon Dicarboxylic Acid as a Function of pH at 25°C

solids, followed by drying and packaging. The aqueous filtrate from the filter is sent to wastewater treatment. It is known that the concentration of the desired product in the filtrate increases dramatically if the reactor pH rises much above 8. This is because the product is a weak organic acid with a pK_a of 7.7. At pH 8 or above, the deprotonated form of the acid will dominate and increase product solubility. In fact, in one instance, the reactor pH was high enough that the majority of the product solids were solubilized and passed right through the filter cloth. These pH excursions have been reduced substantially through better pH control in the reactor.

Second, consider the volatility of volatile organic acids as a function of pH. At a pH below the pK_a for the organic acid, the protonated form, $HA_{(aq)}$, will dominate. In this case, the volatility will be determined by the Henry's law constant for that compound in water. At a pH above the pK_a, however, the Henry's law constant is no longer indicative of the true volatility of the acid. The deprotonated form (A^-), which now dominates, is a nonvolatile, ionic species. As a result, the vapor pressure of HA above the aqueous solution is dramatically reduced at a pH above the pK_a for the acid. The equilibrium reactions for this scenario are

$$HA_{(v)} = HA_{(aq)} \quad (8)$$

$$HA_{(aq)} = A^- + H^+ \quad (9)$$

where K_a and pK_a are given by Equations 3 and 4. An example is given in Figure 13-4 for a common two-carbon carboxylic acid. The basis for Figure 13-4 is the stripping of the organic acid with air at 25°C from a water stream containing 10,000 parts per million by weight (ppmw) organic acid.

13.3 Metals Solubility and pH

The presence of trace metals in the aquatic environment is no longer just the result of natural weathering processes on soils and bedrocks. Because of humankind's ingenuity over the last several centuries, a large fraction of all trace metals in the environment now enter as air and wastewater emissions from industrial sources. In aquatic systems, these trace metals may be present as both insoluble and soluble forms within the water, sediments, and biotic tissues.[2] What distinguishes metals from organic pollutants is that although metals can be transformed by such processes as hydrolysis and oxidation, they cannot be destroyed.

The flux, distribution, and accumulation of trace metals in the aquatic environment present a potential hazard to human, animal, and plant health and aquatic life regardless of the source.[3] More than any other factor, toxicity is driving the need for metals' removal from industrial wastewaters. One noteworthy response to increasing toxicity concerns has been increasingly stringent state and federal regulations in the United States as well as abroad. For example, newer U.S. regulations, such as the federal National Toxics Rule and state water-quality standards, are pushing metal effluent limits to parts per billion (ppb) levels.[4]

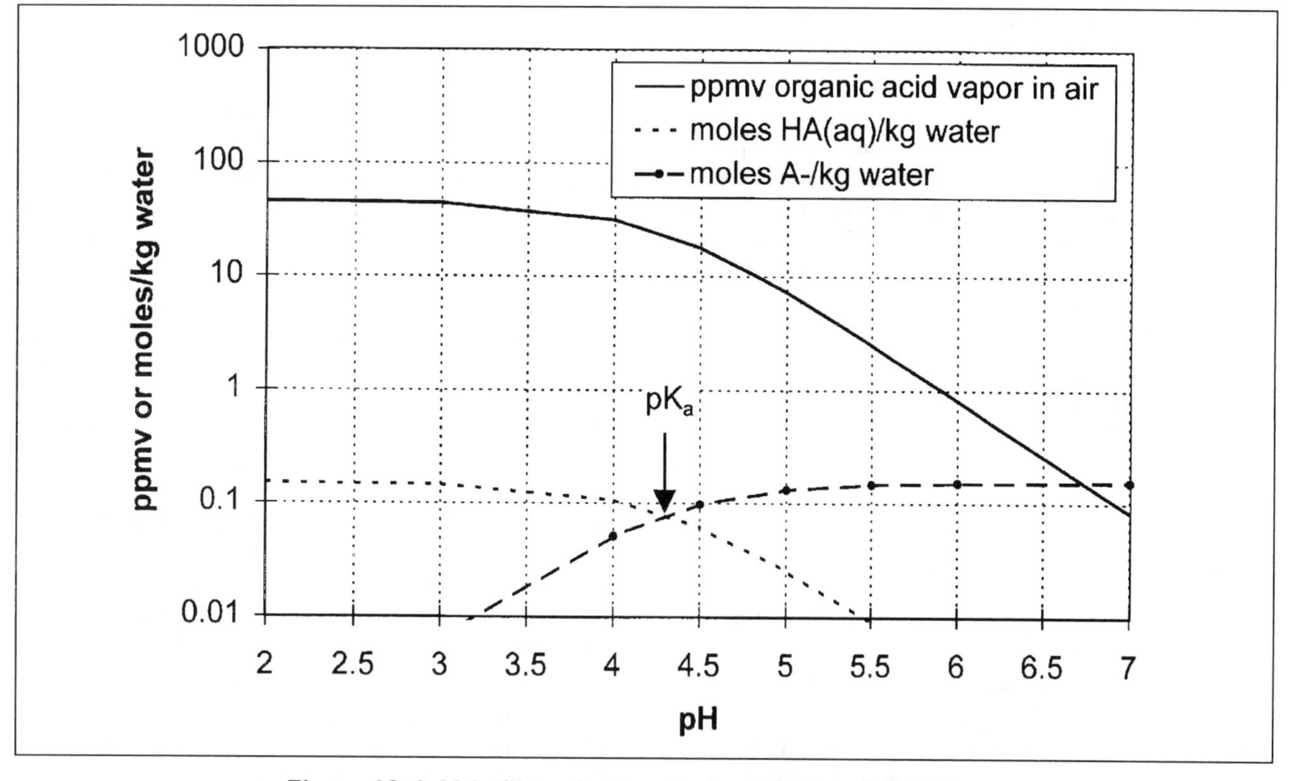

Figure 13-4. Volatility of a Two-Carbon Carboxylic Acid as a Function of pH at 25°C (10,000 ppmw Organic Acid in Water)

In response to these drivers, generators of metal-containing wastewater have new impetus to better understand and predict the behavior of metals in aqueous systems. A necessary ingredient for identifying feasible, cost-effective metal removal technologies (particularly at the source) is the ability to describe the aqueous chemistry of the waste stream. Although not the entire story, the chemical speciation at thermodynamic equilibrium describes the best removal by alkaline precipitation that one can hope to achieve.

Solubility data for metal hydroxides and oxides are available from a variety of sources.[3, 5–10] However, a careful review of these sources by Dyer, Scrivner, and Dentel[11] showed that they do not always agree on the water solubility of a particular metal hydroxide or oxide. In some cases, the difference is several orders of magnitude. This disparity arises from differences in the experimental methods used to collect the data, variations in the size and crystalline forms of the solid phase present, experimental temperatures, pH values, aging times, analytical capabilities, and numerical techniques.

One common source of inaccuracies is the use of solubility data based on the presence of a highly crystalline solid, which may be appropriate for geochemists, but not for engineers seeking to describe apparent equilibria after several hours of equilibration with an amorphous or metastable precipitate. It is also important to understand that the solubility of a given metal will vary significantly with pH. In other words, it is insufficient to know only the solubility product (K_{sp}) of the metal salt. A practicing engineer is often left wondering, "what is the true solubility of the metal and how can I minimize it?"

The answer lies in understanding how metal solubility varies with pH. This section presents the results of a critical evaluation by Dyer et al.[11] of metal hydroxide/oxide solubilities for 12 metals. Overall metal-solubility diagrams are presented to help the reader understand how metal emissions will vary significantly with pH and metal type. In addition, these diagrams emphasize the important role that source reduction and point-source treatment will play in minimizing the cost of metals removal from water waste streams.

13.3.1 Hydrolysis of Metal Cations

The behavior of a metal in aqueous solution is controlled by its chemical speciation; that is, the molecular and ionic species that it forms.[12] If no complexing ligands (e.g., chloride, sulfate, carbonate, organic acids, organic chelating agents, etc.) other than the hydroxide ion (OH^-) are present, the simple metal cation (M^{z+}) will often hydrolyze to form complexes with the hydroxide ion as a function of pH. These complexes can be cations (e.g., $M(OH)_{z-1}^{+}$), neutral molecules (e.g., $M(OH)_z(aq)$), or anions (e.g., $M(OH)_{z+1}^{-}$), where *z* is the oxidation state of the metal. In addition to these mononuclear complexes, polynuclear species containing more than one metal atom (e.g., $M_x(OH)_{xz-2}^{++}$) can also form in some cases. All of these hydrolysis reactions play a significant role in the aqueous chemistry, and hence solubility of metals in water.

The hydrolysis of metal cations has been studied for many years. Baes and Mesmer[12] present a thorough discussion of the kinds of reactions seen in the hydrolysis of a metal cation, the relationship between these various reactions, the definition of the equilibrium expressions for each reaction, and the impact of pH on solubility. Stumm and Morgan[3] also discuss the fundamental chemistry and thermodynamics of metal-ion hydrolysis. The reader may consult these references for more background.

13.3.2 Methodology of Dyer et al.[11] Study

A goal of the study of Dyer et al. was to focus on the concentration of metal hydrolysis species that would be encountered in industrial situations; that is, where the time frame for precipitation is short and aging of the solid phase is minimal. For this reason, metal hydroxides were specifically chosen over the more stable metal oxides, except in the case of Pb(II) where only different crystalline forms of the lead oxide exist.

The OLI Software System from OLI Systems, Inc. (Morris Plains, New Jersey) was used to model the solubility of eleven metal hydroxides and one metal oxide (PbO) in pure water as a function of pH at 25°C. The selected metals were Al(III), Ca(II), Cd(II), Co(II), Cr(III), Cu(II), Fe(II), Fe(III), Pb(II), Mg(II), Ni(II), and Zn(II). Detailed speciation diagrams were generated for each metal showing the concentration of the free metal cation, M^{z+}, and each metal hydrolysis product, $M(OH)_y^{(z-y)+}$, as a function of pH.

Next, an extensive literature review of actual solubility data for each of the twelve metal hydroxides or oxides was completed to validate the model's predictions, and hence thermodynamic database. This included an assessment of how well the predicted solubility curve for each metal agreed with literature data.

Finally, an overall metal-solubility diagram was developed showing the solubility of all twelve metal hydroxides or oxides.

13.3.3 The OLI Software System

The OLI Software System is a commercial simulation package that models aqueous electrolyte equilibria, including chemical speciation and redox reactions; equilibria between aqueous, vapor, organic liquid, and multiple solid phases; biochemical and inorganic reaction kinetics; and ion-exchange, adsorption, and coprecipitation phenomena. The OLI software was chosen for this study because it is commercially available, widely used within the DuPont Company, and supported by a database that is continuously maintained and updated, and that allows for the prediction of thermodynamic properties over a wide range of temperature, ionic strength, and pressure.

The system is built around the OLI Engine, which is the foundation for the Environmental Simulation Program (ESP), Corrosion Simulation Program (CSP), and ProChem. The ProChem facility was used in this study because it is a powerful and flexible tool for modeling single-stage equilibrium at both steady and unsteady state.[11, 13]

13.3.4 Overall Metal Solubility Diagrams

Figures 13-5 and 13-6 present, in a single diagram, the solubility curves at 25°C for the twelve metal hydroxides or oxides. Figure 13-5 presents total metal concentrations in molality (mol/kg H_2O), while Figure 13-6 gives concentrations in ppm by weight. These diagrams are not identical, because the differing atomic weights change the curves' relative positions. It should be emphasized that the solubility diagrams assume that *no complexing ligands other than hydroxide ion are present*, and neglect possible effects that might arise if more than one of these metals were present together in an aqueous system.

The solubility curve for each metal should be viewed as a prediction based on assessment of the most reliable thermodynamic data available to the software supplier at the time. For some metals, this represents the solubility of an active or inactive amorphous precipitate (e.g., $Zn(OH)_2$ and $Cr(OH)_3$); for others, the solubility of an active crystalline modification of the precipitate (e.g., "blue" $Co(OH)_2$); and for still others, the solubility of a stable "inactive" crystalline modification (e.g., $Al(OH)_3$, $Ni(OH)_2$, and PbO). As initially formed precipitates age, more stable crystalline modifications emerge that have a lower solubility.

In most industrial situations, precipitates are formed from supersaturated solutions under relatively short-term conditions. In these cases, the solubility of the incipient active solid rather than the solubility of the aged, stable solid is of interest.[3] Therefore, for some metals, actual solubilities in an industrial precipitation process may be as much as 2 to 10 times higher than those shown in Figures 13-5 and 13-6. For this reason, the solubility "lines" are better viewed as solubility "bands" that extend 200% to 1000% vertically above and below the line, that is, in direct proportion to the variation in the solubility product, K_{sp}.

The benefit of an overall metal-solubility diagram is

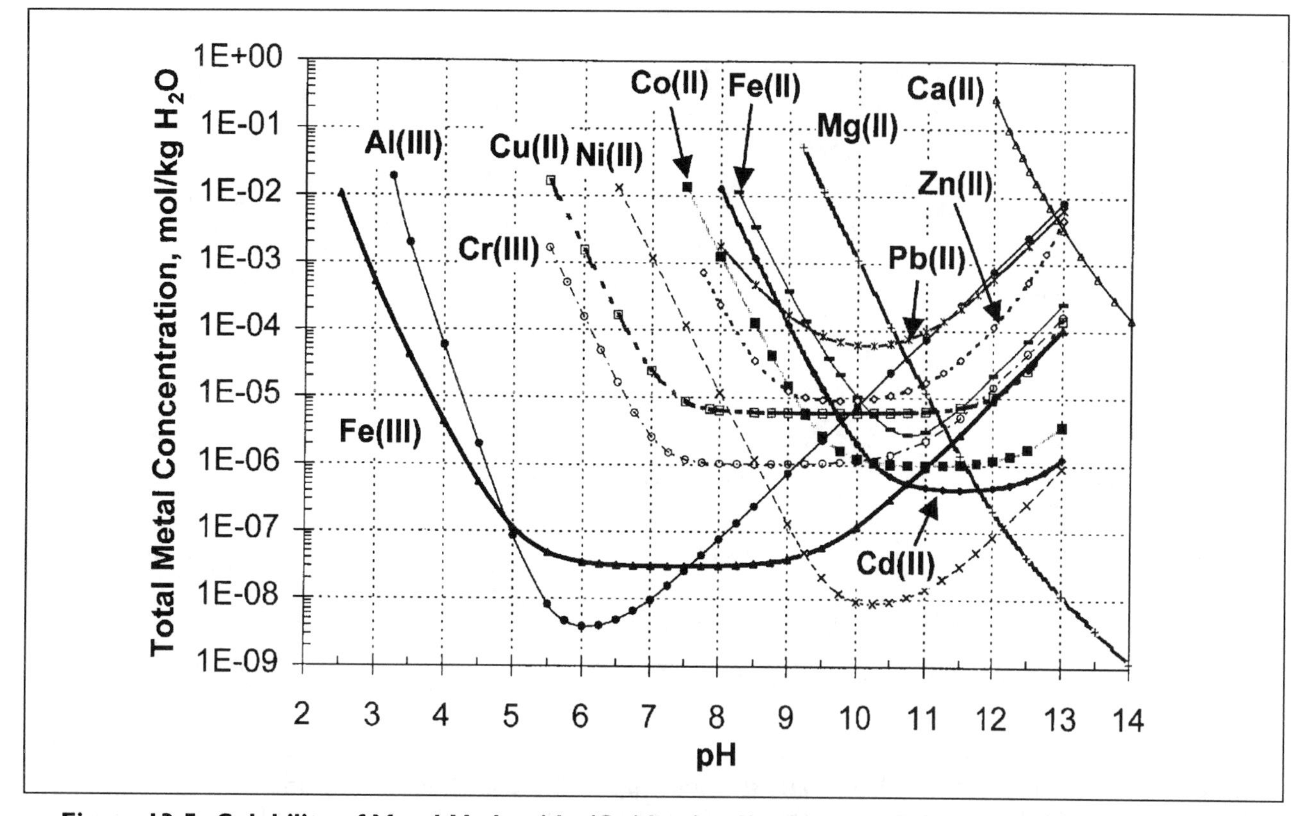

Figure 13-5. Solubility of Metal Hydroxides/Oxides (mol/kg H_2O) at 25°C Based on OLI ProChem Simulation ($M^{z+}(OH)_z$ precipitate present. Exception is Pb which is present as PbO precipitate).

that it can be used as a first-cut screening tool by process and environmental engineers to make decisions. For example, an engineer might want to assess the feasibility of a point-source alkaline precipitation process to remove several regulated metals to a desired level(s). In reality, the solubility curves in Figures 13-5 and 13-6 should be viewed as the best one can achieve.

In many cases, complexing ligands (e.g., organic acids, chelating agents, chloride) and/or solid-phase sorbents (e.g., hydrous ferric oxide, biological solids) that increase the "total solubility" of the metal above the levels shown will be present. Such systems can also be modeled by use of the OLI Software System. Current scientific research is greatly improving the ability of aqueous chemistry speciation models to predict the complexation of metals in solution and the sorption of metals onto solids. These additional "sinks" for metals present a great challenge to engineers trying to limit metals emissions to the environment.

Two case histories that show how to use the diagram are in the next section. In both cases, assume no complexing ligands, such as citrate, ethylenediamine tetraacetate (EDTA), and carbonate ion, are present that would increase or decrease the solubility of the metals.

13.3.5 The Important Role of Source Reduction and Pretreatment

One of the key findings gleaned from the Dyer et al.[11] study is the important role that (1) source reduction and (2) segregation and pretreatment have in minimizing the impact of metals emissions on a wastewater treatment facility. Because regulatory limits for metals are being driven to parts per billion (ppb) levels, once metals-containing waste streams are diluted with the rest of the plant wastewater, alkaline precipitation alone will often be inadequate. Moreover, even though effluent limits may be met, two-stage treatment processes designed for the entire wastewater flow will, in most cases, be exorbitantly expensive to build and operate.

The most attractive alternative, then, will be to minimize metal contamination of the aqueous-waste stream in the first place. For example, this may involve changing from a homogeneous catalyst to a heterogeneous catalyst in the reactor or removing a strong organic complexing ligand from the aqueous system. The next most attractive alternative is to consider pretreatment of the waste stream to remove the bulk of the metals before mixing with the rest of the plant wastewater. This may involve technologies such

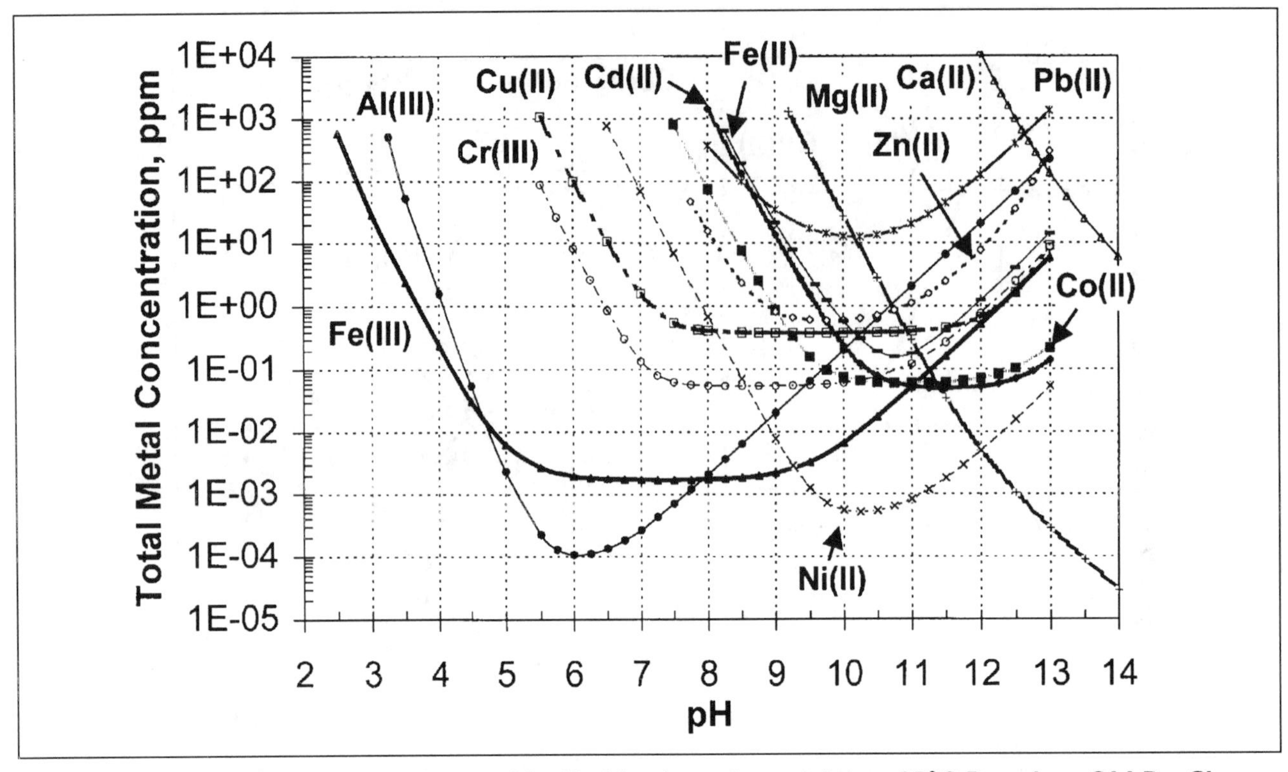

Figure 13-6. Solubility of Metal Hydroxides/Oxides (ppm by weight) at 25°C Based on OLI ProChem Simulation ($M^{z+}(OH)_z$ precipitate present. Exception is Pb which is present as PbO precipitate).

as alkaline precipitation, ion exchange, reverse osmosis, or electrodialysis.

13.4 Case Histories

13.4.1 Plugging of Feed Preheater and Column with Tars

At a DuPont site, tars were plugging a distillation column and feed preheater. The tar buildup resulted in plant shutdowns every 3 months to clean the preheater, and every 9 to 12 months to replace the column packing. This cost the business hundreds of thousands of pounds of lost production each year. A second preheater was installed in parallel with the existing heat exchanger to allow cleaning without shutting the plant down. In addition, lab and plant tests were run to better understand the mechanisms behind the tar formation.

As a result of these tests, the plant discovered that the tar-formation reaction was pH-sensitive. By tightening Standard Operating Procedures and installing alarms on upstream crude-product washers, the tar formation was virtually eliminated. This provided the business with approximately $280,000/yr in additional after-tax earnings.

13.4.2 Tar Formation Due to Acidity in an Incoming Raw Material

A process produces a waste stream of heavy tars that are incinerated. Two types of tars are formed—thermal tars due to the high-temperature reaction and acid tars due to acidity in one of the incoming raw materials. The presence of acids in the reaction step triggers a side reaction that produces the acid tars.[14] A neutralizing agent is added directly to the reactor to try to limit acid-tar formation. This neutralizing agent also becomes part of the waste stream. The ratio of acid tar to thermal tar cannot be distinguished and, historically, this discouraged attempts to reduce waste in the process. The viscous tars entrap significant amounts of product and, over time, reducing this yield loss became more critical to the business' profitability. The solution was for the supplier of the raw material to incorporate better on-line pH control at its own facility. As a result, acid-tar formation and the amount of neutralizing agent added to the reactor have been substantially reduced. By minimizing pH variability in the incoming raw material, the site was also able to better understand the magnitude and causes of thermal-tar formation. The reactor is now operated at a lower temperature. In all, the amount of incinerated waste has been reduced by 60%.

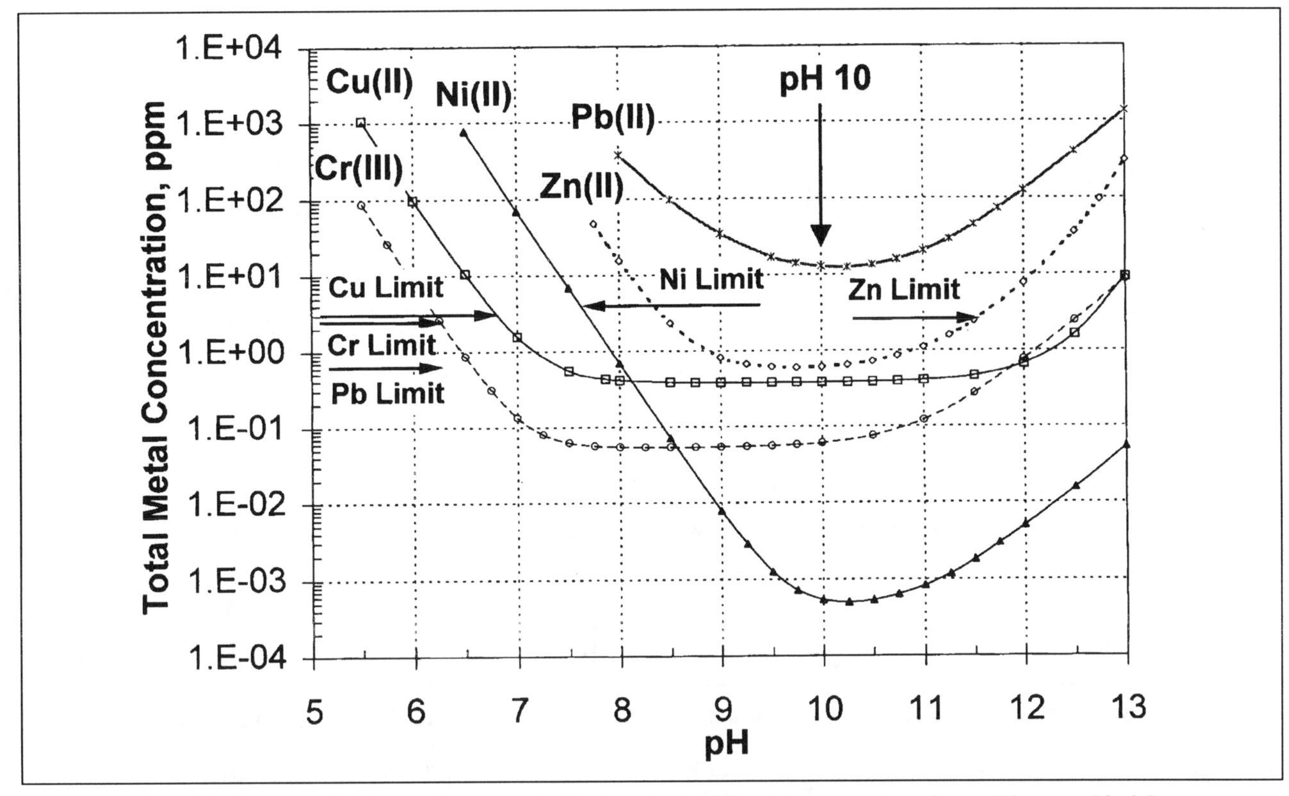

Fig 13-7. Overall Metal Hydroxide/Oxide Solubility Diagram for Case History 13.4.3

13.4.3 Metals Emissions in Wastewater Effluent

At a large chemical manufacturing site, the concentrations of five metals in the site wastewater effluent are regulated by the Clean Water Act's Organic Chemicals, Plastics, and Synthetic Fibers (OCPSF) Guidelines. The targeted metals and their daily maximum discharge limits are 2.77 ppm total Cr [Cr(III) + Cr(VI)], 3.38 ppm Cu; 0.69 ppm Pb; 3.98 ppm Ni; and 2.61 ppm Zn. Will alkaline precipitation alone reduce the concentration of these five metals below the discharge limits? What is the optimum pH to operate the precipitation reaction?

As shown in Figure 13-7, the OCPSF discharge limits could most likely be met with a reasonable margin of safety for total Cr (assuming all Cr is present as Cr(III)), Cu(II), and Ni(II). The margin of safety is defined as the OCPSF discharge limit divided by the predicted metal solubility at the operating pH. At pH 10, the margin of safety for total Cr, Cu(II), and Ni(II) would be about 50, 8, and 8000, respectively. On the other hand, the ability to meet the Zn(II) limit is questionable with a safety factor of only 4. It is clear that an additional treatment step would be needed for Pb(II). The optimum pH would be the pH where close to the minimum solubility is realized for all of the metals. For example, the site might consider a one-stage alkaline precipitation reaction operating at pH 10 to remove the bulk of the metals, followed by an ion-exchange system to remove residual Pb(II) to achieve acceptable levels. Note that the influent metal concentration was not required to assess the feasibility of an alkaline precipitation reaction to meet the effluent limits.

13.4.4 Copper Emissions from Reaction Step

A chemical manufacturing process uses a homogeneous, copper-based catalyst in the reaction step. A 1.8 m^3/h (8 gal/min) aqueous waste stream, containing 1000 ppm Cu(II) at pH 5.5, is discharged from the process and sent to the site wastewater treatment plant. At the treatment plant, the stream is combined with other wastewaters, resulting in a total flow rate of 180 m^3/h (800 gal/min). Can EPA Ambient Water Quality Chronic Freshwater Criteria[4] for Cu (11 ppb total recoverable Cu at 100 mg/L hardness) be met, assuming this is the only source of Cu on the site?

As shown in Figure 13-8, the minimum solubility of copper is 400 ppb between pH 8 and 11. If we take advantage of the 100:1 dilution ratio (180/1.8 m^3/h) in the treatment plant, however, the copper level could be reduced to approximately 4 ppb by pretreating the concentrated stream in the process area. In this case, the more attractive alternative is to segregate and pretreat with alkaline precipitation before dilution with the rest of the site waste-

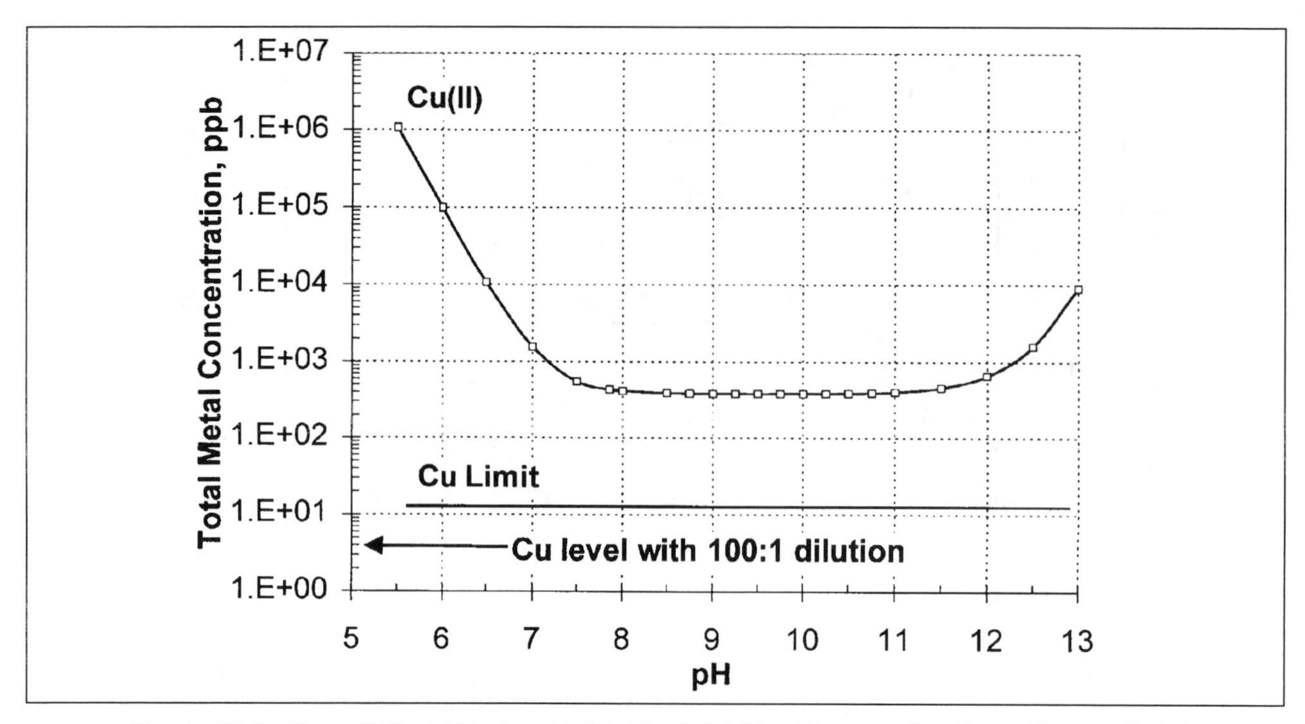

Figure 13-8. Overall Metal Hydroxide/Oxide Solubility Diagram for Case History 13.4.4

water, or, alternatively, to develop a new catalyst that does not create copper emissions in the first place (e.g., less toxic/regulated metal, or heterogeneous catalyst instead of a homogeneous one).

Literature Cited

1. Stumm, W., and J. J. Morgan. 1996. *Aquatic Chemistry*, 3rd ed. New York: Wiley.
2. LaGrega, M. D., P. L. Buckingham, and J. C. Evans. 1994. *Hazardous Waste Management*. New York: McGraw-Hill.
3. Stumm, W., and J. J. Morgan. 1996. *Aquatic Chemistry: Chemical Equilibria and Rates in Natural Waters*, 3rd ed. New York: Wiley.
4. U.S. EPA Office of Water Regulations and Standards. 1996. *Federally Promulgated Water Quality Standards, Code of Federal Regulations, Title 40—Protection of the Environment*, Part 131, Subpart D. Washington, DC: Office of the Federal Register.
5. Dean, J. A. 1992. *Lange's Handbook of Chemistry*, 14th ed. New York: McGraw-Hill.
6. Linke, W. F. 1958–1965. *Solubilities of Inorganic and Metal Organic Compounds*, Vols. I & II, 4th ed. Washington, DC: American Chemical Society.
7. Stephen, H., T. Stephen, and H. L. Silcock, Eds. 1963–1979. *Solubilities of Inorganic and Organic Compounds*, Vols. 1–8. Oxford: Pergamon Press.
8. Lide, D. R. 1992. *CRC Handbook of Chemistry and Physics*, 73rd ed. Boca Raton, FL: CRC Press.
9. Martell, A. E., R. M. Smith, and R. J. Motekaitis. 1995. *NIST Critically Selected Stability Constants of Metal Complexes*, NIST Standard Reference Database 46, Version 2.0, Gaithersburg, MD: NIST Standard Reference Data.
10. Baes, C. F., Jr., and R. E. Mesmer. 1976. *The Hydrolysis of Cations*. New York: Wiley.
11. Dyer, J. A., N. C. Scrivner, and S. K. Dentel. 1998. "A Practical Guide for Determining the Solubility of Metal Hydroxides and Oxides in Water." *Environ. Prog.* 17(1): 1–8.
12. Baes, C. F., Jr., and R. E. Mesmer. 1981. "The Thermodynamics of Cation Hydrolysis." *Amer. J. Sci.*, 281 (Summer): 935–962.
13. Rafal, M., P. Black, S. J. Sanders, P. I. Tolmach, and R. D. Young. 1994. "Development of a Comprehensive Environmental Simulation Program." Paper read at the American Institute of Chemical Engineers' Spring National Meeting, April 17–21, Atlanta, GA.
14. U.S. Environmental Protection Agency. November 1993. *DuPont Chambers Works Waste Minimization Project.* EPA/600/R-93/203. Washington, DC: U.S. EPA, Office of Research and Development.

CHAPTER 14

Pollution Prevention in Vacuum Processes

14.1 Introduction

Volatile organic compounds from vacuum generation are expensive to control and increase the investment and operating cost of the vacuum system. When noncondensibles or inerts are present, the amount of volatile organic compounds emitted to the environment will increase dramatically. Many of us will be challenged in the future to choose and implement technically and economically sound engineering solutions to minimize or eliminate these emissions. In this chapter we address approaches that process, project, and environmental engineers can apply to reduce or eliminate vacuum-system emissions. Actual case histories are included that reflect the range of solutions that are being applied. The discussion that follows is not intended to be a detailed design tool, nor is it a prescription for solving every possible emission problem. Rather, in the chapter we deal with concepts and approaches on how to move toward a zero waste-generation and emissions design.

14.2 Background

A vacuum is primarily used to effect cooling or to vaporize materials at a lower temperature. An example of cooling is freeze crystallization, while examples of volatilization include multiple-effect evaporators; standard distillation; short-path distillation; degassing or solvent removal (such as with polymer melts); and freeze-drying.

A vacuum is generated by one of two devices: vacuum pump or jet ejector. Vacuum pumps are positive displacement devices, the two principal types being sliding-vane and liquid-ring vacuum pumps.

The jet ejector is a device where the kinetic energy of one fluid (the primary fluid) is used to pump another fluid (the secondary fluid). Because steam is readily available and fairly inexpensive, it is often used as the primary fluid to entrain gases from the process, thereby generating the desired vacuum.

14.3 Nature of Emission Sources

When a vacuum pump is used to generate vacuum, two emission sources are possible: the vapor discharge from the vacuum pump, and the seal liquid of a liquid-ring vacuum pump.

In the case of a jet ejector, the motive fluid (e.g., steam) becomes contaminated. The contaminated fluid can be discharged directly as a vapor, or it can be condensed, which results in a liquid-phase emission.

A secondary effect on the environment results from the energy consumed to power the vacuum devices; that is, electricity and steam. Because energy generation produces pollutants, such as carbon dioxide, nitrogen oxides, and sulfur oxides, the vacuum system should be as small as possible.

14.4 Regulatory Outlook

The gaseous emissions from vacuum systems can contain one or more hazardous air pollutants (HAPs) and may be regulated in the United States by the 1990 Clean Air Act Amendments. Typically, regulated emission sources will require greater than 98% control using an end-of-pipe destruction or recovery technology. In addition, the condensed liquids from jet-ejector systems may contain compounds that cause the liquid to be classified as a Resource Conservation and Recovery Act (RCRA) hazardous waste. Work with your site environmental coordinator to determine which environmental regulations are applicable.

14.5 How do Emissions Originate?

For a given system operating at a given pressure, three design parameters will influence the size of the vacuum device and, hence, the potential magnitude of the emissions. They are entrainment, temperature of operation, and the level of inerts.

14.5.1 Entrainment

The first critical design parameter is the surface area of the liquid from which the vapors and gases are being removed. If the space velocity of the gas is too high, excessive entrainment of liquid will occur. Even at low velocities, entrainment of fine mists occurs. If liquid splashing into the vacuum vapor line or vapor condensation can occur, then the vacuum vapor line must be large enough to prevent reentrainment from the liquid draining back to the vessel.

14.5.2 Temperature of Operation

The second key design parameter is the operating temperature of the process unit upon which the vacuum is being drawn. The temperature should be low enough to minimize vaporization of the product or solvents in the system. In the case of polymerization reactions, a compromise must sometimes be made between degassing at a low temperature to minimize vaporization and producing unwanted solids at the lower temperature due to agglomeration of monomers and other particles.

14.5.3 Inert Level

The third key design parameter is the level of inerts in the system. Inerts (e.g., nitrogen or air) have a significant impact on the size of the vacuum system and space velocity of the gases, which, in turn, determines the required surface area of the liquid to minimize entrainment. The inerts arise from gases dissolved in the liquid or leaks from the outside through flanges and other fittings.[1] The dissolved inerts are present due to inert blanketing of storage tanks and vessels, the use of air in oxidation reactions, and/or the addition of inerts for measurement gauges, such as nitrogen-purged dip tubes.

14.6 The Pollution-Prevention Continuum

The continuum depicted in Figure 14-1 shows the rela-

% Pollution Prevention	Moderate Vacuum With Low Vapor Pressure Materials	High Vacuum (< 20 mm Hg) or Deliberate Removal of Volatiles
99.9%	No Inerts	Very difficult
95%	Inline condenser/absorber at all levels of inerts (Figure 14-2) plus vacuum pump with condenser* or jet ejector with noncontact condenser	Inline condenser/absorber at all levels of inerts (Figure 14-2) plus vacuum pump with condenser* or jet ejector with noncontact condenser
50%	Liquid saturated with inerts and an exceptionally tight system**	Exceptionally tight system** and no inerts in liquid
10%	Liquid saturated with inerts plus standard fitting leaks	Standard fitting leaks and no inerts in liquid
<<10%	Saturated liquids plus other inert sources (leaks, gas purged guages, etc.)	Saturated liquids plus other inert sources (leaks, gas purged guages, etc.)

* The replacement of a jet ejector with a vacuum pump and condenser can be greater then 2X the cost of an inline scrubber.

** Requires high level of maintenance of flanges, gaskets, and fittings.

Figure 14-1. The Pollution-Prevention Continuum for Vacuum Systems. Impact of Inert Level and Method of Vacuum Generation on the Degree of Waste Generation

tive merits of options that are available to engineers to minimize or eliminate emissions from vacuum systems. The decision of how far to move toward a zero waste-generation and emissions design will depend on a number of factors, including corporate and business environmental goals, economics, and applicable regulations.

14.7 Source Reduction of Vacuum System Emissions

14.7.1 Entrainment

Entrainment can be minimized by (1) proper design of the vessel being degassed, and (2) proper sizing of the vent line to prevent reentrainment of any liquid flowing down the walls of the vapor line back into the vessel.

Proper Vessel Design. The vessel on which the vacuum is being pulled should have a surface area that satisfies the following relationship:[2]

$$F = \frac{v\,(\rho_{gas})^{0.5}}{A} \leq 1.5 \qquad (1)$$

where

v = gas volumetric flow in ft^3/s

A = vessel cross-sectional area in ft^2

ρ_{gas} = gas density in lb/ft^3

F = rule-of-thumb parameter related to column flooding factor

DuPont experience has shown that an F-factor of less than 1.5 will prevent the entrainment of large droplets; however, there will always be a background level of 30-micron-size entrainment.

Vapor Line Design. The key design parameter for sizing the vapor line from the primary condensers is the level of entrainment. At all vapor flow rates, there is a base load of 30-micron entrainment. For most processes, this small amount of entrainment does not cause any significant problems; however, the larger-sized entrained aerosols could contain solids, monomers, oily materials, and so on. These larger-sized aerosols not only increase the load on the vacuum device and the level of waste, but they also can cause mechanical problems with vacuum pumps, or fouling of noncontact heat exchangers used with steam jets. DuPont experience has shown that the upper-limit F-factor for sizing the vapor line is given by[3]

$$F < 5.4\,(\rho_{gas})^{0.2} \qquad (2)$$

where

ρ_{gas} = gas density in lb/ft^3

F = *F*-factor as defined in Equation 1

If the *F*-factor is greater than the value given by Equation 2, then a portion of any liquid draining back to the primary condenser will be entrained in the vapor.

Vapor-Line Installation. During normal operation, an upset can cause liquid to splash up into the vapor line or the vapor itself can condense in the vapor line. To prevent reentrainment of this liquid, the orientation and design of the vapor line from the vessel on which the vacuum is being pulled (refer to the vapor line between the spray chamber and column reflux condenser shown in Figure 14-2) should have a 15° to 20° slope for drainage of any liquid back to the vessel. The vapor line is designed with a sharp turn or with a T-arrangement with a blind flange. The sharp turn forces the vapor to change direction; any entrained droplets will impact the wall and drain back into the vessel. Do not install a smooth elbow. DuPont's experience has shown that the flow lines in a smooth elbow are sufficiently smooth to carry some entrainment further downstream.

A wall-wash spray oriented countercurrent to the flow of vapor should also be installed in the vapor line. The liquid for the wall-wash spray comes from the liquid in the vessel on which the vacuum is being pulled (see Figure 14-2, where the wall-wash spray is supplied from the column reflux). The liquid is sprayed through a large coarse spray nozzle.

Clean Vapor to a Vacuum Device. A properly sized and installed vapor line will have very few large, entrained droplets entering the vacuum device. However, if

- The inerts level is higher than anticipated,
- The vacuum device is a vacuum pump that cannot handle any entrainment,
- The steam jets use shell-and-tube heat exchangers instead of direct-contact condensers,
- Any process upsets or "burps" can occur,

then a spray-chamber system should be installed in the vacuum vapor line before the vacuum jet or pump.

The spray-chamber system (see Figure 14-2 inside the dashed box) is a recirculation loop with a pump, cooler or chiller, spray chamber with 1 to 4 sprays, and a collection tank. The vessel liquid (column bottoms in Figure 14-2) is used as make-up to the loop. The make-up liquid does not contain any volatile materials (it already has been subjected to the vacuum), and when it is cooled, it will exert minimal vapor pressure in the spray chamber. The recirculating liquid is chilled sufficiently to cool, condense, and absorb material from the vapor line. Moreover, it cleans the vapor going to the vacuum device even if an upset occurs in the process vessel.

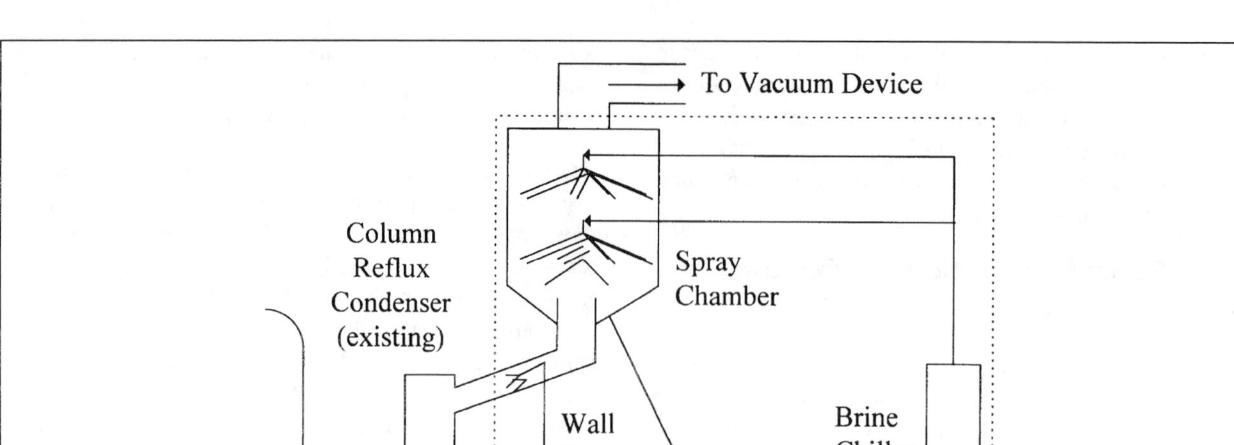

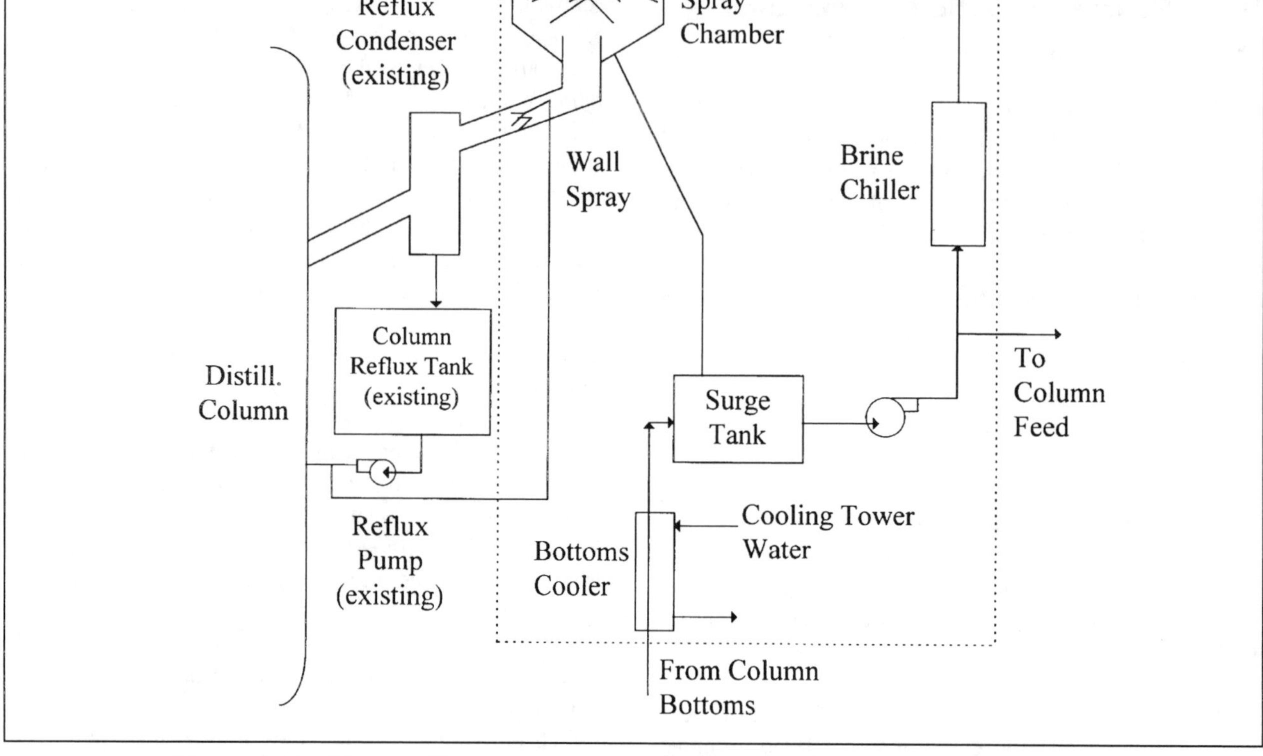

Figure 14-2. Vacuum Vapor Recovery and Cleansing (Vent Scrubber System) for a Vacuum Distillation Column

14.7.2 Temperature of Operation

The effect of temperature is process-specific. If lowering the operating temperature (to minimize the vapor load on the vacuum system) causes freezing, then an in-line filter can be installed in the spray chamber recirculation loop to remove the solids, or a slurry pump can be used to move the material. An economic evaluation is required to balance the added investment for new equipment against the size of the vacuum system and its attendant environmental treatment system.

14.7.3 Inerts Level

The level of inerts in the vacuum system depends on dissolved inerts, including the use of inert-purged gauges and air inleakage. The dissolved inerts originating in storage tanks and vessels can be eliminated by using liquid storage techniques that minimize contact with inerts, such as pressure vessels or vessels with a floating roof. Inert-purged measurement gauges can be replaced by diaphragm gauges or other types of sealed gauges. The leakage of inerts into the vacuum system can be reduced to almost zero by using appropriate gaskets and high-level leak-detection testing.

14.7.4 Vacuum Device

The type of vacuum device also influences the level of emissions to the environment. A jet ejector can use a motive fluid from which the volatile vapors can be easily stripped (e.g., steam when hydrophobic organics are present). A vacuum pump with an exit condenser can recover more than 95% of the volatile vapors. Meanwhile, a dry vacuum pump (sliding vane) is not subject to contamination of a seal liquid, as in a liquid-ring vacuum pump; however, because of the close tolerances required for the sliding vanes, a dry vacuum pump cannot tolerate any entrainment droplets.

14.8 Case Histories and Economics

Three cost factors must be balanced when designing a cost-effective, low-emissions vacuum system: the size of the vacuum device, the cost of end-of-pipe treatment, and the cost to minimize inleakage of gases. The two case histories that follow will illustrate these three cost factors.

14.8.1 Case History #1: Degassing at High Vacuum During Polymer Production

The production of polymer for fiber and film requires a vacuum system to remove the volatile gases to control the degree of polymerization. The vacuum (1 to 16 mm Hg absolute) is produced by a series of steam jets with interstage, water-cooled, direct-contact condensers. The direct-contact condensers require a large volume of water per polymer line. The water from the direct-contact condensers is sent to a cooling tower, where the absorbed volatile organic compunds (VOCs) are stripped from the water by the cooling-tower air.

Two changes in the vacuum system were considered to reduce emissions from these plants:

1. Install vacuum pumps with interstage noncontact condensers. The effluent gas would contain inerts and the most volatile VOCs. The VOCs would then be destroyed in a thermal or catalytic oxidizer, a flare, or other combustion device, such as a boiler.
2. Install spray scrubbers in the vapor line to the vacuum system using a chilled low-vapor-pressure fluid as the scrubbing liquid.

An economic analysis found that the capital investment for the spray-scrubber option was only one-third of that required for the vacuum pumps with interstage condensers, and would give a positive internal rate of return (IRR).

The major problems with the existing vacuum systems were:

- Significant levels of entrainment, which resulted in the coating and plugging of the vacuum jets. In some cases, the jets would plug in less than one month of operation.
- High levels of VOCs in the vacuum system and, ultimately, high emissions from the cooling tower used to cool the jet contact-condenser water.
- A high biochemical oxygen demand (BOD) load in the blowdown from the contact-condenser water system that is sent to wastewater treatment.

The entrainment was caused by a high gas velocity in the vapor line to the vacuum jets. A proper redesign of the vapor line using the techniques described earlier prevented any liquid aerosols from being entrained from the liquid draining back into the source vessel and entering the vacuum device.

To reduce the high levels of VOCs entering the steam jets and the subsequent stripping of these VOCs in the cooling tower, a spray condenser was installed in the vapor line before the vacuum jets. The spray liquid was a low-vapor-pressure compound already present in the process. The spray liquid is chilled and sprayed countercurrent to the gas flow in multiple-spray-bank stages. The spray liquid condenses and absorbs the organics and water from the gas stream. The spray system reduced the VOCs entering the steam jets by more than 90%. As a result, the high BOD load to wastewater treatment was also reduced by more than 98%.

These pollution-prevention actions have resulted in

- Greater than 90% reduction of VOCs emitted to the environment
- Greater than 98% reduction in BOD-load to wastewater treatment
- Improved operability of polymer production through better pressure control of the system
- Recovery of valuable materials, and
- More than 95% reduction in the vapor load to the steam jets, which will ultimately result in steam savings once the steam jets have been modified (i.e., reduced in size due to the smaller vapor/inerts load)

The integrated pollution-prevention system consists of spray condensers, a hold tank, pump, brine chiller and heat exchanger, and coarse spray. Capital investment per polymer line was approximately $1 million, which included extensive dismantlement and rearrangement of existing equipment. Cash flow included savings from ingredients recovery and reduced wastewater treatment load. Table 14-1 summarizes the rate of return for two different project scopes-of-work.

14.8.2 Case History #2: Vacuum Distillation

While manufacturing intermediate monomers that are used to produce a salable polymer product, vacuum distillation columns are employed to purify a monomer. One such column operated at 150 mm Hg absolute pressure using a steam jet that vented to the atmosphere.

The vapor load in the vacuum line depended on the inert load in the system. If no inerts were present, the vapor load was close to zero. The plant recognized the effect of inert loading, and replaced the nitrogen in the dip tubes used for level control with a solvent used in the process. The other sources of inerts—saturation of inerts from storage tanks and inleakage through the fittings—

Scope of Savings	% IRR
Ingredients recovery + reduced wastewater treatment costs	15
Ingredients recovery + reduced wastewater treatment costs + steam savings *	20

* After installation of the spray condenser loop, the vapor load to steam jets was decreased by more than 90%. This allowed for installation of smaller steam jets.

Table 14-1. Internal Rates of Return for Case History #1. Impact of Spray Condenser Loop on Savings Through Ingredients Recovery, Lower Wastewater Treatment Costs, and Steam Savings

have yet to be addressed.

Using as the base case a nitrogen-saturated feed to the column and a standard leak rate through the fittings, an economic analysis of the cost of having inerts shows (see Table 14-2) that considerable capital investment can be spent on projects to reduce the level of inerts in the system while still achieving a reasonable internal rate of return. Any projects with a cumulative capital investment less than that shown in Table 14-2 would result in an IRR greater than 12%. The cash-flow analysis used to calculate these threshold investment numbers accounted for ingredient savings, energy savings, and lower treatment costs if the inleakage and inerts were not present.

If the level of inerts cannot be decreased, either because of operating conditions or safety, then the VOCs could be removed from the vapor system by absorption before reaching the steam jet. Alternatively, the steam jet could be replaced with a vacuum pump and condenser to recover the vapors.

Using the experience from the polymer-production case history presented earlier, and because many vacuum systems use steam jets, an absorption system could be installed to remove the volatiles from the inert carrier gas. The scrubbing fluid would be the bottoms stream from the distillation column. The vent scrubber system (Figure 14-2) would include

- A coarse wall spray directed countercurrent to the vapor flow to keep the walls clean and facilitate drainage in the vapor line
- A scrubbing loop to provide the cooling required to condense and absorb the volatile compounds
- A cooler to remove heat from the bottoms stream, and a chiller to attain the desired temperature for the sprays
- A spray chamber with 2 to 4 spray banks
- Associated pumps, piping, and tanks

A fraction of the bottoms stream would be split to the scrubber system, and a cooling-tower-water cooler would be used to remove the heat from the bottoms stream before addition to the scrubber loop. The chiller would be operated as cold as possible to provide maximum cooling for the sprays. The overflow from the scrubber system would be sent back to the feed of the distillation column. For most distillation columns, there would be only a slight increase (less than 5%) in the liquid traffic in the column and in the heating and cooling loads on the column.

The capital investment for this type of vent scrubber system would be under $250,000. The results of a cash-flow analysis for two levels of inerts are shown in Table 14-3. The first case assumes a process that is exceptionally tight and has no inerts in the feed stream; therefore, there is only a small quantity of volatile organic compounds in the vacuum line. If it is assumed that the "scrubbed" steam can be vented directly to the atmosphere without further treatment, the value of the recovered organic compounds and the reduced steam requirements result in a small positive rate of return on the vent scrubber system investment. On the other hand, if the steam from the vacuum jets was originally condensed and incinerated to destroy absorbed organic compounds, then the return on investment is significantly higher if sufficient material is now captured by the vent scrubber system, so that incineration of the steam is no longer required.

The second case assumes a process that initially had a higher level of inerts (i.e., standard inleakage) that is eliminated through improved maintenance practices. The lower inert level, coupled with the recovery capability of the vent scrubber system, results in a very high rate of

Approach	% Reduction in Waste	Amount of investment[1] that can be spent to give a zero net present value (NPV) at 12% discount rate ($ per lb/hr of feed to the column) Vented*	Abated**
Eliminate inleakage	50	$20 per lb/hr	$120 per lb/hr
Eliminate inleakage and no inerts	100	$40 per lb/hr	$240 per lb/hr

* The vacuum jet steam is vented to the atmosphere.
** The steam is condensed and burned in an incinerator.
1 To estimate the investment that could be justified to eliminate inleakage, multiply the feed rate in lb/hr to the column by the appropriate $ per lb/hr. For example, at 18,000 lb/hr feed to the column where the steam is condensed and incinerated, the investment that could be justified to eliminate inleakage = $120 per lb/hr x 18,000 lb/hr = $2,160K.

Table 14-2. Affordable Pollution Prevention Investment for a Vacuum Distillation Column (Case History #2)

Inert Content of the Process	Vent Scrubber System Internal Rate of Return (IRR) Vented* % IRR	Abated** % IRR
Exceptionally tight system and no inerts in feed	5	100
Standard inleakage eliminated and no inerts in feed[1]	30	200

* The vacuum jet steam that is vented to the atmosphere contains no contaminants.
** The vacuum jet steam no longer needs to be condensed and burned in an incinerator.
1 Assumes that the existing column has standard inleakage, nitrogen-purged gauges, etc. A significant portion of the savings and earnings results from recovered process materials, lower steam and electricity requirements for the vacuum device, and lower treatment costs.

Table 14.3. Internal Rate of Return for Installation of Vent Scrubber System on Vacuum Distillation Column in Figure 14-2 (Assumes Steam-Jet Vacuum System)

return. Both cases assume greater than 95% recovery of volatile compounds, no inerts in the feed to the column, and a $0.40 per pound ingredient cost.

Literature Cited

1. Jackson, D. H. May 1948. "Selection and Use of Ejectors." *Chemical Engineering Progress,* 44(5): 347–352.
2. Fair, J. R., D. E. Steinmeyer, W. R. Penny, and B. B. Crocker. 1984. "Liquid-Gas Systems." *Perry's Chemical Engineers' Handbook,* 6th ed., R. H. Perry and D. W. Green, Eds. pg. 18-6. New York: McGraw-Hill.
3. Alves, G. E. July 1970. "Cocurrent Liquid-Gas Pipeline Contactors." *Chemical Engineering Progress,* 66(1): 60–67.

CHAPTER 15

Ventilation of Manufacturing Areas

15.1 Introduction

Most manufacturing areas require one or more industrial ventilation systems to maintain or improve the workplace environment. In the future, some plants will be required to choose and implement solutions to minimize or eliminate emissions from building or process-area ventilation air. In this chapter we address approaches that process, project, and environmental engineers can consider to reduce contamination of ventilation air at the source and to minimize the volume of air that may ultimately require treatment. Case studies are included that reflect the types of solutions that are being applied throughout industry. This chapter is not a detailed design tool, nor is it a prescription for solving every possible air-emissions problem. Rather, it provides concepts and approaches on how to move toward a zero waste-generation and emissions design.

The three main objectives of industrial ventilation are to:[1]

1. Control contaminants to acceptable workplace exposure levels, for example, comply with the regulations of the Occupational Safety and Health Administration (OSHA)
2. Prevent fires and explosions
3. Control temperature and humidity for worker comfort

All three objectives are important; however, the control of contaminants to acceptable levels is often the primary goal of an industrial ventilation system. In most cases, controlling the contaminant level will also eliminate the risk of fires and explosions; lower explosion limits (LELs) for volatile organics are typically 1 to 2 orders of magnitude higher than OSHA permissible exposure limits (PELs).

In the past, controlling workplace contaminant levels was often achieved using dilution ventilation of the entire manufacturing or process area. Because energy costs were low and volatile organic compound (VOC) and particulate emissions were often not controlled, conservative designs were used that incorporated large volumes of air. In the future, the prospect of having to treat this air will drive businesses to pollution-prevention options that reduce or eliminate contamination in the first place and minimize the volume of air that must be treated.

15.2 The Bottom Line

The high cost of end-of-pipe treatment will drive plants to ventilation-system designs that incorporate source reduction, enclosures, and local exhaust ventilation systems with minimum open area. Case histories show that these approaches can reduce the volume of air requiring treatment by 50–80%. Installed investment for typical enclosures will vary from only $\$40/ft^2$ to $\$120/ft^2$.[2] Invariably, this results in an order-of-magnitude higher savings in investment for the end-of-pipe treatment device.

15.3 Nature and Sources of Emissions

Contamination of ventilation air usually results from the intentional dilution of uncontrolled process emissions or from fugitive emissions around process equipment, such as equipment leaks. Typically, the contaminated ventilation air is very dilute (several hundred parts per million by volume or less), with a volumetric flow rate in the tens or hundreds of thousands of cubic feet per minute. This air is often directly vented to the atmosphere.

Some typical causes of contaminated ventilation air include:

- Warehouses storing polymer pellets/flakes containing residual solvent(s)
- Leaks and openings around hot-air convection dryers in the surface coating industry
- Open-top batch mix tanks in the automotive paint industry
- Openings around parts cleaners
- Leaks from polymer extruders, pumps, packings, and the like
- Solvent cleaning of process vessels
- Solvent off-gassing or dust from open conveying systems
- Dust from crushing and grinding operations
- Chemical spills
- Opening equipment for maintenance

15.4 Regulatory Outlook

Ventilation air contaminated with VOCs or particulates is increasingly coming under regulation. Compliance with these regulations is implemented through comprehensive Title V operating permit programs administered by the states. The need to control ventilation air emissions will depend on a number of factors, including whether the site is in an ozone attainment or nonattainment area, the plant is a major or minor source, the plant is a new or existing source, and whether the vent contains one or more hazardous air pollutants (HAPs). Regulations and compliance options can vary from state to state. For details on applicable state and federal regulations, contact the site environmental specialist.

When controls are mandated, the trend is to require greater than 95% removal efficiency, but there are exceptions. Existing major sources in nonattainment areas are required to reduce VOC emissions through the use of reasonably available control technology (RACT). For example, RACT in New York State is defined as 81% abatement of vents emitting greater than 3 lb/h VOCs. As another example, new VOC sources in the surface-coating industry are faced with the requirements for best available control technology (BACT) in attainment areas, and for lowest achievable emission rate (LAER) technology in nonattainment areas. BACT and LAER both require greater than 95% overall control efficiency, which is defined as the capture efficiency of a room enclosure times the removal efficiency of a control device.

In addition, the 1990 Clean Air Act Amendments require many new and existing major sources of HAPs to achieve emissions limits reflecting the application of maximum achievable control technology (MACT). MACT requires removal efficiencies on the order of 98% for process vents that are subject to control.

15.5 The Pollution-Prevention Continuum

Figure 15-1 gives a hierarchy of options that process and project engineers can consider when addressing VOC or particulate emissions from ventilation systems. The continuum includes minimization of contaminant concentration and air flow as forms of pollution prevention. The decision of how far to move toward a zero-waste and -emissions design will depend on a number of factors, including corporate and business environmental goals, economics, and applicable regulations.

The intent of the continuum is to provide engineers with a perspective on where regulations are driving industry relative to a goal of zero waste and emissions.

15.5.1 Process Modifications to Eliminate the Contaminants

Process modifications to eliminate contaminants represent true pollution-prevention options that prevent contamination of ventilation air in the first place. Often, they require a shift from "open" to "closed" systems. An example would be switching from a batch process with manual addition of powders, solvents, and so on, to a batch process using closed-loop addition of reagents or even to a closed continuous process. In another case, removing the most volatile compound from a wash-solvent mixture could result in a dramatic reduction in VOC emissions. Instituting a good preventive-maintenance program to detect and correct leaks will also go a long way toward reducing contamination of ventilation air.

15.5.2 Minimize the Volume of Contaminated Air

An emerging trend in industry is the concept of a "clean plant."[3] This concept relies heavily on a clean-plant design coupled with general ventilation of the plant and local exhaust ventilation of specific process areas or equipment.[4] By definition, a clean-plant design must incorporate good pollution-prevention practices.

Segregation. An important step in implementing a clean-plant design is to segregate (or isolate) the clean and dirty operations/areas of a facility. For example, cleanroom designers use mini-environments to create a physical barrier to isolate the susceptible part of the manufacturing process from the rest of the room.[5] From an environmental perspective, this is equivalent to isolating the source of emissions from the workplace area. The benefit of this approach is the greatly reduced volume of air that is contaminated. This is in addition to the occupational health and safety benefits that result from reduced employee exposure to harmful emissions.

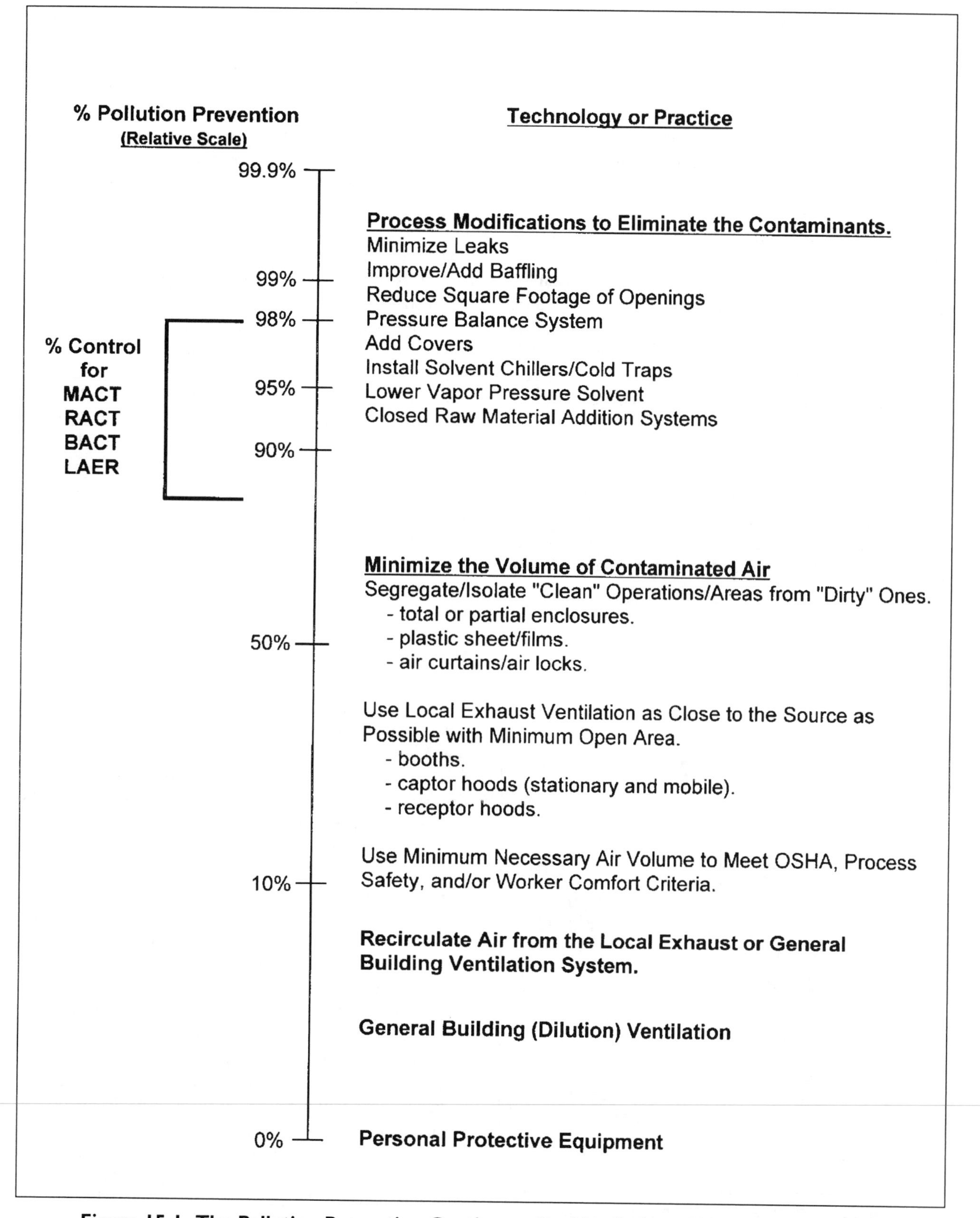

Figure 15-1. The Pollution-Prevention Continuum For Ventilation of Manufacturing Areas

Techniques to segregate the source of emissions from the general work area include total or partial enclosures around equipment, plastic curtains, equipment covers, air curtains, and air locks. A total enclosure (e.g., glovebox, room inside a room) completely surrounds the source of emissions and is preferred over the other segregation techniques, because escape of emissions is limited to leaks through openings.[6-9] A key objective, then, is to minimize the square footage (area) of openings. Enclosures are not always suitable, however, especially in cases requiring ready access to the process.

Air curtains are sometimes used to provide a "seal" to separate environments, while allowing unhindered traffic flow and an unobstructed view through the opening. The degree of seal provided by air curtains is about 70%.[10] They have also been used successfully for fume and dust control as a method to both contain and convey the process emissions to a captor hood or gas cleaning device.

Local Exhaust Ventilation. Local exhaust ventilation (LEV) systems are designed to capture and remove process emissions at the point of generation. When properly designed, LEV offers superior control and is preferred over general building (dilution) ventilation for the removal of many contaminants. LEV systems can effectively control high emissions rates and highly toxic materials, while using less air volume and, thus, less energy than general building ventilation systems.[3] A key advantage of LEV is that the contaminants are concentrated in a small volume of air that can be treated more cost effectively. The goal of the designer should be to minimize the open area between the hood and the emission source.[11]

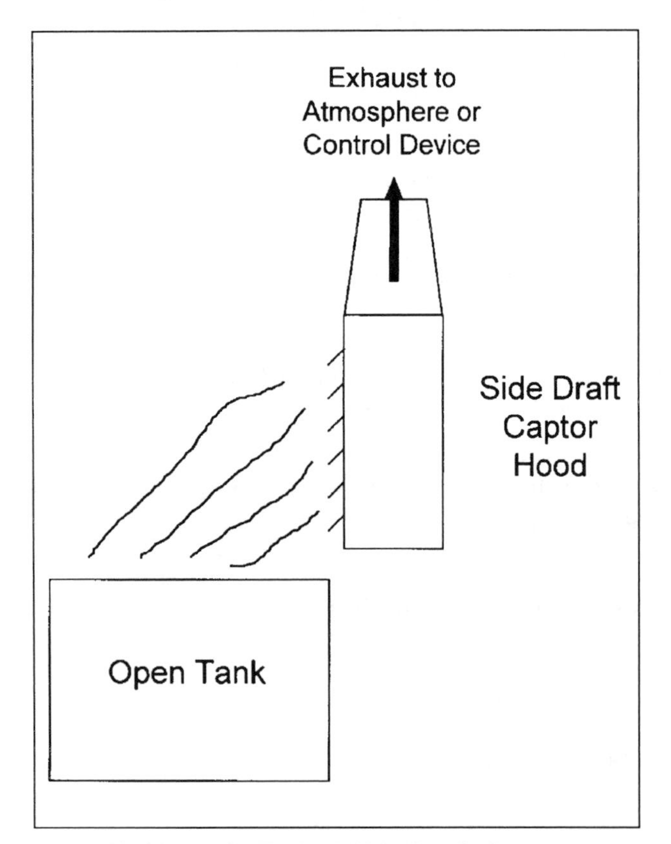

Figure 15-2. Typical Side Draft Captor Hood for an Open Tank

Hood types for LEV systems include booths, captor hoods, and receptor hoods.[12] A booth is nothing more than an enclosure with one wall missing, for example, laboratory hood, paint spray booth. Captor hoods operate at a distance from the emission source and use a fan to draw the contaminants into the hood. They can be stationary or mobile. Examples of stationary captor hoods include side-draft hoods, slot hoods, downdraft hoods, and high-velocity/low-volume hoods. Figure 15-2 shows a typical side-draft captor hood used to control fume or vapor emissions. An example of a mobile captor hood is a flexible-hose trunk. Mobile hoods are best suited for intermittent or transient emission sources, such as some welding operations.[12]

Receptor hoods are like captor hoods, except that the power needed to direct the contaminated air to the hood is supplied by the process (e.g., a hot buoyant plume), not by a fan. Figure 15-3 shows a typical receptor hood.

Minimum Necessary Air Volume. In the past ventilation systems were often designed with healthy safety margins to ensure meeting OSHA and process safety criteria. This was acceptable because energy costs were low and emission controls were usually not necessary. Nowadays, this excess air volume may result in a larger and more expensive control device. Contact a ventilation specialist to help establish the minimum acceptable capture velocity and, hence, air volume for your process.

15.5.3 Recirculation of Ventilation Air

Recirculation of ventilation air is one way to reduce the amount of energy consumed in the conditioning (heating and cooling) of make-up air supplied to a ventilated workplace. When this is done, important safety measures must be taken to ensure that contaminant concentrations remain within acceptable limits and that worker safety is not compromised. Before being returned to the workplace, recirculated air must be cleaned of contaminants or be of acceptable quality to ensure that no person is exposed to hazardous concentrations of any chemical in the exhaust. This approach becomes more feasible as the degree of isolation improves between the general work area and the process equipment emitting contaminants.

15.5.4 Dilution or General Building Ventilation

General building ventilation can be used in a plant for comfort control (temperature and relative humidity), control of contaminant concentration, general housekeeping, and collection of material for reuse. General ventilation is recommended when

- The contaminants have low toxicity
- The workers operate at a distance from the emission source
- The emission rate is low and uniform
- There is no fire hazard
- There are good air-flow patterns and dilution capabilities

General building ventilation can be achieved through three techniques: dilution ventilation, natural ventilation, and displacement ventilation. Dilution ventilation is the most common, so this type of system is discussed further. However, the drivers for pollution prevention apply to all three techniques.

Dilution ventilation systems supply air through vents near the ceiling at a relatively high velocity. Room air becomes entrained, resulting in a nearly uniform, well-mixed distribution of temperature and contaminant throughout the area. If sufficient make-up air is used and adequate mixing is achieved, the contaminants become diluted and provide acceptable air quality in the work area. In most cases, dilution ventilation systems will need to be used in combination with local exhaust systems for toxic contaminants.[3]

Often, the volume of air required to maintain a safe working environment is expressed as the number of room air changes per hour (RAC/h). This measurement is calculated by taking the ratio of the ventilation rate (ft^3/h) to the room volume (ft^3). Typical values for industrial applications range from 8 to 30 RAC/h depending on OSHA requirements and the solvents used.[2] For a 4000-ft^2 by 20-ft high room, this would equate to a ventilation air flow of 11,000 to 40,000 standard cubic foot per minute (scfm). Caution should be exercised before using published RAC/h to design industrial ventilation systems. Ventilation requirements for health-hazard control should be based on a material balance for the contaminants in question, not room volume alone.

The main disadvantage of dilution systems is obvious—contaminants of high toxicity demand large quantities of dilution air to comply with OSHA PELs.

15.5.5 Personal Protective Equipment

If possible, personal protective equipment should not be the sole means of protection against unsafe or unhealthy levels of contaminants because:

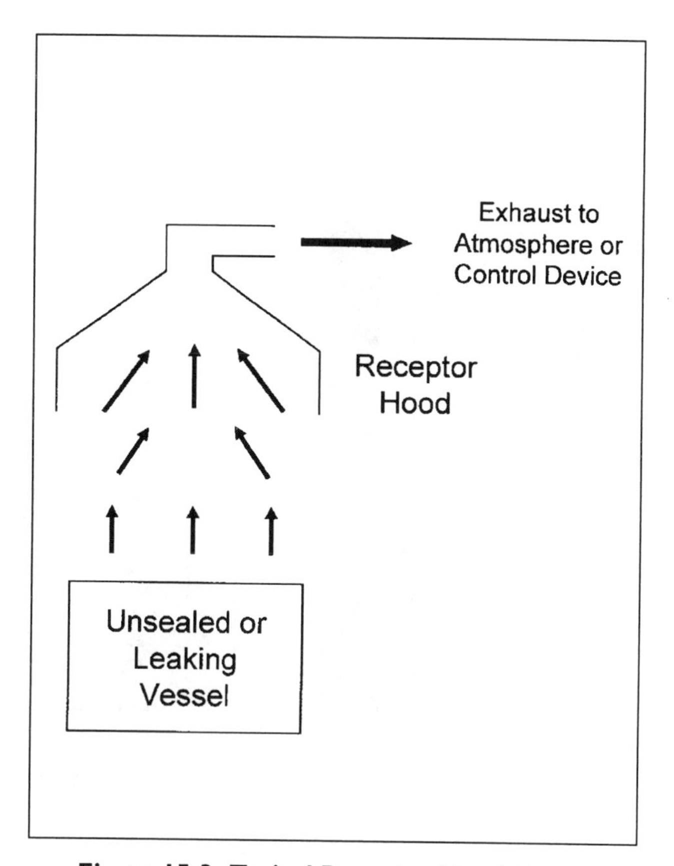

Figure 15-3. Typical Receptor Hood for an Unsealed or Leaking Vessel

- The equipment often presents a secondary hazard by impairing sensory perception, mobility, and communications.
- It is difficult to monitor the equipment's performance.
- It is difficult to ensure the equipment's use at all times.

15.6 The Incentive For Pollution Prevention

Figure 15-4 shows the incentive for preventing contamination of ventilation air in the first place and for minimizing the volume of air that is contaminated. The graph gives the minimum investment for VOC and halogenated VOC (HVOC) end-of-pipe abatement units. The incentive for avoiding particulate contamination of ventilation air is very similar. It is known from previous studies that the cost of end-of-pipe treatment depends primarily on air volume, not concentration. For more discussion on this subject, refer to Chapter 4.

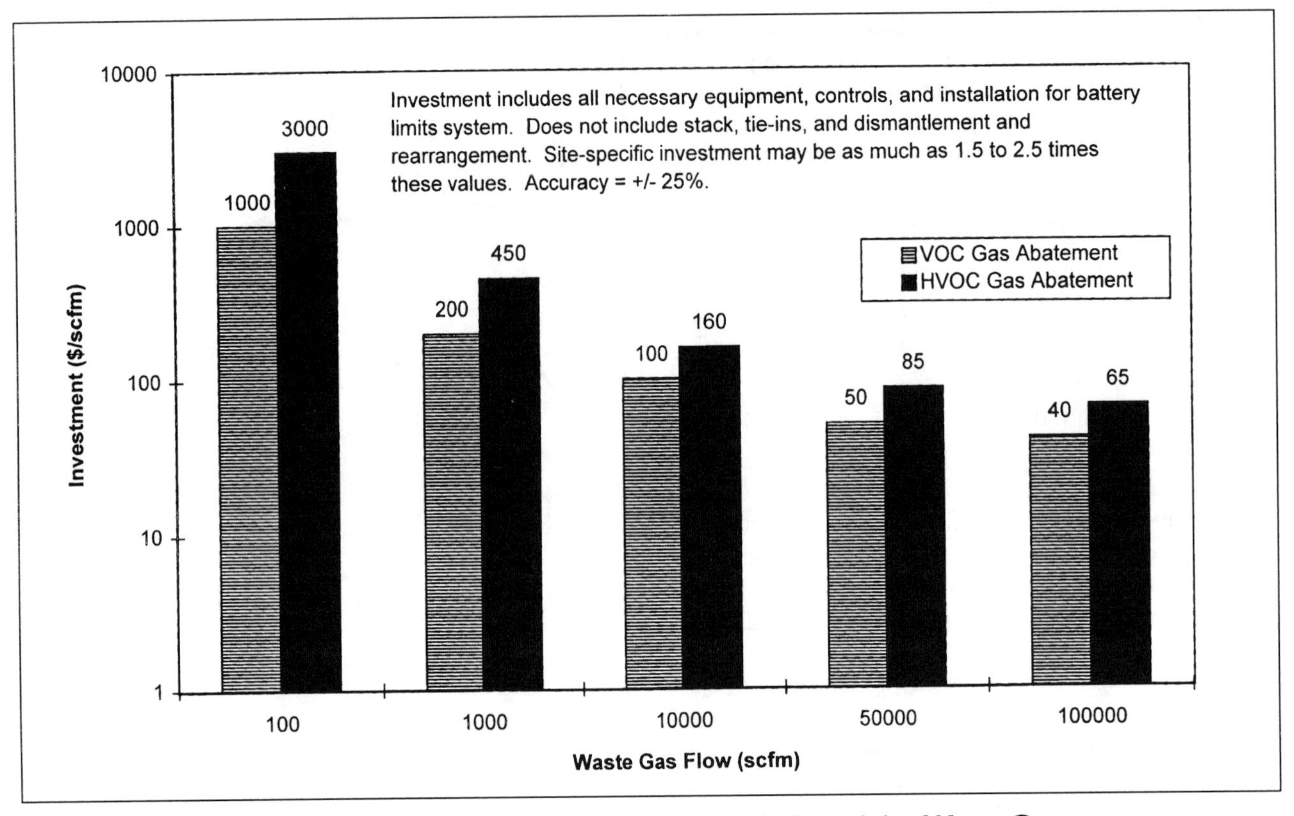

Figure 15-4. Investment For VOC- and HVOC-Containing Waste Gases

15.7 Case Histories

15.7.1 Printing Facility

The Westvaco Folding Carton Division in Richmond, Virginia, uses rotogravure printing presses to produce consumer-products packaging materials. Typical solvents used include toluene, isopropyl acetate, acetone, and methyl ethyl ketone. Driven by the EPA's new source performance standards for the surface-coating industry, the site installed a permanent total enclosure (PTE) around a new press so as to attain a 100% VOC capture efficiency. Leaks from the hot-air convection dryers and other fugitive emissions from the coating operation are captured in the press enclosure and routed, along with the dryer exhaust, to a carbon adsorber for recovery. Overall VOC removal efficiency for the enclosure and recovery system is greater than 95%. While many rotogravure press installations use the total pressroom as the enclosure, Westvaco was one of the first to install a separate, smaller enclosure around the new press. Notable features of the enclosure include

- Quick-opening access doors
- A dryer that serves as part of the enclosure to minimize the enclosure size
- VOC concentration monitors that control air flow to each dryer stage to maintain the dryers at 25–40% of the LEL
- Damper controls that maintain a constant exhaust rate from the enclosure to ensure a slight vacuum within the enclosure

If the pressroom had been used as the enclosure, the amount of ventilation air requiring treatment would have been 200,000 scfm. Instead, the use of the enclosure and the LEL monitors reduced the air flow to the adsorber to 48,000 scfm. This resulted in an investment savings for the carbon adsorber of approximately $5,000,000. The installed cost of the 1700-ft^2 enclosure was only $80,000, or $47/ft^2.

15.7.2 Semiconductor and Pharmaceutical Industries

Both the semiconductor and pharmaceutical industries use cleanrooms to manufacture products. Because cleanrooms are very expensive to build and operate, the trend in the semiconductor industry is to move from large open rooms (where the cleanest air is within the general workplace area) to mini-environments (enclosures) where the cleanest air is maintained within the enclosure and poorer conditions are acceptable within the general workplace area.[5] In the pharmaceutical industry, these enclosures are referred to as isolation chambers because of the need to isolate workers from active/toxic products in addition to supplying clean air.[4] In both cases, segregation through the use of enclosures is a preferred approach. For example, the average cost of a Class 100,000 cleanroom is $100/ft^2$, while a Class 1 cleanroom costs $1000–$4000/ft^2$. A Class 100,000 cleanroom allows 100,000 0.5-micron particles per cubic foot, while a Class 1 cleanroom allows only one 0.5-micron particle per cubic foot.

15.7.3 Parts Cleaner

A DuPont facility was faced with controlling VOC emissions from a solvent parts cleaner. The wash solvent contained a mixture of aromatics and ketones of differing volatility. The exhaust from the parts cleaner was approximately 1000 scfm and contained 1000 parts per million by volume (ppmv) VOCs. A quick analysis of end-of-pipe treatment options identified the incentive for pollution prevention: $200,000 investment and $75,000/yr to operate. Armed with this information, the team identified several opportunities for pollution prevention. They included

- Using a lower vapor-pressure solvent
- Removing the most volatile ketone from the existing blend because it comprised 40–50 vol % of the VOC emissions, while only constituting 10 wt % of the solvent mixture
- Reducing the open area of the parts cleaner hood and lowering the face velocity to the minimum permissible
- Chilling the solvent bath to reduce its vapor pressure

The team is now in the process of further evaluating these options.

15.7.4 Film Facility

A DuPont film facility was required by state air regulations to install reasonably available control technology (RACT) on any process vents emitting more than 3 lb/h VOCs. A RACT analysis showed that the cost to control VOC emissions at the end of the pipe would be a $3,000,000 investment and cost $600,000/yr to operate. With this as the incentive for pollution prevention, a site team was able to identify process modifications that reduced the amount of ventilation air that was contaminated by the process solvent. Changes included improving baffling between dryer zones, minimizing air inleakage, and segregating a quench tank exhaust stream from the surrounding ventilation air. These changes reduced VOC emissions to less than 3 lb/h in each of the vents and eliminated the need to install end-of-pipe treatment.

Literature Cited

1. Goodfellow, H. D. 1985. "Advanced Design of Ventilation Systems for Contaminant Control." *Chemical Engineering Monographs,* Vol. 23: 1–4. Amsterdam: Elsevier.
2. Lukey, M. E. 1992. "Total Permanent Enclosures for the Surface Coating Industry." *Proceedings of the 1992 A&WMA 85th Annual Meeting and Exhibition,* 92-51.07. Pittsburgh, Penn.: Air & Waste Management Association.
3. Grebenc, J., and H. D. Goodfellow. January 1995. "Use Ventilation to Achieve 'Clean' Plants." *Chemical Engineering Progress.* 91(1): 35–41.
4. Moore, S., G. Parkinson, and G. Ondrey. May 1995. "Cleanrooms Go Mainstream: Once the Province of Electronics, Ultrapure Processing is Moving into Other Applications." *Chemical Engineering.* 102(5): 33–37.
5. Whyte, W. September 1993. "Mini-Environments and the Cleanroom." *Professional Engineering.* 6(8): 25–26.
6. Frankel, I., and G. Bisonett. July 1995. "Practical Considerations of Design and Installation of Permanent Total Enclosures." *Journal of Coatings Technology.* 67(846): 63–68.
7. Warlick, T. D. 1992. "Permanent Total Enclosure Technologies Within the Printing Industry: A Company's History from 1986 to Present." *Proceedings of the 1992 A&WMA 85th Annual Meeting and Exhibition,* 92-51.01. Pittsburgh, Penn.: Air & Waste Management Association.
8. Warlick, T. D. 1996. "Permanent Total Enclosures—Their Effect on Manufacturing & Employee Safety: A Ten-Year Study Within the Flexographic Printing Facility." *Proceedings of the 1996 A&WMA 89th Annual Meeting and Exhibition,* 96-TP4C.07. Pittsburgh, Penn.: Air & Waste Management Association.
9. Kashdan, E. R., D.W. Coy, J. J. Spivey, T. Cesta, H. D. Goodfellow, and D. L. Harmon. 1986. "Highlights from Technical Manual on Hood System Capture of Process Fugitive Particulate Emissions." *Chemical Engineering Monographs,* Vol. 24, H. D. Goodfellow, Ed., 497–509. Amsterdam: Elsevier.
10. Goodfellow, H. D. 1985. "Advanced Design of Ventilation Systems for Contaminant Control." *Chemical Engineering Monographs,* Vol. 23, 509–517. Amsterdam: Elsevier.
11. Zhivov, A. M. 1993. "Principles of Source Capturing and General Ventilation Design for Welding Premises." *Proceedings of the 1993 ASHRAE Winter Meeting*, Chicago, Ill. American Society of Heating, Refrigerating and Air Conditioning Engineers.
12. Goodfellow, H. D. February 1987. "Hood Design for Ventilation Systems." *Heating/Piping/Air Conditioning.* 59(2): 60–67.

CHAPTER 16

Volatile Organic-Liquid Storage

16.1 Introduction

Some of us will be challenged in the future to choose and implement sound solutions to minimize or eliminate emissions from storage tanks and vessels containing volatile organic liquids (VOLs). In this chapter we address approaches that process, project, and environmental engineers can use to reduce air emissions of volatile organics from storage tanks and vessels. Case histories are included that reflect the wide range of solutions that is being applied throughout industry. The discussion that follows is not a detailed design tool, nor is it a prescription for solving every possible air emissions problem. Rather, it deals with concepts and approaches on how to move toward a zero waste-generation and emissions design.

Choosing an approach to minimize or eliminate emissions from storage tanks and vessels will depend on many factors, such as tank size, solvent vapor pressure, single tank versus tank farm, retrofit versus new installation, vent header safety, and state and federal regulations. In one situation, a pressure vessel may be the best choice for a new plant design, while in another situation, a floating roof may be the best choice.

In this chapter we specifically address volatile organic compound (VOC) emissions from vents on storage tanks and vessels. Other environmental issues associated with storage tanks and vessels, such as groundwater protection and fugitive emissions, are not covered.

16.2 Nature of Emissions Sources

Vent streams from storage tanks and vessels typically have volumetric flow rates less than 100 standard cubic feet per minute (scfm). These vents can be continuous, such as in the case of a vessel with a continuous nitrogen purge, or they can be intermittent, as is the case for most atmospheric tanks with conservation vents. The diluent or carrier gas can be air or nitrogen, while the VOC concentration will vary with organic type, liquid composition, temperature, pressure, and degree of saturation. Also, water vapor will often be present when air is the diluent. In the United States, many of the VOC(s) will be found on the 1990 Clean Air Act Amendments' (CAAAs') hazardous air pollutants (HAPs) list (e.g., methanol, xylene, toluene, and acetone) and, as such, will be specifically regulated.

16.3 Regulatory Outlook

In this section we provide a snapshot of where the regulatory community stands on VOC emissions from storage tanks and vessels. The discussion is by no means exhaustive. For more detail on applicable regulations, contact your business' environmental specialist.

The regulatory trend is to require at least a 95% reduction in VOC emissions from regulated VOL storage tanks. Applicability of the regulations varies from state to state and is a function of tank size, solvent vapor pressure, organic type, whether the tank is new or existing, whether the facility is a major source of HAPs, whether the site is located in an ozone attainment or nonattainment area, and so on. Compliance requirements will also vary from state to state. For example, Texas prefers flares over refrigerated vent condensers for end-of-pipe emissions control, while most other states list vent condensers as an acceptable means of control.

To ensure compliance with all existing and pending U.S. regulations, it is suggested that consideration be given to emission controls that meet or exceed maximum achievable control technology (MACT) standards as promulgated under the CAAA's Synthetic Organic Chemical

Manufacturing Industry's (SOCMI) Hazardous Organic National Emission Standard for Hazardous Air Pollutants (NESHAP). Table 16-1 summarizes applicability and compliance requirements for the storage-tank provisions of the SOCMI HON. Exemptions from these rules include storage vessels of less than 10,000 gallons and pressure vessels designed to operate at greater than 30.4 psia with no emissions to the atmosphere.

In addition, the Resource Convervation and Recovery Act (RCRA) Phase II VOC Rule regulates certain storage tanks that receive hazardous waste that, at the point of generation or as received from off-site, contain greater than 100 ppm by weight VOCs. Control requirements consist of tank covers (e.g., roofs) equipped with a 95% efficient control device or, for enclosed combustion devices (i.e., thermal oxidizers versus flares), a reduction of the total organic content of the vapor stream to a level less than 20 ppm by weight on a dry basis corrected to 3% oxygen.

16.4 How Do Emissions Originate?

16.4.1 Fixed Roof Tanks

The two main sources of emissions from fixed roof tanks are breathing losses and working losses. Breathing

Vent Classification	Applicability		Compliance Requirements
	Existing Source	New Source	
Group 1	> 40,000 gallon and vapor pressure > 0.75 psia* or 20,000-40,000 gallon and vapor pressure > 1.9 psia*	> 40,000 gallon and vapor pressure > 0.1 psia* or 10,000 to 40,000 gallon and vapor pressure > 1.9 psia*	Internal floating roof with double seals, single liquid-mounted seal or single metallic shoe seal; OR External floating roof with double seal; OR Internal floating roof converted from external floating roof; OR Closed vent system and 95% control device; OR Include in emissions average**
Group 2	All others	All others	Reporting and recordkeeping requirements only.

* Vapor pressure is at the temperature equal to the highest calendar-month average of the VOL storage temperature for VOL's stored above or below the ambient temperature or at the local maximum monthly average temperature as reported by the National Weather Service for VOL's stored at the ambient temperature.

** A facility-wide equivalent level of emissions reduction may be achieved through a combination of less effective controls on some vents and more effective controls on other vents.

Table 16-1. MACT Standards for Storage Tanks per the 1990 Clean Air Act Amendments SOCMI HON

loss is the expulsion of vapor from a tank vapor space that has expanded due to daily changes in temperature and barometric pressure. This loss occurs without any liquid level change in the tank and is impacted by the size of the tank's vapor space, tank location (i.e., direct sunlight or shaded), tank color, atmospheric conditions, whether the tank is insulated or not, and method of inerting (blanket or continuous purging).

Working losses result from filling and emptying operations. Filling losses are caused by an increase in the liquid level in the tank. Vapors are expelled from the tank when the pressure inside the tank exceeds the set pressure of the relief device. Emptying losses occur when air or nitrogen that is drawn into the tank during liquid removal saturates with VOCs and expands, thus exceeding the pressure setpoint of the vent device. Working losses are mainly a function of the filling/emptying rates and turnover frequency, and tend to be larger than breathing losses, except for very large tanks.

16.4.2 Floating Roof Tanks

Emissions from floating roof tanks are the sum of standing storage losses and withdrawal losses. Standing storage losses include rim-seal and roof-fitting losses for external floating roof tanks and rim-seal, deck-fitting, and deck-seam losses for internal floating roof tanks. In the case of an external floating roof tank, the rim-seal losses can be wind induced. Withdrawal losses are evaporative losses that occur as the liquid level, and thus the floating roof, is lowered. A liquid film remains attached to the tank wall and is exposed to the atmosphere. Standing storage losses are the major source of emissions from floating roof tanks.

16.4.3 Stratification

Stratification of the tank's vapor space will occur in large tanks with little turnover or agitation of the liquid surface. In these cases, the tank vent will not be saturated with organic but, instead, will be at some fraction of saturation, typically 10–25%. This may also be true of tanks with a continuous nitrogen purge if the purge rate is high enough. Otherwise, it is often assumed that the vent is 80–100% saturated with organic at the tank operating temperature and pressure. This issue becomes particularly important when designing a refrigerated vent condenser for 95% control efficiency versus a maximum mass emission rate. For example, consider a vent that, at saturation, contains 10 lb/h VOC. If saturated, 95% control of this vent would equate to 0.5-lb/h maximum VOC discharge concentration. However, if the vent is only 25% saturated due to stratification, then the maximum discharge concentration would drop to 0.125 lb/h to achieve the same 95% control efficiency.

16.4.4 Estimating Emissions

The most widely accepted method for estimating emissions from fixed roof and floating roof tanks is EPA Method AP-42.[1] This method allows one to calculate an annual mass emission rate, but it does not provide a procedure for calculating a design basis (maximum mass emission rate and volumetric flow rate) for end-of-pipe control devices, such as vent condensers.

In general, reaching agreement with state regulators on the design basis (flow rate and concentration) for an end-of-pipe control device can be difficult, especially for vent condensers. In the case of vent condensers, the control efficiency will vary from winter to summer as the vent inlet concentration and composition change with atmospheric conditions. Therefore, maintaining 95% control of a wintertime vent stream of 100 scfm and 100 parts per million by volume (ppmv) requires a larger refrigeration unit than achieving 95% control of a 100 scfm summertime vent at 700 ppmv.

16.4.5 Fugitive Emissions (Equipment Leaks)

This chapter does not specifically address fugitive emissions from storage tanks and vessels. Sources of fugitive emissions include equipment leaks from pumps, flanges, valves, and other tank seals.

16.5 The Pollution-Prevention Continuum

The continuum depicted in Figure 16-1 shows the range of options that are available to engineers to minimize or eliminate VOC emissions from storage tanks and vessels. The decision of how far to move toward a zero-waste and -emissions design will depend on a number of factors, including corporate and business environmental goals, economics, and applicable regulations. The intent of the continuum is to provide the user with a perspective on where regulations are driving industry relative to a goal of zero waste and emissions.

16.6 Discussion

16.6.1 Source Reduction of Breathing Losses

Source reduction approaches that typically reduce breathing losses between 10 and 95% are described below. One or more of these should be considered even when an end-of-pipe treatment device will be installed to help reduce the size and/or cost of the treatment unit.

Reduce the Volume of the Vapor Space in the Tank. A large fraction of breathing losses is caused by thermal expansion of the VOC-laden vapor in the tank head space. Therefore, operate five tanks completely full rather than ten tanks half full.

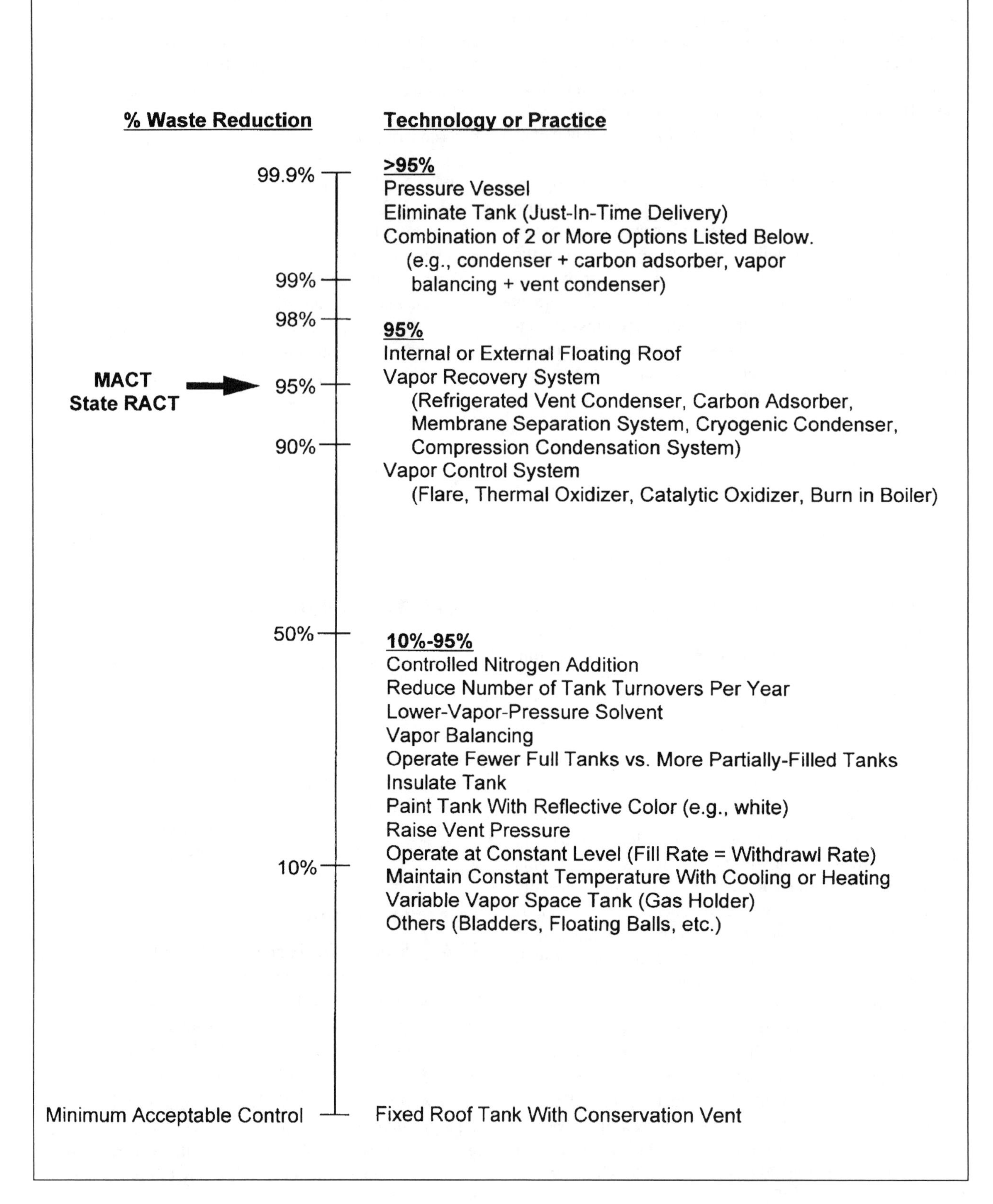

Figure 16-1. The Pollution-Prevention Continuum For Storage Tanks and Vessels

Controlled Addition of Inerts. Whenever possible, eliminate continuous nitrogen addition for padding. Consider a design based on pressure control rather than flow control. If continuous addition is necessary, use a rotameter or other flow control device.

Keep the Tank at Constant Temperature. Options include painting the tank white to reflect more sunlight, insulating the tank to reduce heat transfer between the tank contents and the surrounding air, and operating the tank at a constant temperature with supplemental cooling and/or heating. The economic feasibility of the last two options, in particular, would need to be evaluated on a case-by-case basis.

Raise the Vent Pressure. Examine the possibility of raising the pressure at which the conservation vent opens to reduce emissions caused by thermal expansion of the tank vapor space. However, do not exceed the tank design pressure! Tanks that operate at 2 1/2 psig or higher will often have little or no breathing loss.

16.6.2 Source Reduction of Working Losses

Source reduction steps that can reduce working losses between 10 and 95% are described below. One or more of these should be considered even when an end-of-pipe treatment device will be installed to help reduce the size and/or cost of the treatment unit.

Vapor Balancing. Consider installing vapor return lines to send the displaced vapor from the tank being filled back to the liquid source (e.g., tank truck, barge). This approach is used at gasoline stations during Stage I (vapor balancing between unloading tank truck and underground storage tank) and Stage II vapor recovery (vapor balancing between car gas tank and underground storage tank). Eventually, however, the displaced vapor will have to be treated. In the case of gasoline stations, this occurs at the tank truck loading terminal. A corollary approach would be to equalize the vapor space between tanks in a tank farm using a common vent header. This would apply to tanks containing the same solvent or where cross-contamination is acceptable. Vapor displaced while filling one tank would fill the vapor space of a tank being emptied. When considering vapor balancing techniques, vent-header safety must be addressed to avoid creating a situation where fire could propagate between tanks.[2]

Maintain Constant Tank Level. If the fill rate equals the withdrawal rate, then working losses are eliminated.

Reduce the Number of Tank Turnovers. The annual number of tank turnovers (tank volumes/yr) is equivalent to the annual solvent consumption (gal/yr) divided by the liquid storage volume of the tank (gal/tank). For a fixed tank volume, reducing the number of tank turnovers per year will reduce emissions. In addition, the more frequently a tank is "turned over," the more likely the tank vapor space will be saturated with the VOC.

16.6.3 Other Source Reduction Approaches

Source reduction options that reduce both breathing and working losses are discussed below.

Eliminate Tank. Consider just-in-time delivery from a tank car or tank truck.

Alternative Solvent. Particularly in the design of new processes, it may be feasible to choose a solvent with a lower vapor pressure.

Other Approaches. Install a gas holder (variable vapor space tank) to contain displaced vapor during tank filling and thermal outbreathing. The vapor from the gas holder can then be returned to the liquid storage tank during tank emptying and inbreathing. Another option is to place Styrofoam balls or a fabric cover over the liquid surface to minimize evaporation. This option, however, has had some negative reviews. A third approach is to place an expandable bag or bladder inside the tank to contain the vapor; however, bag tears would be difficult to detect, particularly in large tanks.

16.6.4 Floating Roofs

Floating roofs are a source-reduction option that has traditionally been used in the petroleum industry to minimize VOC emissions from large storage tanks.[3] In recent years, however, vendors have upgraded their technology to make floating roofs a more viable option for the smaller tanks found in the chemical industry.

Floating roofs minimize solvent evaporation by reducing or eliminating the vapor space above the liquid surface. The liquid surface is completely covered by the floating roof except for a small annular space between the roof edge and the tank wall. The roof rises and falls with the liquid level, while a sliding seal attached to the roof fits against the tank wall and covers the exposed annular space. Both single- and double-seal systems are available, although only certain types meet MACT Standards. Improper or loose-fitting seals can significantly increase emissions. Figure 16-2 shows a comparison between a fixed roof solvent storage tank and a pan-type external floating roof tank.

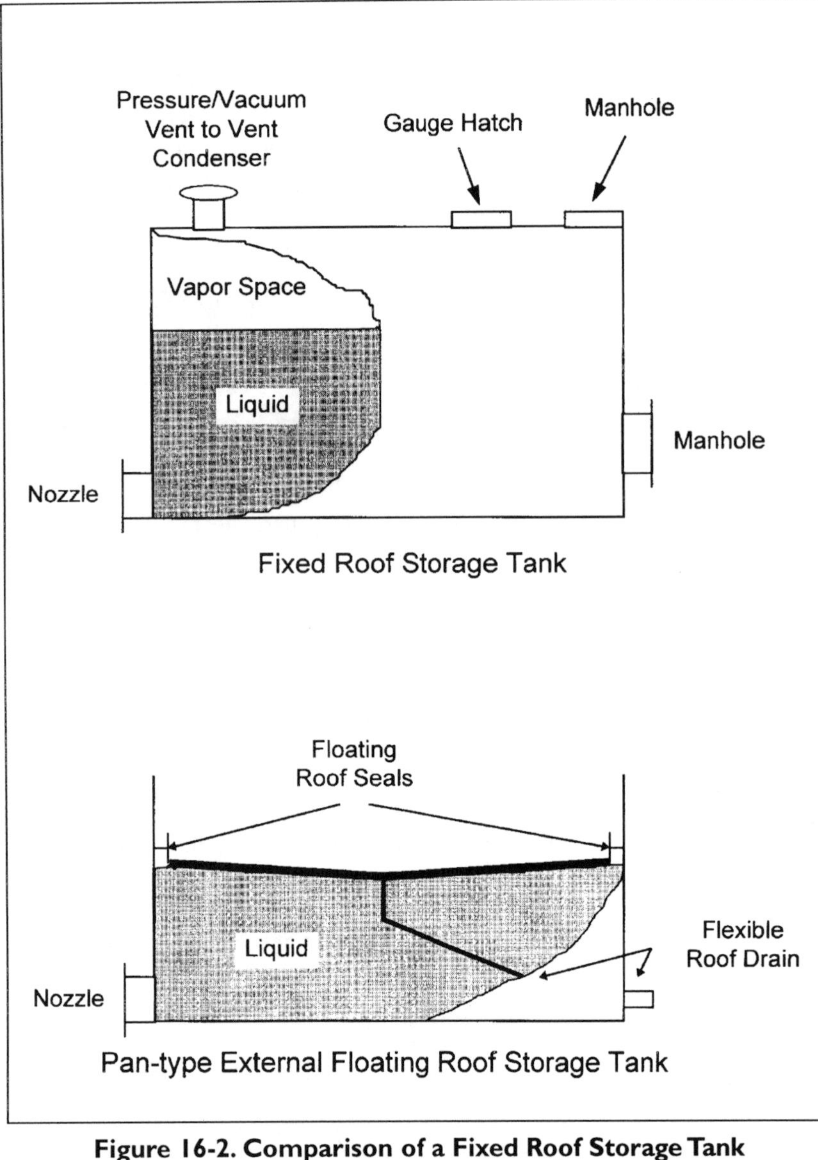

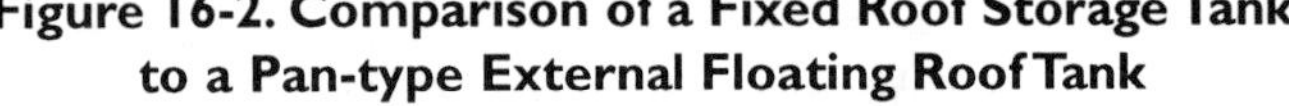

Figure 16-2. Comparison of a Fixed Roof Storage Tank to a Pan-type External Floating Roof Tank

External floating roofs are constructed of welded steel plates and are of three general types—pan, pontoon, and double deck. The present trend is to use pontoon and double-deck types. The four basic types of internal floating roofs are a full-contact steel pan, a full-contact aluminum honeycomb panel, full-contact buoyant FRP panels, and a noncontact aluminum skin and pontoon. For more detail, refer to the EPA Control Techniques Guideline Series, *Control of Volatile Organic Compound Emissions from Volatile Liquid Storage in Floating and Fixed Roof Tanks.*[4]

Issues to consider before installing a floating roof include:

- Floating roofs are difficult to install in tanks less than 10–14 feet in diameter.
- Retrofitting existing fixed roof tanks with an internal floating roof needs to be evaluated on a case-by-case basis. For example, the cost of cleaning and degassing the tank before installation should be included in the economic evaluation. In addition, structural modifications and special vents on the fixed roof may be necessary.
- Refer to appropriate state and federal regulations to ensure that the roof and seal designs meet applicable standards (e.g., MACT). Consider monitoring and recordkeeping requirements also.

DuPont Company estimates of installed investment for some typical internal floating roof retrofit systems are summarized in Table 16-2.

16.6.5 Pressure Vessels

Although pressure vessels are not an option in retrofit applications, they should be considered for new installations. The pressure vessel is enlarged in volume to allow compression of the vapor space during tank filling, thermal expansion of the vapor space, and solvent vaporization. Breathing and working losses are practically eliminated (the tank "breathes" only when the setpoint of the pressure/vacuum vent is exceeded). Pressure vessels are currently used for storing volatile compounds such as ammonia, propane, and butane. Savings realized by using a pressure vessel in place of a fixed roof atmospheric tank with an end-of-pipe control device include reduced maintenance, less emissions monitoring and recordkeeping, and lower utility costs. However, safety and design issues need to be carefully considered.

Economic evaluations in DuPont have shown that for a design pressure less than 200 psig and a tank volume less than 75,000 gallons, a single pressure vessel is often more cost effective than a single atmospheric fixed roof tank with an end-of-pipe control device. On the other hand, when more than three tanks are being installed (e.g., a tank farm), it will usually be more economical to install atmos-

Diameter (ft)	Volume (gallons)	Solvent	Type of Internal Floating Roof	Installed Investment ($1000)
10	5,900	Acetone	Aluminum Skin and Pontoon	21,000
15	23,900	Toluene	Aluminum Skin and Pontoon	23,000
30	127,000	Xylene	Aluminum Honeycomb Full Surface Contact	42,000
48	500,000	Methanol	Fiberglass Full Surface Contact	100,000
70	1,000,000	Methanol	Fiberglass Full Surface Contact	170,000
100	2,800,000	Xylene	Aluminum Honeycomb Full Surface Contact	220,000

Table 16-2. Installed Investment for Internal Floating Roof Retrofit Systems (ENR=5700)

pheric tanks with floating roofs or atmospheric tanks headered to a common control device.[5]

16.6.6 End-of-Pipe Controls

Vapor Recovery Units. Alternatives for recovering the VOCs include refrigerated condensation, compression condensation, cryogenic condensation, carbon adsorption, and membrane separation. In those instances where a greater than 95% reduction in emissions is required, two of these technologies may need to be used in series (e.g., condensation followed by carbon adsorption). Although there are a few exceptions, a fully installed vapor recovery system (≤ 100–200 scfm) will usually require a capital investment of $100,000 to $200,000.[6]

Vapor Control Units. Abatement of the VOCs can be accomplished using catalytic oxidation, thermal oxidation, flaring, and burning in an on-site boiler. Although there are a few exceptions, a fully installed vapor control system (≤ 100–200 scfm) will usually require a capital investment of $100,000 to $200,000.

16.7 Technology Selection Criteria

As mentioned in the Introduction to this chapter, the choice of what approach to use to minimize or eliminate emissions from storage tanks and vessels will depend on a number of factors, such as retrofit versus new installation, tank size, single tank versus tank farm, solvent type, required level of emissions reduction, and preferences of the regulatory authority. The technology selection criteria given below are based on economics and *current* industry experience for organic solvents containing carbon, hydrogen, and oxygen. The criteria assume that a 95% or higher reduction in emissions is required. Your situation may be different. For example, the selection criteria do not apply to highly toxic and highly corrosive materials, such as HF, phosgene, and ammonia. Also, be cautious when applying the selection criteria to halogenated organics. Halogenated organics are a special class of compounds that sometimes require a more complex end-of-pipe control device and/or specialized materials of construction.

16.7.1 Floating Roofs

A floating roof is often the economic choice for *single tanks* with a diameter *greater than* 10–14 feet (tank volume larger than 6000-30,000 gallons) in both *new and retrofit* applications. In both *new and retrofit tank farm* applications (more than 3 tanks), floating roofs may be competitive with a headered vent system using a common control device when each tank's diameter is *greater than* 10–14 feet (individual tank volume larger than 6000–30,000 gallons).

16.7.2 Dedicated Control Device

A dedicated control device is often the economic choice for *single tanks* with a diameter *less than* 10–14 feet (tank volume smaller than 6000–30,000 gallons) in *retrofit* applications.

16.7.3 Pressure Vessel

A pressure vessel is often the economic choice for *single tanks* with a diameter *less than* 10–14 feet (tank volume smaller than 6000–30,000 gallons) in *new* applications.

16.7.4 Headered Control Device

A headered control device is often the economic choice for *tank farms* (more than 3 tanks) where the tanks have a diameter *less than* 10–14 feet (tank volume smaller than 6000–30,000 gallons) in both *new and retrofit* applications. In both *new and retrofit tank-farm* applications, a headered-vent system using a common control device will be competitive with floating roofs when each tank's diameter is *greater than* 10–14 feet (individual tank volume larger than 6000-30,000 gallons).[5]

16.7.5 Source Reduction Options

Do not underestimate the level of emissions reduction that can be achieved by combining one or more of the source reduction approaches discussed previously. In some cases, vapor return lines alone may be the most cost-effective solution that satisfies the regulatory requirements.

16.8 Case Histories

The case histories that follow will provide a flavor of the types of approaches that are currently being implemented or considered within the chemical industry. They support the technology selection criteria in most cases.

16.8.1 Refrigerated Vent Condenser

At a DuPont site, a prototype direct-expansion, refrigerated vent condenser was installed to condense and recover acetone vapors from a single 30,000 gallon (13-ft-diameter) acetone storage tank. The vent condenser was installed on the tank roof and was designed for ambient conditions encountered at the site: from – 23°C in the winter up to 43°C in the summer, and for relative humidities from 10% to 95%. The unit was sized to achieve 95% recovery of acetone-saturated streams at 15°C ambient temperature (–40°C condensing temperature) and 35°C ambient temperature (–15°C condensing temperature). The system was designed with a dehumidification section to remove water vapor from the humid air during tank inbreathing, a defrost cycle to remove water ice from the condensation coils, and an electrical classification of Class I, Divisions 1 and 2, Group D. Actual removal efficiencies have ranged from 90% to 93% due to the absence of acetone recovery during the defrost period. Installed investment for the prototype unit was under $50,000. Retrofitting the tank with an internal floating roof was not pursued because the tank diameter (13 ft) was on the borderline of where floating roofs are technically feasible (10–14 ft diameter).

16.8.2 Headered Vent System

In a tank farm at a DuPont site, vent lines from multiple tanks containing dimethyl terephthalate (DMT) were tied together and routed to a common abatement device. Operating experience had shown that some tanks were being emptied, while other tanks were being filled. The combined vapor displacements were such that net outbreathing occurred infrequently, and only a small abatement device was needed for the entire tank farm. This approach was possible because all the tanks contained the same material, and the vapors were compatible. In this case, a combination of vapor balancing and a headered control device proved to be the most cost-effective solution. Remember, however, that vapor compatibility, product cross-contamination, and vent-header safety are three factors that must be considered before trying to implement this technique.[2]

16.8.3 Floating Roofs

At a DuPont site, state and federal regulations were driving the plant to reduce VOC emissions from methanol storage tanks by 1997. Originally, there were nine 500,000-gallon tanks and one 2,500,000-gallon tank located in a tank farm remote from the operating plant. An engineering evaluation completed in the mid-1990s showed that the best approach to reduce emissions was to (in order of preference):

- Reduce the volume of vapor space by operating each tank as full as possible and/or operate with fewer full tanks
- Install a floating roof on the 2,500,000-gallon tank
- Header the active tanks to a common vent condenser
- Header the active tanks to a common water scrubber
- Install a pipeline connecting the storage tanks to the barge area

Subsequent to this evaluation, the site became aware of a spare 2,500,000-gallon tank. They then concluded that the most cost-effective approach was to replace the nine 500,000-gallon tanks with the spare 2,500,000-gallon tank. The two 2,500,000-gallon tanks were then retrofitted with an internal floating roof (or they could also have been headered to a common control device). By going to two tanks with less total volume (5,000,000 gallons instead of 7,000,000 gallons), the volume of vapor space was reduced, and the cost to retrofit the two tanks with floating roofs was less than the cost to retrofit ten tanks.

16.8.4 Pressure Vessels

At a DuPont site, a new manufacturing facility will utilize 10,000-gallon pressure vessels to store a volatile organic raw material, instead of low-pressure tanks with conservation vents. The air permit for the new facility limits its VOC emissions to less than 5 ton/yr (including fugitives). Storage tanks at the existing manufacturing facilities are designed for 5 psig and are operated with conservation vents set at 12 inches water column. The tanks are purged with nitrogen to exclude oxygen from the tank vapor space, and use nitrogen-purged level indicators. The new tanks, on the other hand, will be designed for 100 psig, use diaphragm level indicators, add nitrogen to maintain a minimum pressure of 0.5 psig during liquid withdrawal, and vent at pressure to a minus 40°C refrigerated vent condenser followed by a headered flare. Seventy-five percent VOC recovery is expected in the vent condenser, with the flare destroying 98% of the remaining VOC emissions. Other options considered, but ruled out, included floating roofs and venting to a water scrubber followed by biotreatment of the aqueous waste stream.

Literature Cited

1. U.S. Environmenal Protection Agency. July 1993. *Compilation of Air Pollutant Emission Factors (AP-42)*, Vol. I: *Stationary Point and Area Sources*, 4th ed., Chap. 12 and Supplement F. Research Triangle Park, NC: U.S. EPA, Office of Air Quality Planning and Standards.

2. Clark, D. G., and R. W. Sylvester. January 1996. "Ensure Process Vent Collection System Safety." *Chemical Engineering Progress*. 92(1): 65–77.

3. TI-3 Petroleum Committee Air Pollution Control Association. May 1971. "Control of Atmospheric Emissions from Petroleum Storage Tanks." *Journal of the Air Pollution Control Association*. 22(5): 260–268.

4. EPA Emissions Standards Division. October 1993. *EPA Control Techniques Guideline Series–Control of Volatile Organic Compound Emissions from Volatile Liquid Storage in Floating and Fixed Roof Tanks*. Research Triangle Park, NC: U.S. EPA, Office of Air and Radiation, OAQPS.

5. Chetty, A. S., J. A. Dyer, and K. L. Mulholland. 1993. "Reducing the Cost of Treatment Through Standardization." *Proceedings of the 1993 A&WMA 86th Annual Meeting and Exhibition*, 93-TA-31A.01. Pittsburgh, Penn.: Air & Waste Management Association.

6. Dyer, J. A., and K. L. Mulholland. February 1994. "Toxic Air Emissions: What is the Full Cost to Your Business?" *Environmental Engineering: A Special Supplement to February 1994 Chemical Engineering*. 101(2): 4–8.

CHAPTER 17

Separation Technology Selection

17.1 Introduction

Most chemical manufacturing processes are composed of two major sections, one that transforms feed materials or reactants into products (e.g., a chemical reactor), and a second that consists of unit operations that separate the resulting mixed streams into their individual constituents (Figure 17-1). The separation unit-operations purify the feeds to the reactor and separate the desired products from any byproducts, solvents, and unreacted feed materials. The general types of separation processes are:[1]

1. Energy-related separations, such as distillation and crystallization
2. Mass-separation agents, including absorption and extraction
3. Transport-related separations, such as carbon adsorption and membranes
4. Combinations, including extractive distillation and azeotropic distillation

In this chapter, separation technology selection and application is discussed from a pollution-prevention perspective.

17.2 Pollution-Prevention Strategies

To approach the ideal of a minimum waste-generation process (where energy transforms feed materials into only salable products), separation technologies must be selected that utilize component property differences most efficiently and are sequenced in an optimum manner. Figure 17-2 shows an idealized chemical manufacturing process with minimum waste generation.

17.2.1 Selection Heuristics

Table 17-1 lists general heuristics for selecting and sequencing the separations methods.[1] Each item in the table is described in more detail below.

Separation Method Selection. The separation methods chosen depend on the differences in component properties[2] as illustrated in Table 17-2.

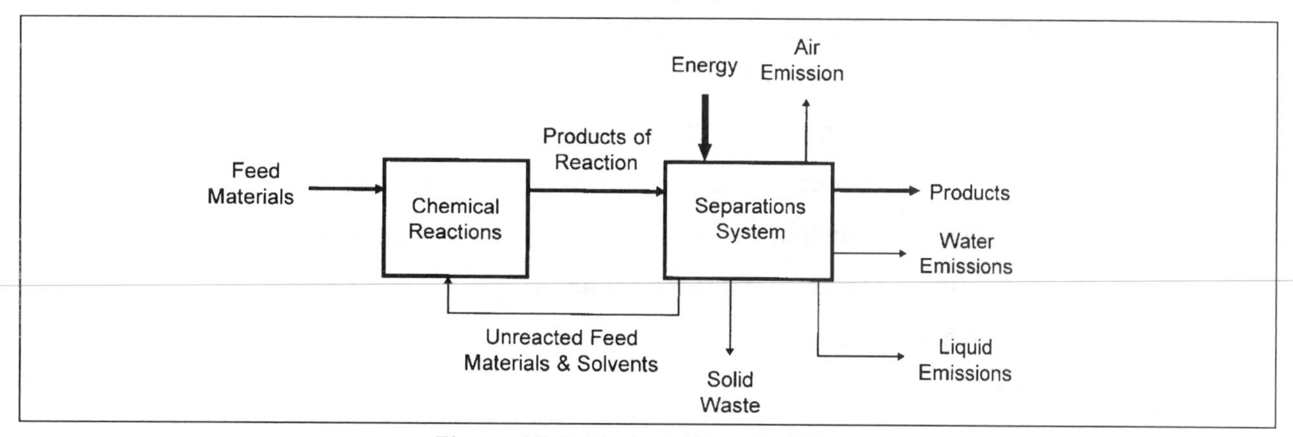

Figure 17-1. Typical Chemical Plant

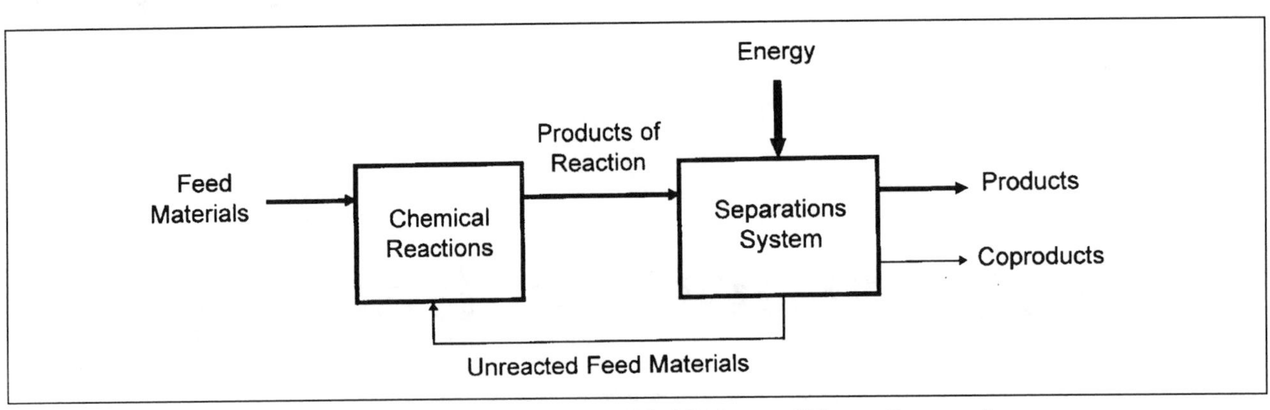

Figure 17-2. Chemical Plant With Minimum Waste Generation

Separation Load. Examine the streams to determine if some could be combined to reduce the number of unit operations, for instance, streams of similar composition.

Corrosive and Unstable Materials. Removing corrosive materials minimizes investment and the generation of undesirable trace metals. Removal of unstable compounds minimizes the formation of undesirable waste products, such as tars.

Highest-Volume Components. Removing the highest-volume components minimizes downstream equipment investment, and minimizes the addition of new materials required for processing that could become another source of process waste.

Difficult and High-Purity Separations. If component properties, such as vapor pressure, are close to each other, or if high purity is required, the fewer the number of additional components in the mixture, the easier and more efficient the separation.

Smallest Number of Products. The sequence with the smallest number of products will normally be the lowest investment.

Minimize the Addition of New Components. Raw materials, intermediates, and products have value to the business, while other materials added to the process, such as extraction solvents, only increase investment, operating cost, and waste load. Avoid using separation technologies that require adding new materials to the process.

1. Select separation methods.
2. Combine similar streams to minimize the number of separation units.
3. Remove corrosive and unstable materials early.
4. Separate highest-volume components first.
5. Do the most difficult separations last.
6. Do high-purity recovery fraction separations last.
7. Use a sequence resulting in the smallest number of products.
8. Avoid adding new components to the separation sequence.
9. If a mass separating agent is used, recover it in the next step.
10. Do not use a second mass recovery agent to recover the first.
11. Avoid extreme operating conditions.

Table 17-1. Separations Heuristics to Minimize Waste Generation

Recover New Components Quickly. If a separating agent, such as a solvent for liquid–liquid extraction, is used, the separation agent must be recovered immediately in the next separation step. Because the addition of new components increases cost and waste, do not use another separating agent to recover the first separation agent.

Operating Conditions. Avoid pressures and temperatures far from ambient conditions. When compared to below-ambient conditions, temperatures and pressures above ambient require lower operating and investment costs per unit of energy consumed.

17.2.2 Separation Unit-Operation Selection

Proper selection of separation technology involves consideration of differences in component properties (see Table 17-2), design reliability, energy requirements, and the number and amount of low-value materials required for processing, such as solvents and mass-transfer agents.

The differences in component properties dictate which unit operations are feasible. After feasible unit operations have been selected, design reliability becomes the next most important criterion. The plant must work properly, not only to make a profit, but also to minimize waste generation due to poorly realized separations and excessive startups and shutdowns. The energy requirements pertain not only to the quantity of energy, but also to the type of energy. For example, the use of efficiently produced on-site boiler steam for distillation requires much less energy than an electrical heater that includes losses in electrical generation, transmission, and resistance-heat generation.

17.3 Unit Operation Parameters

The traditional approach to process design has been to first engineer the process and then to engineer the treatment and disposal of waste streams. With increased regulatory and societal pressures to eliminate waste emissions to the environment, however, the total system must now be analyzed—process plus treatment—to find the most economic option. Experience in all industries teaches that processes that minimize waste generation at the source are the most economical. The proper selection, design, and operation of separation processes influence both the generation and introduction of waste materials.

- *Minimize Generation.* Select the separation process that minimizes the formation of waste constituents, such as tars and fines, in all chemical and physical separation steps.

- *Minimize Introduction.* Select the separations process that minimizes the addition of materials to the process that must be recovered and recycled and that are transformed to make waste. This implies minimizing the introduction of materials that are not essential ingredients in making the final product. Examples of introducing nonessential ingredients include: (1) using water as a solvent when one of the reactants, intermediates, or products could serve

Unit Operation	Property Differences
Distillation	Vapor pressure
Crystallization	Melting point or solubility
Extraction	Distribution between immiscible liquid phases
Adsorption	Surface sorption
Reverse osmosis	Diffusivity and solubility
Membrane gas separation	Diffusivity and solubility
Ultrafiltration	Molecular size
Ion exchange	Chemical reaction equilibrium
Dialysis	Diffusivity
Electrodialysis	Electric charge and ionic mobility
Liquid membranes	Diffusivity and reaction equilibrium
Electrophoresis	Electric charge and ionic mobility
Chromatographic separations	Depends on type of stationary phase
Gel filtration	Molecular size and shape

Table 17-2. Unit Operations and Required Property Differences Between Species[2]

the same function; and (2) adding large volumes of nitrogen gas, because air is used as the oxygen source, heat sink, diluent, or conveying gas.

The strengths and weaknesses of the major separation processes are discussed briefly in the following sections.

17.3.1 Distillation

Distillation is the principal technology used for the separation of a mixture into its individual components. Distillation uses the difference in boiling points and heat to separate the components. There are over 40,000 distillation columns in the United States (90 to 95% of the separation units in use).[3]

The two key design parameters for distillation columns are efficiency—how well the component separation occurs—and capacity—hydraulics and the rate at which materials can be processed without loss of efficiency. Factors favor distillation when the relative volatility is greater than 1.5:1, the products are thermally stable, the continuous process production rate is 10,000 lb/day or more, and high corrosion rates and explosive reactions do not occur at distillation conditions.

Weaknesses of distillation that impact energy use and waste generation include the following:

- When the relative volatility is less than 1.5 but greater than 1.15, distillation is feasible; however, it consumes large amounts of energy (that is, it becomes much more costly).
- Distillation is not energy efficient when attempting to separate a large volume of liquid from high boilers, for example, boiling water overhead to collect a small quantity of low-volatility contaminants in the bottoms stream.
- Poor liquid or gas distribution.
- Distillation cannot separate the components of an azeotropic mixture.
- Distillation can degrade thermally sensitive materials.
- Small continuous-distillation columns lose their investment advantage, for example, a 6-inch-diameter column is only 20 to 25% less expensive than a system having a 12-inch-diameter column.

Because distillation is the most widely used technology for component separations, extensive studies have been done to determine the major sources of poor operation.[4] Based on the analysis of a series of case histories, Kister[4] developed a hierarchy of distillation column malfunctions as shown in Table 17-3.

Chemical-process simulators, such as Aspen Technology's ASPEN PLUS (Cambridge, MA), Simulation Sciences' PRO/II (Brea, CA), and Hyprotech's HYSIM (Calgary, Alberta, Canada), can be used to model distillation columns. However, the accuracy of the simulation is dependent on the quality and availability of vapor–liquid-equilibrium data and the relationships chosen for the mixture components and their interactions. For example, an accurate definition of an azeotrope depends on accurate binary activity coefficient information. Other information, such as liquid- or vapor-phase dimerization that occurs with compounds such as hydrogen fluoride and acetic acid, will impact not only the vapor–liquid–equilibrium relationships but also the heat balance for the system.

17.3.2 Crystallization

Crystallization is similar to distillation in that (1) energy is used to achieve a separation change by forming a

Cause	Percent of Reported Cases
Instrument and control problems	8
Troublesome column internals	17
Startups and shutdowns	16
Operational difficulties	13
Design, foaming, installation, etc.	36

Table 17-3. Distribution of Distillation Column Malfunctions

solid from a saturated solution, and (2) the heat of fusion is generally much lower than the heat of vaporization, making crystallization a more energy-efficient process.

Factors favoring crystallization include the presence of thermally sensitive materials (e.g., pharmaceuticals), vapor pressures or relative volatilities that are too low for distillation to be feasible, reaction systems where a solid is one of the products, and cases where a competitive way of concentrating aqueous systems (e.g., freezing out water) is needed.

Weaknesses of crystallization are that the process design cannot easily be predicted from phase-equilibrium and physical-property data, and bench-scale and pilot-plant tests are normally required.

The principal sources of malfunction in crystallization systems are

- Poor temperature control due to heat exchanger fouling or plugging
- Poor solid–liquid separation
- Poor control of slurry density
- Encrustation or unwanted buildup of crystals in the crystallizer

DuPont has several double-draw-off (DDO) processes that can reduce or eliminate the problems associated with poor control of slurry density, encrustation, and solids buildup.[5]

17.3.3 Liquid–Liquid Extraction

Liquid–liquid extraction separates components of a mixture based on their different solubilities in a solvent. While there are 40,000 distillation columns in the United States, there are only 1000 or so extractors.[3] Two types of liquid–liquid extractors are shown in Figure 17-3. Two requirements must be met for extraction to work: the component(s) to be separated from the feed phase must preferentially distribute themselves in the solvent phase, and the feed and solvent must be substantially immiscible.

Factors that favor liquid/liquid extractors over distillation columns are

- Azeotropes or low relative volatilities are involved
- Low to moderate processing temperatures are needed
- Solvent recovery is easy
- Energy savings can be realized

An example of energy savings is the extraction of an organic from water (heat of vaporization of about 1000 Btu/lb) with a solvent having a lower heat of vaporization than water.

One weakness of extraction is that it requires a solvent, which brings with it added complexity in use, that is, storage, distribution, recovery, and reuse. Another weakness is that the use of a solvent often increases waste generation both from solvent losses themselves and lost salable products.

The principal sources of malfunction in extraction systems include:

- Poor liquid distribution
- Plugging of the packing or trays—agitated columns avoid this problem (see Figure 17-3)
- Surfactants that impact mixing and mass transfer
- Stable emulsions that form layers in the extractor
- Unknown contaminants in the feed that affect the surface chemistry of the system. To ensure proper design, testing of the liquid–liquid extraction system in a pilot plant is required.

Supercritical Fluid Extraction. A supercritical fluid, such as carbon dioxide, is a fluid operating above its critical point (for carbon dioxide, this corresponds to a pressure of 70 to 350 atm, depending on the temperature). The biggest single advantage of supercritical fluid extraction is the nontoxic nature of supercritical solvents. For processes and products where there are concerns about the environment, worker exposure, or purity of products for human consumption, supercritical carbon dioxide can play an important role. However, the higher costs associated with high operating pressures must be compared against the cost of using other solvents or methods of separation.

17.3.4 Adsorption

In adsorption, one or more solutes are transferred from a gas or liquid mixture to the surface of a solid adsorbent. Desorption is the reverse process, whereby the solute is removed from the surface of the adsorbent by a change in temperature or pressure (or both), or the addition of desorption fluid, such as steam, an organic solvent, or a mineral acid.

Adsorption involves the selective concentration of a species in a fluid at an interface, usually a solid–fluid interface. Solid adsorbents are used in fixed beds, moving beds, and slurries, and include zeolites, alumina, silica, activated carbon, polymeric materials, and ion-exchange resins. Adsorptive separations require high specificity to provide an economic advantage, for example, desiccant drying, ion exchange, and carbon adsorption. Adsorptive separations are normally utilized to remove the last traces of impurities.

Factors that favor adsorption are:

- A high degree of solute removal is needed, especially when at low contaminant concentrations

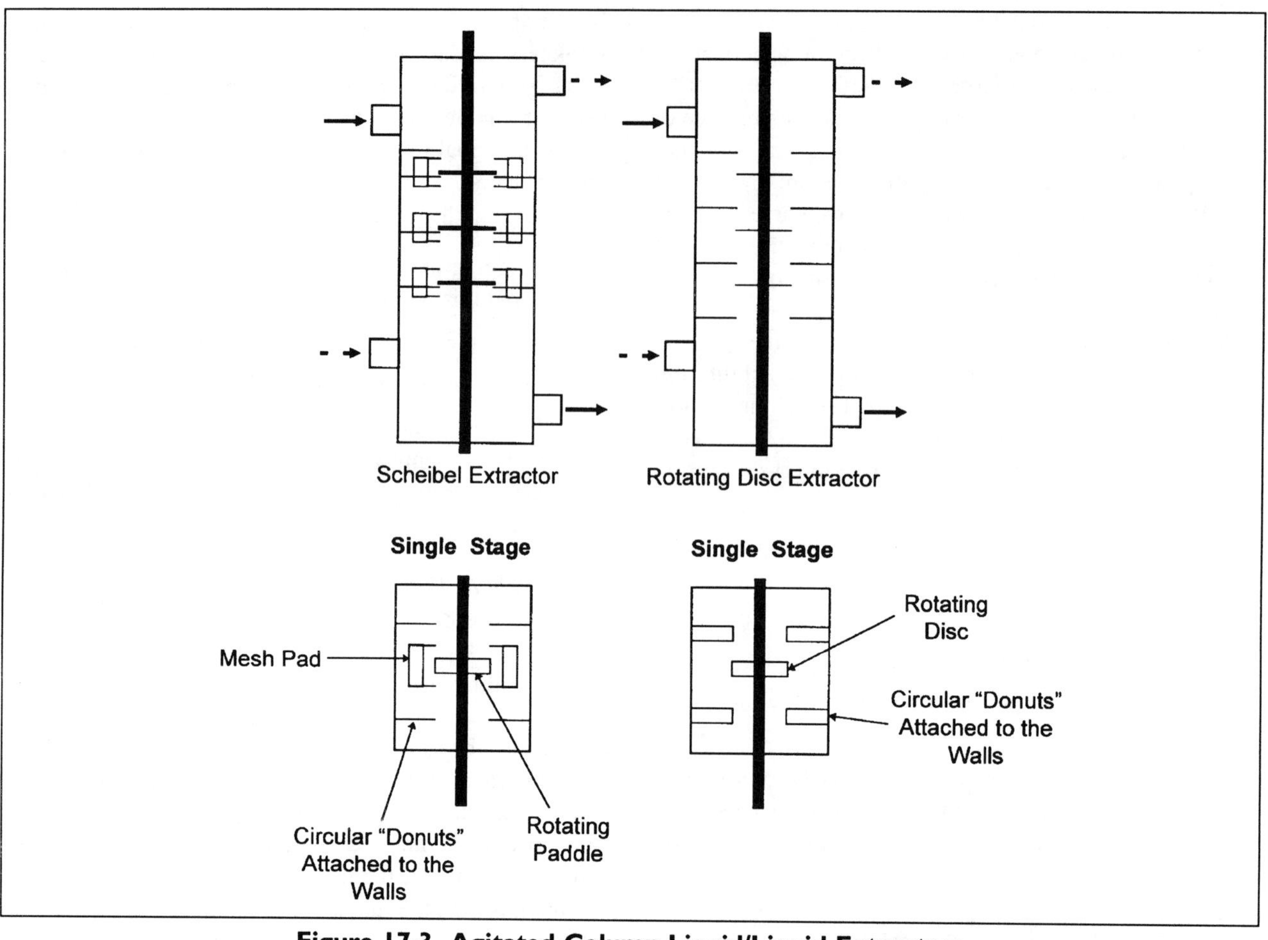

Figure 17-3. Agitated-Column Liquid/Liquid Extractors

- Low-to-moderate operating temperatures are desired
- The spent bed(s) can be regenerated easily
- The adsorbent is not susceptible to fouling or attack by the feed components[6]

One weakness of adsorption systems is the need to switch from one bed to the other. This results in an unsteady-state process and the inherent disadvantages associated with a continually changing operation. A second weakness is that the performance of an adsorbent can be affected significantly by poisons or solids present in the feed. A third disadvantage is that adsorbent regeneration by solvent or chemical stripping usually creates a waste stream that must be further treated. Finally, the large heat effects during adsorption and desorption must be considered in the design and operation.

The principal sources of malfunction in adsorption systems are blinding, chemical attack, or deactivation of the adsorbent; incomplete regeneration; high humidity in the inlet gas; and change in the feed composition, such that a different type of adsorbent is required.

17.3.5 Membranes

Membrane processes separate the components of a mixture on the basis of their relative permeation rates through a membrane material. A key requirement of this technology is finding a membrane that selectively permeates the components in the mixture.[7]

The factors that favor membrane technology are when bulk rather than precise separations are sufficient, relatively low flow rates (a few hundred lb/h) are involved, the membrane is found to be resistant to fouling by the feed components, energy cost is a key consideration, and no added materials, such as solvents, are necessary to achieve the separation.

Weaknesses of membrane systems include susceptibility to fouling, lack of durability of membrane materials, high capital-investment costs, and the impact of bulk rather than precise separation.

The principal sources of malfunction are usually fouling, membrane rupture, or membrane compaction.

17.4 Case Histories

17.4.1 Distillation: New Design Criteria

Once a plant is running, the temptation is to leave the process alone. In a DuPont process, the plant had been "running" for 20 years; however, after a particularly difficult production campaign, the column operation was reviewed, with the expectation that the column internals would need to be replaced. Instead, the review indicated that the column feed location was incorrect.

The conventional wisdom is to locate the feed stream at the tray on which the mixture composition matches that of the feed stream. The better method is to locate the feed on the tray that results in minimum energy consumption; this results in a smaller capital investment and lower operating cost. Relocating the feed to the latter tray reduced the loss of product to waste from 30 lb/h to 1 lb/h, increased column capacity by 20%, and decreased the refrigeration cooling load by 10%. The net benefit was a greater than $9,000,000/year increase in revenue to the business.

17.4.2 Distillation: Recovery of Waste

Water-soluble solvents from a solution polymerization process were water scrubbed from an air stream. Recovery of the solvents from the water stream was considered to be too expensive, so the water stream (now a Resource Conservation and Recovery Act (RCRA) hazardous waste) was incinerated. An extensive review of the vapor–liquid equilibrium data and a pilot-plant test showed that the solvents could be separated from the water stream by distillation followed by extraction. The distillation step separated the three solvents, with one solvent going overhead into the distillate with the water, and the other two solvents remaining in the bottoms. The solvent in the water stream was then extracted from the water with a low-boiling immiscible hydrocarbon, and the solvent was recovered from the hydrocarbon by an azeotropic distillation column. By recovering more than 10 million lb/yr of solvents and reducing the waste incineration load by more than 4 million lb/yr, the new capital investment had only a two-year payback.

17.4.3 Crystallization: New Technology

A solution-spun fiber process generated a byproduct sulfuric acid stream of 8.5% concentration at a flow rate of 150,000 lb/h. DuPont Engineering developed a DDO neutralization process using an aragonite form of limestone ($CaCO_3$) that produced a salable gypsum ($CaSO_4$) centrifuge cake that was low enough in moisture that expensive drying was not needed. The DDO feature gave a superior crystal habit or shape, and the increased solids concentration eliminated encrustation, which gypsum processes were notoriously known to experience.[5] The business needed a 95% uptime, which was achieved. In addition, 600,000 tons of toxic waste were avoided by this process, while the pure CO_2 that evolves from the gypsum crystallizer was sold as liquid CO_2.

This DDO process, which was based upon DuPont Engineering's crystallization technology, competed with a commercial process, saving $5,000,000 in investment ($10,000,000 vs. $15,000,000 for the commercial process). Two other DDO processes have been developed and built to neutralize dilute HF streams to produce salable CaF_2, which is the raw material for HF manufacture.

17.4.4 Extraction: Thermodynamic Review

In distillation, the conventional wisdom is to remove the low-boiling material first. After the low-boiling material was removed in a DuPont batch process, the remaining mixture was very difficult to separate, because of azeotropes and pinch points formed by the remaining compounds. The separation difficulties resulted in about one-third of the production run having to be incinerated.

The vapor–liquid equilibrium data for the compounds was reexamined, especially the binary interaction parameters. This reexamination revealed that the low-boiler could extract the product from the remaining compounds. A pilot-plant test confirmed the concept, and a continuous extraction process was designed and constructed. The new process reduced the lost product from 200,000 lb/yr to less than 2000 lb/yr, and the impurities in the final product were decreased from 500 ppm by weight to less than 1 ppm by weight.

17.4.5 Adsorption: Technology Selection

The toxicity of Cu^{2+} toward aquatic life is well-documented, and, as a result, a number of separation methods have been investigated for the removal of this heavy metal from water streams. Some of these methods include chemical precipitation, reverse osmosis, ion exchange, and activated carbon adsorption. The relative merits and economics of ion exchange and activated-carbon adsorption are compared for the removal of 10 ppm Cu^{2+} from 100 gallons per minute (gpm) of water at a pH of 4–5.

Relatively few studies have been done using granular activated carbon (GAC) for the removal of cations from water due to the poor intrinsic affinity of charged species for the primarily nonpolar surfaces of carbon. A notable exception is the work of Dr. C. P. Huang at the University of Delaware,[8] who published the equilibrium loading of Cu^{2+} onto a number of activated carbons at various pH values. At pH 4, Dr. Huang's best GAC loaded with Cu^{2+} at a level of approximately 0.4 wt %. Halving the equilibrium loading as an estimate of dynamic capacity, it is easily calculated that about 6000 lb/day of GAC would be required using this approach. Typical strong cation-exchange resins, on the other hand, possess capacities on the order of 2.0 equivalents per liter. This translates to an approximately 20-fold increase above the carbon loading on a weight basis, and about 45-fold on a volume basis. Regeneration of either system would involve elution of the copper with acid; normal off-site thermal reactivation techniques employed by carbon vendors would not remove the copper from the carbon. Considering that the cost per mass of adsorbent is essentially the same, and assuming the same capital costs would be involved, ion exchange is in almost all cases a clear winner over GAC treatment for this particular separation problem.

17.4.6 Membrane: Process Simplification

A toxic surfactant was removed from a gas stream with a water scrubber, and then removed from the water stream by means of ion exchange. The regenerate from the ion-exchange bed was further concentrated using a reverse-osmosis membrane process. The concentrated stream with the surfactant was then returned to the supplier for further purification. Over time, the ion-exchange resin capacity was reduced and significant amounts of the expensive surfactant were lost.

To further reduce emissions of the surfactant, the collection system was expanded to include gas streams from other process units. A review of the recovery process showed that the ion-exchange system, and its attendant costs and wastes, was not required to purify the water stream. The ion-exchange system was replaced with a new reverse-osmosis system to concentrate the surfactant-containing stream from the water scrubber to 1000 ppm by weight. This 1000 ppm-by-weight stream was then further concentrated to 20% by weight in the existing reverse-osmosis system. The recovered water could then be recycled back to the water scrubbing system. The 140,000 lb/yr of recovered surfactant was returned to the supplier at a credit of $20/lb or $2,800,000/yr.

Literature Cited

1. Kelly, R. M. 1987. "General Processing Considerations." *Handbook of Separation Process Technology*, R. W. Rousseau, Ed.: pp. 197–222. New York: Wiley.
2. Null, H. R. 1987. "Selection of a Separation Process." *Handbook of Separation Process Technology*, R. W. Rousseau, Ed.: pp. 982–995. New York: Wiley.
3. Humphrey, J. L. October 1995. "Separation Processes: Playing a Critical Role." *Chemical Engineering Progress.* 91(10): 31–41.
4. Kister, H. Z. 1990. *Distillation Operation*, 2. New York: McGraw-Hill.
5. Randolph, A. D., S. Mukhopadhyay, B. C. Sutradhar, and R. E. Kendall. 1990. "Double Draw-Off Crystallizer: Major Player in the Acid Rain Game?" *Crystallization as a Separations Process*, ACS Symposium Series No. 438, Chap. 9: 115–129. Washington, DC: American Chemical Society.
6. Keller, G. E. October 1995. "Adsorption: Building Upon a Solid Foundation." *Chemical Engineering Progress.* 91(10): 56–67.
7. Koros, W. J. October 1995. "Membranes: Learning a Lesson from Nature." *Chemical Engineering Progress.* 91(10): 68–81.
8. Corapcioglu, O. M., and C. P. Huang. 1987. "The Adsorption of Heavy Metals onto Hydrous Activated Carbon." *Water Research.* 21(9): 1031–1044.

CHAPTER 18

Equipment Leaks: Regulations, Impacts, and Strategies

18.1 Fugitive Emissions

Fugitive emissions—defined as unintentional air emissions resulting from the effects of malfunctions, age, lack of proper maintenance, operator errors, improper equipment specifications, use of inferior technology, and externally caused damage—account for about one-third of the total organic air emissions (385,000 ton/yr) from chemical plants in the United States.[1] EPA test programs indicated that pumps and valves accounted for 72% of these fugitive volatile hazardous air pollutant emissions as shown in Figure 18-1.[2] The information in this chapter should be of interest to anyone dealing with emissions caused by leaks from valves and pumps. (Equipment leaks are a specific subcategory of fugitive emissions.)

18.2 New Regulations in Mid-1994

The 1990 Clean Air Act Amendments (CAAA) have provisions likely to drive many plants toward adopting programs to deal with equipment leaks. These include Title I restrictions in nonattainment areas, Title III regulations on Hazardous Air Pollutants (HAPs), and Title V federal permits.

Regulations intended to reduce emissions from equipment leaks will apply to many organic products and processes. Such programs are often referred to as monitor and restore equipment seals (MARES), also known as leak detection and repair (LDAR) programs. These regulations are "work practices" and require monitoring according to specified methods and corrective actions when certain conditions are exceeded.

The enforcing regulatory agency requires documentation as proof that you are complying with the regulations. Therefore, the setup and recordkeeping is an important part of any equipment-leak program.

18.2.1 Impact and What to Do

Businesses are moving toward pollution prevention and reduced waste generation to avoid costly proscriptive control requirements. Also, tight capital budgets are forcing businesses to renew existing investment.

The emissions from equipment leaks could trigger high-cost emission reduction through hardware replacement. One apparently simple way to avoid complications caused by equipment leaks is to install so-called leakless valves and pumps. This will generally be the most costly approach and frequently will not achieve an appreciably lower emission rate than less costly approaches.

The new regulations require a MARES program. This is a work practice that can be based on quality-control principles and does not restrict equipment choice. A well-structured maintenance program, with modest upgrades in packings and seals, will be the most cost-effective way to deal with leaks and leakage regulations. Some form of recordkeeping will be necessary regardless of the reason for the program.

18.3 Recordkeeping Setup

One of the first and most important tasks is to set up the recordkeeping system. This requires identifying the valves, pumps, and flanges in the program, with some way to track performance. Identification can be as simple as color coding hardware in the field and recording data on a clipboard. When such a manual task gets too large to handle, more complex systems may be desirable. To provide a cost-effective way to leverage with other piping-integrity programs, various software vendors as well as valve and pump manufacturers have developed or are developing software for proper recordkeeping.[3]

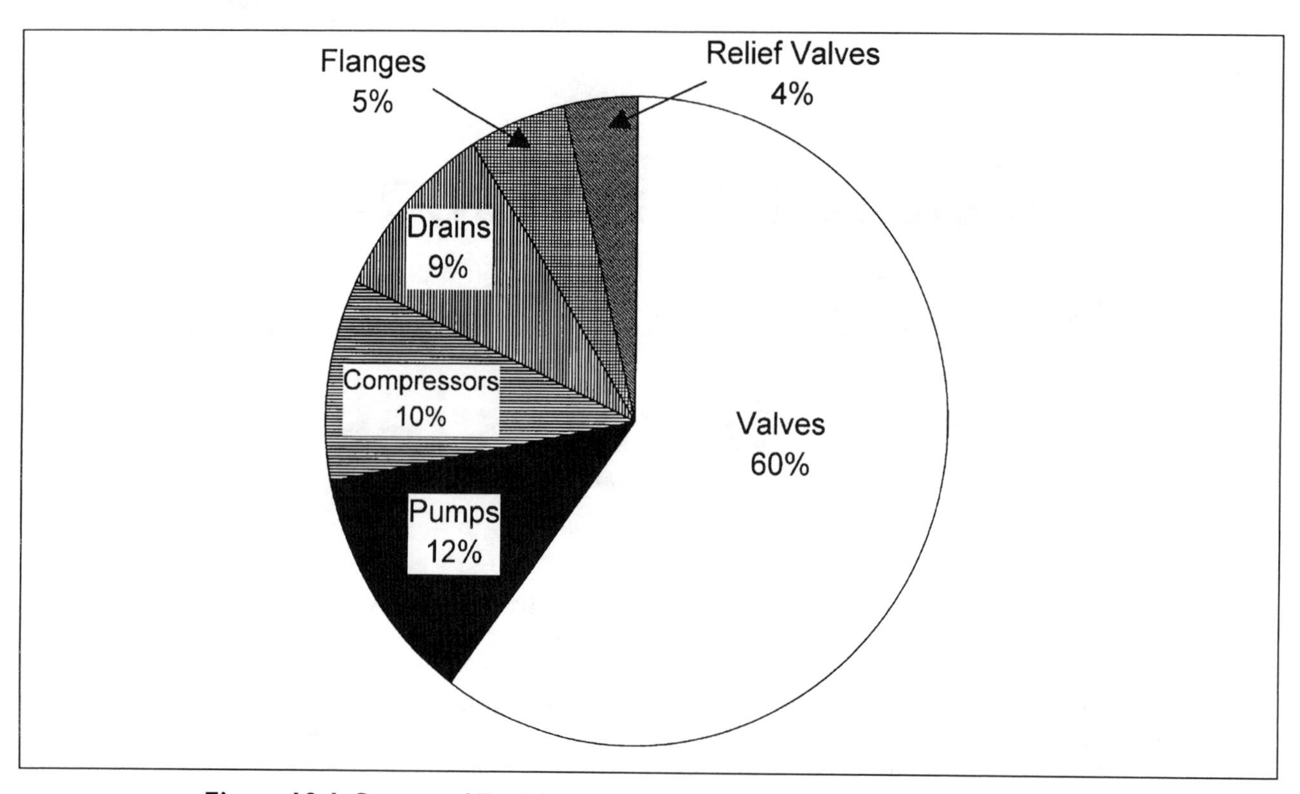

Figure 18-1. Source of Fugitive Volatile Hazardous Air Pollutant Emissions

18.4 Strategy for Valves

The suggested strategy for valves is as follows:

1. Do not jump into a replacement program until monitoring has given you data for rational decision making. Monitoring will identify any particular areas that might warrant a change in hardware or work practices. In addition, it can indicate the improvements and cost-effective changes.
2. If monitoring identifies a problem, equipment modifications should follow this sequence:
   (a) Upgrade packing with an appropriate low-leak material, such as the new DuPont Kalrez polymeric packing.[4,5]
   (b) Replace full-turn valves with quarter-turn valves.
   (c) Consult with a valve expert before considering replacing existing valves with new expensive technology, such as bellows seal valves.[6,7]
3. For a new process, it will be necessary to predict the equipment leak rates before the process is built. However, if EPA factors are used to predict leak rates, the resulting emission estimations will cause difficulties (new permits, emission taxes, regulatory attention, etc). Rather than accept greatly inflated emission estimates from EPA factors, two possible alternatives for valves and pumps are appropriate:
   (a) Use data showing low leak rates for a sister process.
   (b) Use data showing low leak rates for similar elements and maintenance practices at the same site.

18.5 Learnings From a Study of Pump Seals

The number of pumps in a process is much less than the number of valves: 12 to 25 pumps per 1000 valves. Average pump life in the DuPont Company is from 6 to 12 months, and 50% to 80% of the repairs are due to seal failure. Repair costs are high.

A study of the effect of seal arrangements, wear faces, secondary seals, pump speed, shaft size, and several properties of the fluids being pumped indicated that:[8]

- Single mechanical seals appear to provide an extremely cost-effective way to deal with most requirements.
- Actual double-seal flush-fluid leak rates are not much lower than single-seal leak rates. The material leaking from dual seals is the barrier fluid and may or may not be a hazardous air pollutant (HAP).
- 8.3% of inspected seals of all types in the petroleum and hydrocarbon processes (over 400 screened) were above a screening value of 1000, compared to

a target of less than 10% in the regulation. (A screening value of 1000 or greater is the trigger for corrective action in some regulations.)

- Modest upgrades in seal face and secondary seal materials will often improve performance at relatively low cost.
- Emphasis on the maintenance issues can ensure good performance without major retrofits or replacement of equipment.
- The lowest leak rate observed for single seals was from a combination of rotating and stationary wear faces of silicon carbide/carbon.
- Teflon, Viton, and Kalrez were all found to be effective materials for secondary seals when properly used. More demanding service justifies the more expensive materials.
- Cartridge seals gave lower average screening values than noncartridge seal designs.

18.6 Strategy for Pumps

The suggested strategy for pumps is as follows:

1. When the seals need to be repaired, consider replacing them with cartridge seals. A cartridge seal is assembled in the shop by the seal manufacturer and is more precisely aligned. Some pump modifications may be necessary.
2. Determine if a different secondary seal material is appropriate.
3. Determine if a better combination of rotating and stationary wear faces is appropriate.
4. If the single mechanical seal is not sufficient, then replace it with a double mechanical or tandem seal (more expensive solution).
5. Finally, if necessary, consider a sealless pump (most expensive solution).

Literature Cited

1. Adams, W. V. 1991. "Control of Emissions From Rotary Equipment in the Pharmaceutical Industry." Proceedings of the Bioprocess Engineering Symposium. 21: 49–57. New York: American Society of Mechanical Engineers.
2. Lipton, S. June 1989. "Fugitive Emissions." *Chemical Engineering Progress.* 85(6): 42–47.
3. Konrad, K. K., and K. W. Wilkerson. May 1995. "Tough Clean Air Rules Boost Fugitive Emissions Technology." *Intech.* 42(5): 48–53.
4. Buckley, G. J. September 1995. "Valve Users Get Help on Fugitive Emissions." *Control Engineering.* 42(10): 117–122.
5. Wolz, D., and L. Winkel. April 1995. "Controlling Fugitive Emissions by Retrofitting Valves." *Chemical Processing.* 58(4): 81–86.
6. Wright, J. B. May 1995. "Curbing Fugitive Emissions." *Chemical Processing.* 58(5): 70–77.
7. Miller, M. December 1991. "Fugitive Emission Regulations Change Valve Selection/Care." *Pollution Engineering.* 23(13): 80–84.
8. Adams, W. V. August 1991. "Control Fugitive Emissions from Mechanical Seals." *Chemical Engineering Progress.* 87(8): 36–41.

APPENDIX A

A Collection of Pollution-Prevention Opportunities

A.1 Introduction

Pollution prevention encompasses a broad range of activities. Some require a long lead time and major investment, while others involve procedural and operational changes requiring little investment and resulting in immediate savings to the business. In this Appendix, we have compiled a number of pollution-prevention ideas that can reduce the cost of doing business. Looking at it another way, pollution prevention offers the opportunity to improve business performance through lower variable cost and improved after-tax operating income (ATOI) and cash return on investment (CROI).

Focus on opportunity! Look at the cost sheet, then give thought to the impact of your process on the variable cost of utilities. Put your effort into opportunities that represent hundreds of thousands of dollars, not hundreds of dollars.

A.2 Did You Know That...

- Treating the incremental air flow from a 1-ft^2 opening with a 100 ft/min face velocity will cost a minimum of $5000 investment and $1000/yr to operate. This is financially equivalent to a one-time cost of $12,500.
- An additional 10 gallons/minute (gpm) of pure water in your wastewater will increase the investment for a new wastewater treatment plant by at least $30,000 and its operating cost by $3000/yr. This is financially equivalent to a one-time cost of $55,000.
- If you add 1 pound/hour (lb/h) of an organic into your wastewater, it will increase the investment for a new wastewater treatment plant by at least $6000 and its operating cost by $2000/yr. This is financially equivalent to a one-time cost of $21,000.
- If you add 1 lb/h of an organic plus 10 gpm of pure water into your wastewater, the effects on your new wastewater treatment plant are additive: $36,000 of new incremental investment and $5000/yr extra to operate.
- One pound of organics in your wastewater will generate about two pounds of biosludge (15% solids). At a processing/disposal cost of $100/ton, the biosludge processing and disposal costs alone come to $0.10 for that one pound of organic.
- It costs $50 to $1000 per cubic yard to remediate soil. Ex situ remediation may cost at least $300 per yd^3.
- If you landfill 10,000,000 pounds of packaging waste in a year, the disposal cost will be $400,000 at $0.04/lb.
- The minimum incremental cost to treat 1000 gallons of wastewater containing no organics is $2.00. This compares with a typical cost-sheet cost for filtered water of $0.15 per 1000 gallons. You must consider *both* the cost of the water and its subsequent treatment!
- A typical pump seal flush creates about 1/2 gpm of wastewater, which will add at least $1500 investment and increase the operating cost by $150/yr for a new wastewater treatment plant. This is financially equivalent to a one-time cost of $2750.

A.3 Pollution-Prevention Opportunities List[1-3]

A.3.1 Process/Product Conceptualization and Development

- *Pollution prevention.* Include a "zero-waste discharge" option in all alternative process evaluations. This increases waste awareness, which can lead to valuable cost-savings ideas.

- *Raw materials.* Consider buying purer raw materials or removing the impurities before they enter the process. Many times this is best accomplished by the suppliers, because they may already have the infrastructure in place to remove and treat the impurities.

- *Improve reaction kinetics.* To reduce byproduct and coproduct generation, consider reactions with higher selectivity and lower conversion over reactions with higher conversion and lower selectivity. For reversible reactions, consider recycling byproducts back to the reactor to minimize production of byproducts. Modify reaction sequences to reduce the amount of intermediates and byproducts, or change their composition.

- *Look closely at the molar excess for minor ingredients in batch reactions.* Understand the relationship between yield and cost and avoid excess use of inorganic compounds, such as mineral acids and bases.

- *Consider solvent selection.* Historically, solvent selection was based only on functionality, availability, and purchase cost. Now, solvent selection must also be based on hazard and toxicity, regulatory impact, and cost of recycle or destruction. Table A-1 contains a proposed solvent selection hierarchy (listed in order of decreasing desirability) based on environmental, toxicity, and process-hazard concerns.

- *Use of air.* Use oxygen instead of air for oxidation reactions to minimize the introduction of noncondensibles. Ultimately, these must be purged from the process and will likely require treatment.

- *Catalyst selection.* Frequently, heterogeneous heavy-metal catalysts are more easily retained within the process than homogeneous (soluble) catalysts. Noble-metal catalysts can often be recycled by both on-site and off-site reclaimers.

- *Minimize residual solvent in polymer pellets and films shipped to customers.* This may be a more cost-effective solution for a customer's waste problem and will result in a greener product.

- *Lower solvent emissions from paints.* Emissions from solvent-based coatings can be reduced by using extrusion-, powder-, CO_2-, or water-based coatings.

- *Use solventless separations processes.* Evaluate unit operations or separation technologies that do not require the addition of solvents or other nonreactant chemicals, such as membranes and melt crystallizers.

- *Develop new packaging material.* Make packaging material out of the end product. For example, sell polymer resin in bags made of that resin so the customer can grind the bags and use them as a feedstock.

- *Optimize plant siting.* Locate the plant next to a consumer to reduce raw-material costs, eliminate packaging, and maximize product recycle.

A.3.2 Plant Design

- *Look at the streams leaving the flowsheet.* What costs do the streams represent in replacement solvent or lost product? In waste treatment? In utilities consumption? For gas streams, what can be done to reduce the volume of air going to pollution-control devices? The cost to treat an incremental 100 standard cubic foot per

Water	
Nonpolar Organics	**Polar Organics**
Aliphatics (flash point above 140°F)	Alcohols
Aliphatics (flash point below 140°F)	Organic acids
Aromatics	Other oxygenates
Halohydrocarbons	Nitrogen-bearing compounds
1. HCFCs	Halogenated organics
2. Chlorocarbons	Chlorinated organics
3. CFCs	
Listed or acutely toxic compounds	Listed or acutely toxic compounds

Table A-1. Proposed solvent selection hierarchy.

minute (scfm) of gas is $1000/yr or more. For wastewater, one lb/h of oxygen demand costs about $2000/yr to treat, in addition to the lost organic value. A gallon/minute of water costs $1000/yr to get to the process and then discharge.

- *Recover and recycle solvents.* Many processes stage the use of wash solvents to reduce make-up solvent and waste-disposal costs.
- *Reuse high-quality water.* Consider recovery of condensate as boiler feedwater make-up, process water, or cooling-tower make-up. Condensate is often economical in place of soft water for applications such as diluting sodium hydroxide.
- *Strip volatile-organic-compound (VOC) containing wastewater and recover the volatile organics.* This approach is frequently more attractive than stripping followed by abatement of the overhead stream, for two reasons. First, you have more flexibility in designing the treatment system, and second, you recover value from the recycled material.
- *Improve solvent displacement from solid products and intermediates.* Solvents can be more completely removed by improved solid–liquid separators, such as pressure filters, before contact with water. Also, washing can be staged to reduce dilution and make recovery more economical.
- *Improve liquid–liquid separations.* More stages, colder operating temperatures, and improved interface control can improve decanter performance. Also, consider additional mass-transfer stages in distillation and extraction operations.
- *Eliminate the hazardous air pollutant.* No one approach can be defined to accomplish the elimination of hazardous air pollutants. However, brainstorming can often identify opportunities (e.g., using unlisted heptane in place of regulatory-listed hexane).
- *Do not let air pollution become water pollution.* Avoid water scrubbing as a means of solving air-pollution problems if this, in turn, creates a water-pollution problem.
- *Control inert addition.* Use controlled (flowmeters, rotameters, etc.) instead of uncontrolled addition of inerts for blanketing, purging, and pressure transfer.
- *Consider water as an expensive solvent.* Treat water as any other solvent; that is, minimize its introduction into the process and reuse directly or treat and recycle.
- *Segregate air streams.* Segregate process-contaminated air from heating and ventilation air; that is, stage ventilation from cleaner air not requiring abatement to a smaller volume of dirtier air requiring abatement.
- *Optimize dryer operation.* Follow the following steps for minimizing emissions from indirectly heated air convection dryers using once-through air.
  (a) Recycle air
  (b) Recycle air plus minimize inleakage (improved baffling, seals, air locks, etc.)
  (c) Recover organics from the purge stream
  (d) Recover energy value from the purge stream
  (e) Abate organics from the purge stream
- *Consider inerted dryers* operating "fuel-rich" with condensation for solvent recovery.
- *Consider dense-phase conveying* instead of dilute-phase conveying.
- *Replace or modify contact condensers to make recovery more economical.* Before installing direct-contact heat exchangers to condense or absorb vapors, consider the downstream processing and treatment consequences. Contact condensers can sometimes be economically replaced with absorbers (replacing the contacting fluid with a higher boiling or immiscible fluid) or refrigerated noncontact condensers. The more concentrated volatile organic is typically cheaper to recover. Also, consider reboilers in place of live steam injection.
- *Optimize process conditions.* Determine optimum process conditions for phase separation, for example, colder decanter temperatures.
- *Segregate waters.* Segregate process wastewater from stormwater runoff by diking and covering areas. Installing roofs and curbs may be less expensive and more environmentally sound than building larger treatment facilities to handle the additional flow.
- *Optimize plant layout* to shorten piping runs and to minimize the number of emission points—particularly fugitive emission points. In other words, minimize the number of valves, flanges, and pumps. Minimize the number of outlets for waste streams, for example,

minimize hoods, drains, and tankage.

- *Use dedicated shipping containers, bulk packaging and supply, or ship by pipeline.*

- *Consider pressure vessels or floating roof tanks* instead of traditional storage tanks with conservation vents.

- *Use of fewer pumps* leads to fewer seal leaks, so consider pressure or gravity feed.

- *Minimize product degradation.* Use staged heating (i.e., multiple heat exchangers in series, using progressively hotter heating media) to minimize product degradation and undesirable side reactions due to high tube-wall temperatures.

- *Replace a solvent.* Use the actual product or a process intermediate as a carrier for stabilizers and inhibitors instead of introducing additional solvents to the process.

- *Well-maintained mechanical seals* generate less waste than packings with seal flushes.

- *Consider using vacuum pumps* instead of steam jets.

A.3.3 Plant Operation

- *Do not overtreat wastes.* Significant energy has been saved by lowering the temperature of thermal oxidizers (typically $3000/yr for every 1000 scfm), by reducing the pressure drop across particle scrubbers (typically $500/yr for every 1000 scfm), and by turning off surplus aerators in wastewater treatment ($27,000/yr for every 100 hp).

- *Are aerators run to meet an occasional peak demand?* If so, can one or more be shut down except during turnarounds or other peak demands?

- *Understand whether your wastewater treatment plant adds base or acid to control pH.* Look for the opportunity to reduce pH-control chemical demand. For example, the cost to neutralize hydrochloric acid is about $200/ton acid. Question all acid additions to your process. Consider using waste acid from another part of the plant or an acid that is generated as a byproduct within the process. This strategy saves money twice, through reduced ingredients cost and reduced use of a base to neutralize the acid.

- *Avoid excessive carbon dioxide scrubbing in caustic scrubbers.* Scrubbers can often be operated at pH less than or equal to 8 and still remove chlorine, bromine, sulfur dioxide, and hydrochloric acid. Higher pH wastes caustic by scrubbing carbon dioxide. For example, at pH 10 and typical carbon dioxide partial pressures found in oxidizer off-gases, the incremental caustic consumption for scrubbing carbon dioxide is $100–$700/yr per scfm.

- *Look closely at equipment cleaning practices.* High waste loads resulting from equipment cleaning set the peak demand on many wastewater treatment plants and represent product down the drain. If the wash fluid is compatible with the process, consider recovery of the first wash and return it to the process during startup. Also, work with the wastewater treatment plant to generate ideas to reduce peak loads. If cleaning must be done frequently, save the last stage of rinse fluid for the initial rinse the next shutdown. Or make the first rinse with mother liquor. Water washing can sometimes be made much more efficient by adjusting pH.

- *Recycle samples to the process.* This is particularly cost effective if excess samples are hazardous waste. The dead volume of sample lines should be eliminated and the volume of samples should be reduced to actual analytical requirements. Do not underestimate the impact of discarded laboratory samples on your wastewater effluents. There are cases where the dominant source of heavy-metal contamination in the wastewater was from laboratory samples. In general, minimize the number and size of samples.

- *Consider operating with fewer full tanks instead of more partially filled tanks.* Operate tanks at a constant level. Reducing the number of tanks in service and operating them at a constant level reduces ingredient losses.

- *Control product purity.* Customers want consistent quality. Variable purity—including intermittent over-purification—reduces capacity and increases waste without providing value to the customer. Examples are tighter control of near-boilers in refining trains, tighter control of filtrate conductivity, avoiding excess solvent wash, and measuring the moisture level in front of a drying step. Better control of product purity can result in substantial capacity increases for sold-out processes.

- *Modify production schedules.* Are waste-generation costs considered when scheduling a multitude of similar products on one production line? Can a "mixed" product be sold to a lower-quality market, eliminating the need for equipment cleaning and reducing waste?

- *Seal up the process.* Several processes have substantially reduced air exchange between process equipment and the work place, achieving volatile organic compound (VOC) concentration ratios >1000 inside the process versus in the workplace. This saves money by reducing heating, ventilation, and air-conditioning (HVAC) costs and the operating cost of emissions control equipment.

- *Routinely monitor for fugitive emissions.* Leaks of organics into utility water should be detected and corrected early. Typical sources of leaks include heat exchangers and pump seals. Both can be economically corrected with proper maintenance.

- *Review cleaning practices.* Use improved cleaning technology to minimize solvent and wash-water wastes, such as pipe-cleaning pigs, rotating spray heads in tanks, high-pressure water jets, antistick coatings on vessel walls, better draining equipment, mechanical cleaning/sweeping, and multiple small rinses instead of filling and draining. Minimize or eliminate water washdowns, and even consider reducing the number of water-hose stations.

- *Minimize storage* of raw materials, intermediates and products.

- *Minimize startups and shutdowns.* When optimizing production schedules, consider the waste generated during startups and shutdowns.

- *Minimize air leakage into vacuum systems.* Air inleakage will eventually become contaminated with organics or particulates and will require treatment.

- *Turn off or shutdown idle equipment* such as steam jets, cooling-tower water to heat exchangers, and water flushes.

- *Equalize tank and reactor vents* during filling, unloading, and transfer operations.

- *Return packaging, impurities, and unused raw materials* to suppliers.

- *Increase efforts to find markets for byproducts and coproducts.*

Literature Cited

1. U.S. Environmental Protection Agency. November 1993. *DuPont Chambers Works Waste Minimization Project.* EPA/600/R-93/203. Washington, DC: U.S. EPA, Office of Research and Development.
2. Chemical Manufacturers Association. May 1993. *Designing Pollution Prevention into the Process: Research, Development and Engineering.* Chemical Manufacturers Association.
3. U. S. Environmental Protection Agency. May 1992. *Facility Pollution Prevention Guide.* EPA/600/R-92/088. Washington, DC: U.S. EPA, Office of Research and Development.

APPENDIX B

Description of Screened Ideas from Case Study

Use Superheated Steam (as the Inert) and Pure O_2 Instead of Air in the Reactor

Description: Air supplies nitrogen as a heat sink to prevent hot spots from developing in the fixed catalyst bed during the course of the reaction. The nitrogen could be replaced with dry steam and pure oxygen or, alternatively, enriched air could be used. The gas exiting the reactor could then be cooled, producing a low-flow-rate stream consisting of oxygen and any gaseous byproducts of reaction. The condensed steam could be sent to (1) a stripping column, where the water would be separated from the reactant and product, or (2) the existing benzene extractor.

Concerns: What is the effect of the steam on the catalyst? Do any of the reactants or products react with the steam? What are the safety hazards when the steam is condensed and the oxygen-rich gas is sent to an abatement device?

Investment and Costs:

Case (1): Heat exchanger for condensation.
Flare to abate the off-gas.
Fired superheater to superheat the steam.
\$6,700,000 investment and \$3,500,000 annual operating cost.

Case (2): Heat exchanger for condensation.
\$2,000,000 investment and \$800,000 annual operating cost.

Benefits: Avoided cost of abating the reactor off-gas.
Eliminates need for benzene-extraction system.
Smaller compressor and gas handling system.

Case (1): Reduces end-of-pipe investment by \$800,000.
Reduces annual operating cost by \$300,000.
Reduces wastewater treatment load from 35 gallons per minute (gpm) to 1 gpm.

Case (2): Reduces end-of-pipe investment by \$800,000.
Reduces annual operating cost by \$200,000.
Reduces wastewater treatment load from 35 gpm to 17 gpm.

Resources: Bench-scale testing to understand the effect of steam.
Process hazards evaluation.
Engineering analysis of steam addition, condenser, and modification of distillation system to handle a small amount of water.

Timing: One person-week of experimentation.
Two person-weeks for engineering evaluations.

Probability of Success: 30%

Recycle Vent-Gas Stream and Add Additional O_2 or Enriched Air as Needed

Description: Recycling the vent gas with a purge stream would reduce the size of the off-gas abatement device. The system would also be run at high pressure.

Concerns: What is the effect of the higher impurity concentration on the catalyst?

Investment and Costs:

Piping for recycle.
\$200,000 investment and \$500,000 annual operating cost.

Benefits: Reduces size of abatement device.
Reduces size of compressor.
Reduces end-of-pipe investment by \$700,000.

Reduces annual operating costs by $70,000/yr.

Resources: Experiments to test the effect of contaminants on the catalyst.
Timing: Four person-weeks of experimentation.
Probability of Success: 90%

Install Condenser to Remove Product

Description: The Product (PROD) freezes at about 60°C. A cooler operating at about 65°C would condense the PROD with little condensation of the reactant (REAC).
Concerns: What is the split between PROD and REAC? Are higher-molecular-weight solids formed?
Investment and Costs: $500,000 investment for a condenser and receiver.
Benefits: Removes product from the system sooner and changes the nature of the downstream unit operations.
Resources: Model evaluation of concept.
Timing: One person-week for engineering evaluation.
Probability of Success: 90%

Fluidized-Bed Reactor

Description: The fixed-bed reactor requires a large volume of air to act as a heat sink to prevent hot spots. A fluidized bed would permit the use of a lower volume of air, because only a small amount of oxygen is actually required for the reaction. Also, the improved temperature control of a fluidized bed could reduce tar formation.
Concerns: What is the effect on catalyst life due to mechanical abrasion of a fluidized bed? What is the effect of the oxygen-to-inert-gas ratio on the reaction yield and selectivity? Does the reaction work at higher reactant concentrations as would be present in the fluidized bed (10 times the fixed-bed reactor concentration)?
Investment and Costs:
New reactor and catalyst make-up system.
Fines collection system, such as a cyclone or bag filter.
$500,000 investment and $500,000 annual operating cost.

Benefits: Reduces size of abatement device.
Reduces size of compressor.
Better reaction control without hot spots.
Reduces end-of-pipe investment by $800,000.
Reduces annual operating costs by $400,000.

Resources: Experiments to test the effect of fluidizing the catalyst bed.
Timing: Four person-weeks of experimentation.
Probability of Success: 80%

New Catalyst with Better Selectivity/Conversion

Description: Develop a catalyst that produces less tars and emissions.
Concerns: Comparable yield and selectivity.
Investment and Costs: New catalyst.
Benefits: Increases overall yield.
Reduces waste generation by 50% for a 5% increase in yield.

Resources: Experiments to test the new catalysts.
Timing: Six person-months of experimentation.
Probability of Success: 50%

Change the Air-to-Feed Ratio in the Reactor to Reduce COS Generation

Description: Add air above the reactor bed to oxidize COS to SO_2.
Concerns: Will the level of COS be reduced sufficiently to eliminate the need for a thermal oxidizer? Will the increased level of SO_2 require a caustic scrubber? Will the technology work?
Investment and Cost: Air feed above the fixed catalyst bed. May require a larger thermal oxidizer.
Benefits: Reduces COS generation.
Resources: Experiments to test feasibility.
Timing: Four person-weeks of experimentation.
Probability of Success: 50%

Replace the Water Scrubber with a Solvent Scrubber

Description: Using a high-boiling solvent, the product and reactants can be scrubbed from the gas stream. This would eliminate the water/benzene system. The product stripper would be replaced with an extraction system. In fact, the ideal solvent would be the reactant, because REAC boils at 120°C and freezes at –30°C.
Concerns: What would be the yield loss of REAC in the off-gas stream from the scrubber? The recycle rate needs to be more than 80% to prevent excessive yield loss. Can the product-stripper overhead stream, which contains the REAC and water, be recycled directly, or does it need to be sent to the REAC purification system?
Investment and Costs:
Replace packed-bed water scrubber with a spray tower using solvent.
Install a brine-chiller recirculation system.
Decommission the benzene extraction system.
Lost REAC in the exiting off-gas stream from the scrubber.

$1,300,000 investment and $100,000 annual operating cost.

Benefits: Eliminates benzene extraction and water scrubbing systems.
Reduces annual operating costs by $700,000.
Reduces wastewater treatment plant load from 35 gpm to 1 gpm.

Resources: Modeling study to determine the design parameters.

Timing: Two person-weeks for modeling.

Probability of Success: 90%

Use Freeze Crystallization to Separate REAC, PROD, and Water

Description: Replace the extraction step with a series of freeze-crystallization steps to separate the product from the reactant and water.

Concerns: What is the split between the product and reactants? Do the salts have any effect and where do they go? Because the PROD freezes first, can it be removed by filtration or is it soluble with the REAC in the water?

Investment and Costs:
Replace the benzene extraction step with freeze crystallization technology.
$3,500,000 investment and $1,500,000 annual operating cost.

Benefits: Eliminates benzene extraction system.
Reduces annual operating costs by $700,000.

Resources: Experiments required to test feasibility.

Timing: Two person-months of experimentation.

Probability of Success: 40%

Recycle Benzene from the Steam Stripper Overhead Stream

Description: Cool and pipe the stripper overhead stream to the benzene extractor.

Concerns: Are there any materials, such as trace byproducts, that would accumulate in the system and require further separation?

Investment and Costs:
Cooler and piping.
$100,000 investment and $10,000 annual operating cost.

Benefits: Saves approximately 32 lb/h benzene.
Reduces annual operating costs by $90,000.

Resources: Modeling study to determine the appropriate design parameters.

Timing: One person-week for modeling.

Probability of Success: 90%

Use the Steam Stripper Bottoms as the Source of Water for the Water Scrubber

Description: Cool and recycle the steam stripper bottoms to the water scrubber.

Concerns: What level of inorganic salts can be tolerated in the system?

Investment and Costs:
Cooler and piping.
$300,000 investment and $10,000 annual operating cost.

Benefits: Reduces water flow to wastewater treatment by about 90%.

Resources: Modeling study to determine the appropriate design parameters.

Timing: One person-week of modeling.

Probability of Success: 90%

Purify REAC and PROD Directly From the Water Phase, Rather Than Using Extraction

Description: Use pressure-swing absorption to separate the REAC and PROD from the water.

Concerns: How much REAC will distill overhead with the water? Will the inorganic salts leave with the PROD?

Investment and Costs:
Three distillation columns and steam usage.
$10,500,000 investment and $4,500,000 annual operating cost.

Benefits: Eliminates benzene extraction step.
Reduces annual operating costs by $700,000.

Resources: Modeling study to determine the appropriate design parameters.

Timing: Two person-weeks of modeling.

Probability of Success: 70%

Recover and Recycle Tars from the REAC PROD Splitter

Description: Use a batch-still pot or wiped-film evaporator to recover any PROD in the tars. Also, consider using a small reactor to convert the tars back into the reactants or into another useful product.

Concerns: Will a sufficient quantity of PROD be extracted from the tars to justify the investment and operating cost? What is the nature of reactions that convert the tars to other compounds?

Investment and Costs:
Batch-still or wiped-film evaporator and reactor.

Benefits: Recovers PROD.
Increases overall yield by converting the tars to useful product.

Resources: Modeling study to determine the appropriate design parameters for the batch still or wiped-film evaporator. Experimental studies to determine the reactions to convert the tars to useful products.

Timing: Two person-weeks of modeling.
Six person-months of experimentation.

Probability of Success: 80% for the batch still/wiped-film evaporator; 70% for new reactor.